INTRODUCTION TO BUILDING

Derek Osbourn   Dip Arch [Hons] RIBA MSIAD

Longman
Scientific &
Technical

Longman Scientific & Technical,
Longman Group Limited,
Longman House, Burnt Mill, Harlow,
Essex, CM20 2JE, England
and Associated Companies throughout the world

First published 1985
Revised reprint 1989
Reprinted 1991,
Reprinted by Longman Scientific & Technical 1993, 1994, 1995

ISBN 0 582 23159 0

British Library Cataloguing in Publication Data
A CIP record for this book is available from the British Library

Produced by Longman Singapore Publishers Pte Ltd
Printed in Singapore

PART A

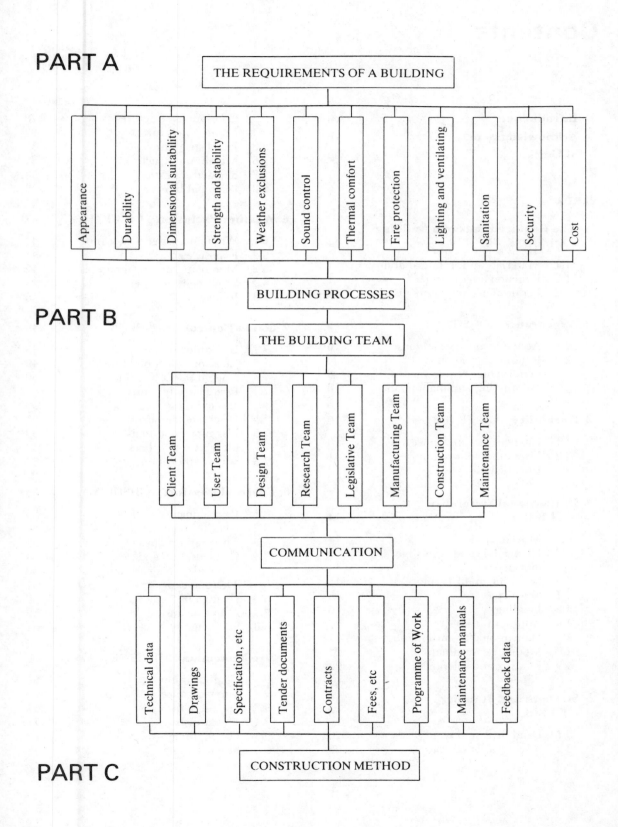

THE REQUIREMENTS OF A BUILDING

- Appearance
- Durability
- Dimensional suitability
- Strength and stability
- Weather exclusions
- Sound control
- Thermal comfort
- Fire protection
- Lighting and ventilating
- Sanitation
- Security
- Cost

BUILDING PROCESSES

PART B

THE BUILDING TEAM

- Client Team
- User Team
- Design Team
- Research Team
- Legislative Team
- Manufacturing Team
- Construction Team
- Maintenance Team

COMMUNICATION

- Technical data
- Drawings
- Specification, etc
- Tender documents
- Contracts
- Fees, etc
- Programme of Work
- Maintenance manuals
- Feedback data

CONSTRUCTION METHOD

PART C

Contents

CONTENTS

Introduction

This volume is intended for those who are commencing a serious study of the various mental and physical processes which are involved during the creation of a building. It is, therefore, primarily for students of building, architecture and interior design. It provides a basic introduction to building: each section is designed to stimulate interest and encourage further reading from the other more advanced volumes of Mitchell's Building Series. A bibliography is provided for the purpose at the end of each chapter which also includes relevant BRE Digests, obtainable from the Building Research Establishment, Garston, Watford WO2 7JR.

The factors involved in the creation of a building are complicated, numerous and varied. To understand them in detail it would first be necessary to clearly identify each, and after placing in a sequential order of dependence, provide a thorough analysis of the role they play. This would be an onerous task, since the factors are closely interdependent and generate 'sub-factors' which make a structured sequential order somewhat arbitrary. However, as an *introduction* to the subject, the various factors can be combined and simplified, an order established which relates to purpose and elementary knowledge, and a brief description given to clarify some of the considerations.

The main divisions which have been chosen for this book are as follows:

PART A: an analysis of a building in terms of what it is expected to do – its functional and performance requirements.

PART B: an analysis of a building in terms of the processes required, the building team which implements them, and the methods used for communicating information.

PART C: an analysis of a building in terms of one typical construction method.

Acknowledgment

The publishers and I acknowledge with thanks the Controller of Her Majesty's Stationery Office for permission to quote the Building Regulations 1985 and the Directors of the Building Research Establishment, the British Standards Institute, and the research and development associations for permission to quote from their publications.

The performance requirement headings derive from *Principles of Building* published by HMSO; figure 5 is based on drawings in Mitchell Beazley's *Great Architecture of the World*; and several drawings orginate from the Brick Development Association, *Architects' Journal*, and the RIBA (see captions). I am grateful to RIBA Publications Ltd for figures 115(a) and (b), and also to them, as well as the CSD, for permission to reproduce standard contract forms in Appendix C. The computer drawings of figures 105 and 112 were generously supplied by GMW Computers Ltd in co-operation with Diamond Redfern and Partners, and the information for figures 41, 42, 43, 44 and 46(b) was readily given by J E Moore. Permission was given by Professor Peter Burberry to include figures 47, 54 and 176 from *Environment and Services*, and by Jack Stroud Foster to include figures 13, 166 and 168 from *Structure and Fabric* Part 1 and figures 11 and 82 from Part 2.

David Clegg of RIBAS Ltd gave invaluable guidance on the CI/SfB classification system. Teaching in the subject of this book, it is inevitable that I have also drawn upon much accumulated data supplied by others as documents or during discussions over many years. I apologize to those not credited and can only offer my sincere thanks.

George Dilks, assisted by Jean Marshall, prepared the final drawings and I owe thanks for the skilful and tolerant way in which sketches were interpreted. I am especially grateful to Nori Howard-Butôt for not only typing and retyping the draft text on many occasions, but also for making many important suggestions regarding content. Lastly, of course, my thanks go to the publishers and particularly to Thelma M Nye for her enthusiasm, patience and extremely able editorial assistance.

London 1984   DO

SI units

All quantities in this volume are given in SI units, which have been adopted by the United Kingdom for use throughout the construction industry as from 1971.

Traditionally, in this and other countries, systems of measurement have grown up employing many different units not rationally related and indeed often in numerical conflict when measuring the same thing. The use of bushels and pecks for volume measurement has declined in this country but pints and gallons, and cubic feet and cubic yards are still both simultaneously in use as systems of volume measurement, and conversions between the two must often be made. The sub-division of the traditional units vary widely: 8 pints equal 1 gallon, 27 cubic feet equal 1 cubic yard; 12 inches equal 1 foot; 16 ounces equal 1 pound; 14 pounds equal 1 stone, 8 stones equal 1 hundredweight. In more sophisticated fields the same problem existed. Energy could be measured in terms of foot pounds, British Thermal Units, horsepower, kilowatt hour, etc. Conversion between various units of national systems were necessary and complex, and between national systems even more so. Attempts to rationalise units have been made for several centuries. The most significant stages being:

The establishment of the decimal metric system during the French Revolution.

The adoption of the centimetre and gramme as basic units by the British Association for the Advancement of Science in 1873, which led to CGS system (centimetre, gramme, second).

The use after approximately 1900 of metres, kilograms, and seconds as basic units (MKS system).

The incorporation of electrical units between 1933 and 1950 giving metres, kilograms, seconds and amperes as basic units (MKSA system).

The establishment in 1954 of a rationalised and coherent system of units based on MKSA but also including temperature and light. This was given the title *Système International d'Unites* which is abbreviated to SI units.

The international discussions which have led to the development of the SI system take place under the auspices of the Conference General des Poids et Mesures (CGPM) which meets in Paris. Eleven meetings have been held since its constitution in 1875.

The United Kingdom has formally adopted the SI system and it will become, as in some 25 countries, the only legal system of measurement. Several European countries, while adopting the SI system, will also retain the old metric system as a legal alternative. The USA has not adopted the SI system.

The SI system is based on six basic units:

Quantity	Unit	Symbol
Length	metre	m
Mass	kilogramme	kg
Time	second	s
Electric	ampere	A
Temperature	degree Kelvin	°K
Luminous intensity	candela	cd

The degree Kelvin will be used for absolute temperatures, for customary use the degree Celsius (°C) will still be used and for temperature intervals (difference between two temperatures) degrees Celsius (°C) will also be used. (273.15°K is 0°C and °K and °C are identical in terms of temperature intervals.) In addition to the basic units there are two supplementary units:

Quantity	Unit	Symbol
Plane angle	radian	rad
Solid angle	steradian	sr

Degrees °, minutes ' and seconds " will also be used as part of the system.

From these basic and supplementary units the remainder of the units necessary for measurement are derived, eg:

Area derived from length/m^2.

Volume derived from length/m^3.

Velocity derived from length and time/m/s.

Some derived units have special symbols:

Quantity	Unit	Symbol	Basic units involved
Frequency	hertz	Hz	1 Hz = 1/sec (1 cycle per sec)
Force, energy	newton	N	1 N = 1 kg m/s^2
Work, quantity of heat	joule	J	1 J = 1 Nm
Power	watt	W	1 W = 1 J/s
Luminous flux	lumen	lm	1 lm = 1 cd sr
Illumination	lux	lx	1 lx = 1 lm/m^2

Multiples and submultiples of SI are all formed in the same way and all are decimally related to the basic units. It is recommended that the factor 1000 should be consistently employed as the change point from unit to multiple or from one multiple to another. The following table gives the names and symbols of the multiples. When using multiples the description or the symbol is combined with the basic SI unit eg, kilojoule kJ.

Factor		Prefix	
		Name	Symbol
one million million (billion)	10^{12}	tera	T
one thousand million	10^9	giga	G
one million	10^6	mega	M
one thousand	10^3	kilo	k
one thousandth	10^{-3}	milli	m
one millionth	10^{-6}	micro	μ
one thousand millionth	10^{-9}	nano	n
one million millionth	10^{-12}	pico	p

It will be noted that the kilogram departs from the general SI rule with respect to multiples, being already 100 g. Where more than three significant figures are used it has been United Kingdom practice to group the digits into three and separate the groups with commas.

This could lead to confusion with calculation from other countries where the comma is used as a decimal point. It is recommended therefore that groups of three digits should be used separated by spaces, not commas. In the United Kingdom the decimal point can still, however, be represented by a point either on or above the bottom line.

PART A

An analysis of a building in terms of what it is expected to do – its functional and performance requirements

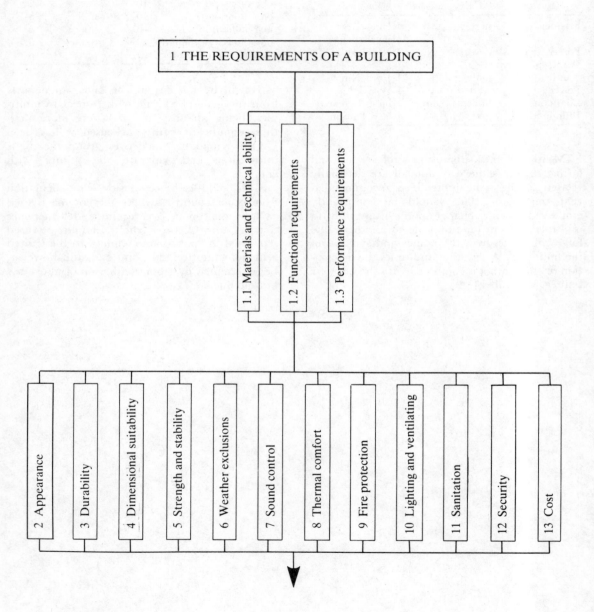

1 THE REQUIREMENTS OF A BUILDING

- 1.1 Materials and technical ability
- 1.2 Functional requirements
- 1.3 Performance requirements

- 2 Appearance
- 3 Durability
- 4 Dimensional suitability
- 5 Strength and stability
- 6 Weather exclusions
- 7 Sound control
- 8 Thermal comfort
- 9 Fire protection
- 10 Lighting and ventilating
- 11 Sanitation
- 12 Security
- 13 Cost

1 The Requirements of a Building

1.1 Materials and technical ability

When a building is constructed two main physical resources are involved. These are, *materials* necessary to form the various parts, and *technical ability* to assemble the parts into an enclosure (figure 1). Initially, the materials employed were those which could most easily be obtained from the accessible areas of the surface of the earth. The technical ability was mostly simple, having evolved from the convenient methods of economically working the rudimentary characteristics of these available materials. The gradual widening in means of communication and corresponding development in attitudes, led to an increased range of these resources becoming available.

The current uses of particular construction methods no longer need to rely on locally available materials or traditional technical ability. Continued investigation has resulted in the enormous range of materials now becoming available which may be used singly, in combination with one another, or even to form new materials. Technological developments are inter-related with this range and use of materials, and enable virtually anything to be constructed.

Nevertheless, there are certain considerations which have always exerted some control on the indiscriminate use of resources. These controls remain, and now that the range of resources is wider, and attitudes towards the function of a building are more complicated, the selection of appropriate construction method becomes much more difficult. For this reason, it is first necessary to understand precisely what is required of a building before selecting an appropriate method of construction.

1.2 Functional requirements

Elaborate shelters have been, and still are, made by most species of insects, reptiles and animals capable of using the readily available materials (earth, stones, branches and leaves, etc) with the aid of the inherent manipulative skills (technology) of their arms, legs, wings, claws, beaks and jaws. Early man also required shelter which provided security for him, his possessions and activities. However, he developed his inherent manipulative skills by inventing tools which led to less indiginous construction methods and also ways of changing the natural state of materials so that they could be used to greater advantage. Each innovation devised usually

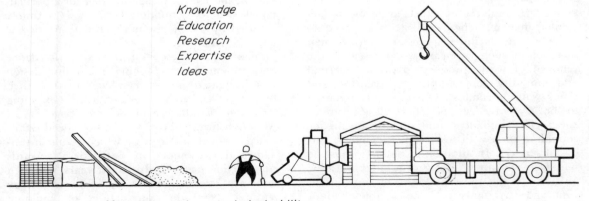

Knowledge
Education
Research
Expertise
Ideas

Materials   +   technical ability

Figure 1   Construction method

resulted in shelters, which, although initially providing a good standard of comfort and convenience, eventually became substandard accommodation as requirements became more elaborate. But, regardless of technical developments, the provision of a physically comfortable shelter was not the only or even the principal reason for building. From early times, a building was also required to give an established place of social and religious *identity*: it must indicate culture, status and mood, whilst creating the humanized space in which to learn, experience and carry out normal daily functions in comfort.

Sir Henry Wooten, a fifteenth century humanist who adapted the writings of Vitruvious for his book, *The Elements of Architecture* (1624), wrote that a good building must satisfy three conditions:

Commodity (comfortable environment conditions);
Firmness (stability and safety); and
Delight (aesthetic and psychological appeal).

These *functional requirements* are implicit in the provision of a shelter which is also a building fit for human habitation. *A well constructed building reflects contemporary attitudes towards environmental control, structural concepts and aesthetic excellence.* And the materials and technical ability used throughout history have normally provided the means of achieving these particular ends.

1.3 Performance requirements

A modern building is expected to be a life-support machine (figure 2). It is required to provide the facilities necessary for human metabolism such as clean air and water, the removal of waste produce, optimum thermal and humidity control, privacy, security, and visual/acoustic comfort. It is generally required to be a source of (perhaps self-generating) energy for appliances, and provide means for communication with television, telephones, and postal services. In addition, a building must be safe from collapse, fire, storm, and vermin; resistant to the physical forces of snow, rain, wind and earthquakes, etc; be capable of adaption to various functions, external landscaping or internal furniture arrangements. It must also

be easily, economically, quickly and well constructed; and allow easy maintenance, alterations and extension. All this must be accomplished in the context of providing a building which has character and aesthetic appeal.

Criteria of this nature form today's interpretation of the basic functional requirements for a building quoted earlier. In order for them to be conveniently considered, it is necessary to divide a building into the various related duties to be fulfilled and establish the precise *performance requirements* for each. When these duties are incorporated into a building where the *functional requirements* have been clearly defined, the selection of a suitable construction method (materials and technology) can be achieved by using the criteria given by the *performance requirements* under the following headings:

Appearance
Durability
Dimensional suitability
Strength and stability
Weather exclusion
Sound control
Thermal comfort
Fire protection
Lighting and ventilation
Sanitation
Security
Cost

Performance requirements cannot be placed in order of importance because any one of them may be more critical than another for a particular element of a building. Priority is normally dictated by the precise function and location of a specific building.

The use of these inter-related performance requirements in establishing a building design was instigated many years ago by the then Building Research Station (now Building Research Establishment – see page 122). Although firm principles have now been established, critical factors arise which result in fundamental changes in attitudes towards construction methods. In recent years, one such influence concerns the availability of energy resources for the production of building materials, and for the heating and lighting of buildings.

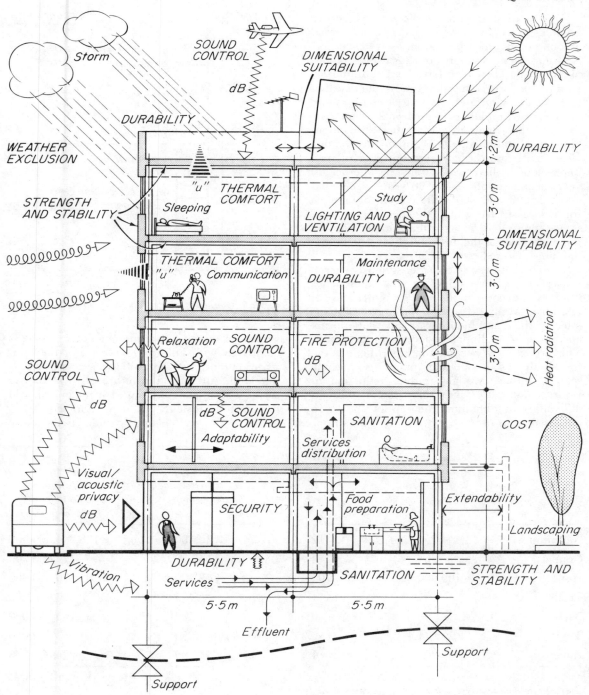

Figure 2 Performance requirements for a building

During the formation of the earth some five million years ago, only a relatively small amount of hydrocarbon atoms were incorporated and these now form our fossil fuels of gas, oil and coal. These fuels have been continuously used in one form or another during the development of mankind and the rate of consumption has increased rapidly over the last 40 years. As a result of this situation, even if the world population was stabilised, the requirements for fuel were static, and the poorer countries remained undeveloped, there would only be another 40 years of gas supply, 25 to 30 years of oil, and between 200 and 300 years of coal obtainable through easy access. Of the net energy consumed, more than 50% is subject to decisions by those involved in the design and construction of buildings (figure 3).

The current need to consider seriously energy conservation is resulting in new approaches towards the construction and maintenance of buildings. Many prejudices and preconceptions must be shed, and buildings of different appearance to those traditionally accepted must result. This premise will become more pertinent as the commonly used building materials become scarcer, either through gradual reduction in world availability or (more likely) through difficulties in obtaining them from the more remote regions of the earth. Figure 4 illustrates

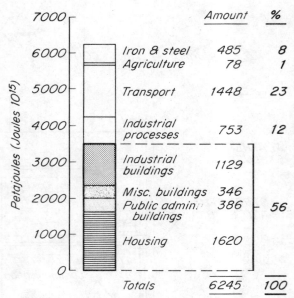

Figure 3   Amount of energy consumed by the building industry (measured in petajoules)

the rate of annual consumption for certain metals compared with known reserves, and the total natural occurrences beneath the oceans as well as dry land. Although the future of these metals seems guaranteed for many year to come, it will become increasingly difficult and costly to establish the technology to mine in the more

Metal	Known reserves	Current annual consumption	Total natural occurrence in oceans including the seabed	Total natural occurrence in the first mile of the earth's crust (under dry land)
Aluminium	4,250	12	15,500	138,244,000,000
Copper	364	8	4,650	119,000,000
Iron	109,000	700	12,400	84,630,000,000
Lead	94	4	465	27,400,000
Nickel	90	0.5	3,100	135,960,000
Tin	4.4	0.25	4,650	67,760,000
Zinc	306	4.5	15,500	224,000,000

(Million Tonnes)

Source of data on reserves: Commodities Research Unit London

Figure 4   Mineral content of oceans and the earth's crust related to present consumption

inaccessible regions. It is also highly probable that any future winning of these hitherto inaccessible minerals will result in dramatic ecological changes in the earth's surface and atmosphere. Although aggregates and cements for concretes, clays for bricks, and timbers are similarly likely to remain in worldwide abundance for many years, their retrieval for conversion into building materials is already becoming limited by environmental conservation issues. On a positive note as far as timber is concerned, world-wide attempts are being made to reduce indiscriminate de-forestation, particularly the equatorial rain forests. Efforts are also being made to replenish the already depleted timber stocks in Europe through re-planting, and, in future, timber may well become the critical building material.

Nevertheless, the development of substitutes for the traditional building materials are likely to increase and these will result in the need to re-assess construction methods. These changes will be paralleled by changes in the performance requirements for the whole building. As already inferred, research and development is continually taking place in these areas and it is the duty of those involved in the design and erection of buildings to be aware of current trends. The achievement of this knowledge cannot be solely through any published building law or contemporary code of good practice, as by the time they have been established other discoveries and experiences may already have taken place.

Further specific reading

Mitchell's Building Series

Environment and Services: Chapter 1 *General*
Materials: Introduction
           Chapter 1 *Properties generally*
Structure and Fabric Part 1: Chapter 1 *The nature of buildings and building*
Components: Chapter 1 *Component design*

Building Research Establishment publication

Principles of Modern Building, Vols 1 and 2, HMSO

2 Appearance

The appearance of a building is initially determined by the activities to be accommodated, as these strongly influence the scale and proportion of the overall volumetric composition (figure 5). The shape of the individual spaces forming the collective volume are defined by 'boundaries' which become the walls, floors, roofs, etc. These are ultimately required to conform with precise *aesthetic and technical* criteria, and the materials employed for these purposes are as numerous and varied as the methods which can be adopted for their use. However, the underlying principle remains that both aesthetic and technical criteria are affected by the composition, form, shape, texture, colour and position of the materials employed. To this must be added the skill with which they are placed in a building (craftsmanship) and the cost, since these factors often provide a deciding role.

2.1 Aesthetic aims or fashions

In a well designed building, appearance is the reflection of a balance between *aesthetic aims or fashions* and the *construction method* derived from the desire for optimum environmental control, structural stability and logical techniques of instigation. Building designers of an earlier and simpler time than today had available only a comparatively limited choice of technical resources, which through the influences of socio-cultural attitudes, led to particular 'styles' in the appearance of buildings. In contrast, the current range of resources, more complicated performance requirements, and widening of cultural influences have resulted in multifarious 'styles'. Some of these evolve from an overwhelming bias towards 'high technology' and the absolute economies of industrial forms. However, if technology is allowed to dominate the aesthetics of a building without compromise, there is a loss of human understanding, scale and proportion and, perhaps, colour, form and texture. By way of comparison, some 'styles' endeavour to imitate the psychological appeal associated with the less complicated require-

ments of the past. Sometimes this may be done on the premise that a building of today, with its highly technological requirements, is beyond the aesthetic comprehension of man. The technological advantages may be accepted, but they are combined with efforts which deny their visual influences as well as reduce their performance efficiency.

A building design will invariably be unsuccessful if it relies on purely technology for aesthetic appeal, or purely out-moded conventions of a past era. A skilful building designer, whatever his/her aesthetic leanings, resolves conflicts through detailed understanding and sympathetic consideration of *all* the performance requirements. None must be ignored or denigrated.

2.2 Relationship to other performance requirements

Although appearance is only an aspect of the total aesthetic quality, it is generally *the* one on which most first impressions of a building are formed. The majority of users have definitive ideas about what a building should look like, and apart from rules concerning 'correctness', their satisfaction or otherwise may be entirely subjective. In this respect, therefore, the requirement for 'appearance' to some extent contrasts with other performance requirements – although from a designer's point of view it is the factor which unites them all.

For this reason, reference is made in most of the chapters of the book to the effects on appearance arising from decisions about other performance requirements. This serves in reinforcing the view that appearance and function are inseparable in a building although the degree to which this may be recognised depends upon the skill of the designer. Even certain forms of 'decorative motifs' may be based on a requirement to provide solar shading devices, or perhaps even form the structural tie between two separated flank walls. The appearance of a modern building should not rely on 'functional'

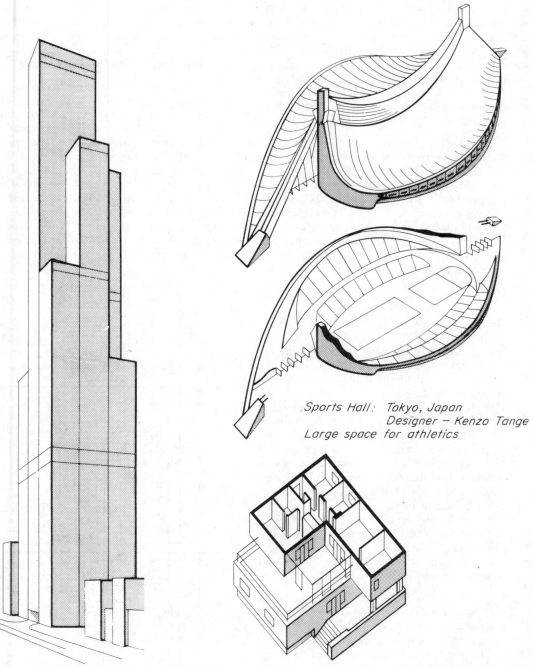

Sports Hall: Tokyo, Japan
            Designer – Kenzo Tange
Large space for athletics

Office block: Sears Building, Chicago, USA
            Designer – B. Graham
Cellular spaces on confined urban site

House: Dessau, East Germany
            Designer – Walter Gropius
Spaces for living

Figure 5   The effect of building function on appearance

or 'decorated functional' aspects alone. Properly controlled and sensitively located non-functional items (in terms of technical performance) – decorations, murals and sculptures – can be incorporated as an essential part of the overall aesthetic achieved by a building.

Some of the basic areas of consideration affecting the appearance of a building, both *externally and internally*, are as follows:

- the aesthetic objectives of the designer in terms of preferred form, shapes, pattern, texture and colour, etc;
- the effects of location and siting on the design and construction methods adopted with particular reference to Town Planning and Building Regulation requirements;
- the design as part of the larger composition of the area – harmony with adjacent buildings and/or specific features including landscaping;
- the 'viewing distances' applicable to the design;
- the use of a particular structural organisation;
- the use of materials which are suitable for particular application which enhance, modify or even change the appearance of the design as it ages;
- the relationship of window/door openings and the creation of a rhythm in the design;
- architectural detailing used to reinforce the character required by the design and a location eg the use of particular types of window/door lintol construction, the creation of shade and shadows through location of components;
- the positioning of ductwork, service pipes etc and their contribution towards aesthetic character;
- the effects of maintenance on the initial design and subsequent use of a building (see 2.3).

It is important that the aesthetic achieved by the smaller parts of a building is a reflection of the same design philosophy applied to its larger parts. The whole building, internally and externally (including approaches and landscaping) should display a similar character, taste, interest, wealth and asperation. If sympathetically conceived and constructed, a homogenous environment is created which is understandable to users, establishes interest and develops taste for good design and construction generally.

2.3 Weathering and maintenance

More specific reference must be made about the necessity to anticipate the effects of future weathering and maintenance on the appearance of a building. Well designed and constructed, a building should accommodate the progress of time without causing a lessening of any functional requirement (see chapter 3 *Durability*). Furthermore, the inevitable effects of weathering should make a positive, rather than a negative, contribution to the appearance of a building. This relies on carefully detailed constructional solutions which derive from a thorough understanding of the behaviour of

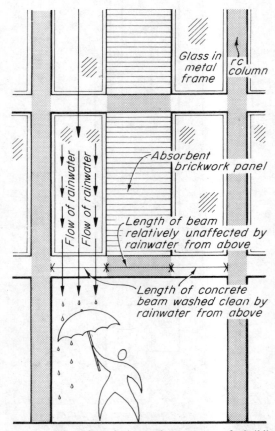

Figure 6   Effect of weathering on the appearance of a building

materials to be used in a particular manner. Figure 6 illustrates a common design 'fault' which has marred the appearance of a building. The impermeable glass and metal frame surface within the opening between brick panels allows water to be caught, run over and clean the concrete beam immediately below. A lesser amount of rainwater will flow over the part of the beam occurring below the brick panels because of the greater absorption properties of the bricks. The result is a striped staining of the concrete beam which changes the appearance of the building in a manner presumably never intended by the designer. The provision of an adequate cill and drip below the opening to prevent the free flow of water over the beam face would have helped to overcome this problem. Alternatively, a continuous gutter could have been provided along the top of the beam, or the beam faced with a material less susceptable to staining.

Further specific reading

Mitchell's Building Series

Materials: Chapter 1 *Properties generally* (Appearance)
Structure and Fabric Part 1: Chapter 1 *The nature of buildings and building*
Components: Chapter 5 *Doors* 5.2(a) Appearance
Chapter 6 *Windows* 6.2(a) Appearance
Chapter 7 *Glazing* 7.2(a) Appearance
Chapter 11 *Demountable partitions* 11.2(a) Appearance
Chapter 12 *Suspended ceilings* 12.2(a) Appearance
Chapter 13 *Raised floors* 13.2(a) Appearance
Chapter 14 *Roofings* 14.2(a) Appearance
Finishes: Chapter 2 *Polymeric Materials* 2.1 Texture and colour
Chapter 3 *Ceramic Materials* 3.1 Internal and external tiling
Chapter 4 *Composites* 4.1 Floorings
Chapter 5 *Metallic Materials*
Chapter 6 *Endpiece*

Building Research Establishment Digests

BRE Digest 45: *Design and appearance* Part 1
BRE Digest 46: *Design and appearance* Part 2
BRE Digest 176: *Failure patterns and implications*
BRE Digest 226: *Thermal visual and acoustic requirements in buildings*
BRE Digest 269: *The selection of natural building stone*
BRE Digest 280: *Cleaning external surfaces of building*
BRE Digest 286: *Natural finishes for exterior timber*
BRE Digest 296: *Timbers: their natural durability and resistance to preservative treatment*

3 Durability

Apart from daily wear and tear by its users, a building is subjected to the constant influences of climate (wind, rain, snow, hail, sleet, sunlight); perhaps attack from vandals and vermin, or even damage by fire, explosions and structural movements. Both the inside and the outside of all buildings are, therefore, subject to forces which can cause deterioration during their life. Durability is the measure of the rate of deterioration resulting from these and other forces.

3.1 Changes in appearance

When related to external climatic or environmental factors, the durability aspects of a building are known as *weathering*. The action of frost, temperature variations, wind and rain on the materials of a building can cause *changes in appearance* by gradual erosion and/or the transportation of atmospheric pollutants which cause staining. Unless the building is carefully designed and detailed, these visual changes seldom enhance appearance, and, therefore, result in disfiguration. The degree to which a building will suffer from erosion and/or staining depends upon many inter-acting variables. One involves the relationship between type and amount of atmospheric pollution with the exposure of a building to wind, rain, frost, snow and solar radiation. Their effects will be determined by the characteristics of the materials used in a building and include their capacity or otherwise for moisture absorption, as well as their surface profiles, orientation, texture, and colour.

In an urban situation it is normal to see dark bands of staining below most horizontal projections (mouldings and cills, etc) of a surface permeable masonry wall. These projections provide a shield against direct rainfall or rainfall run-off from above, and dirt deposits are therefore left relatively undisturbed when compared with the lower regions of the wall (see also examples in figure 6). It is interesting to note that the design of projected mouldings for buildings of the past generally ensured an even weathering to the surrounding vertical surfaces. Nevertheless, although today's buildings are often devoid of mouldings and decorative devices, it is possible for designers to allow water to be guided down a specific route by means of ribbed projections or recessed grooves, and, therefore, create 'controlled' areas of weathering which enhance rather than detract from the appearance of the building. Porous surfaces may accumulate dirt and dust; vertical surfaces stain according to roughness and absorptivity; surface colour affects visual density of staining; but carefully detailed designs can also use these properties to advantage.

It is first necessary to establish the degree of *climatic exposure* which is likely to affect a building (see chapter 6 *Weather Exclusion*). Generally, disfiguration will be slight where there is a moderate rainfall and a moderate rate of pollution, and constructional detailing needs to be far less 'bold' than for areas suffering greater amounts of rainfall and pollution. However, freak weather conditions can result from a group of tall buildings which create a weather pattern between them contrary to accepted predictions. Also, as buildings receive more direct rainfall on their upper storeys, high buildings are generally cleaner at the top and dirtier towards the lower storeys. This is caused by dust being washed down the face of the building and being retained at the point where the porosity of the surface absorbs (or constructional detailing collects and channels) a major part of the water. Low buildings, lacking exposure, will obtain an overall covering of pollution which will be unrelieved by washing. This is preferable to the random streaking to be seen on elevations which are subject to strong dust accumulation and which do not receive direct rainfall in sufficient quantity to provide overall cleansing.

The prevailing rain-carrying winds in the British Isles blow mainly from the south-west and the north-west quarters (figure 7). In non-polluted areas the rain will be clean and will wash the building down. In polluted areas the rain will absorb the dust in the air and thus will

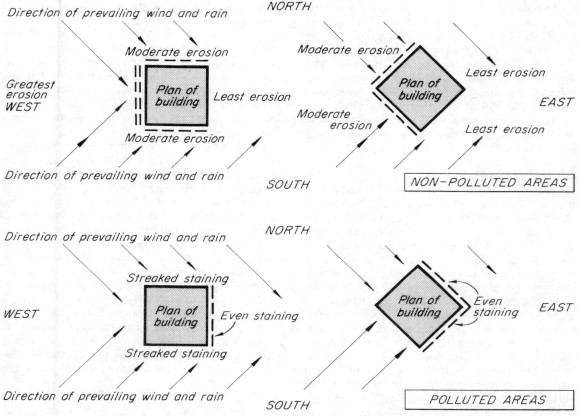

Figure 7   The effect of south-west and north-west rain carrying winds on a building

be dirty when striking the surface. It will also carry dirt from the upper surface to the lower surface and cause staining. South-west and north-west aspects of buildings will generally remain cleaner than the north-east and south-east aspects. North and north-east aspects are particularly liable to severe accumulation of dirt. Thus for a building to be cleaned, the volume of clean rain water must exceed the volume of dirty water passing over the surface – snow has no beneficial cleaning effect.

3.2 Physical deterioration

Apart from causing the staining of a building, certain forms of weathering *also* result in chemical actions which cause physical deterioration by decay. From the time they are placed on a building site, all materials commence on a path of deterioration which could continue after a building has been constructed until a point is reached when they are not fulfilling a useful purpose. Again, the precise form of attack depends upon exposure conditions, and on the susceptibility of the materials used. (For further comment see chapter 4: *Dimensional Suitability*, 4.1 *Movements*.) Corrosion, erosion, and disintegration of materials and construction details can follow from the effects of changes in moisture content, frost, sunlight, soil and ground-water action, atmospheric gasses, electrolytic action, fungal and insect attack, or domestic and industrial wastes, etc. Critical conditions are influenced by the selection of materials appropriate for their design function, detailing for their application, and workmanship for their installation. These provide the 'control' on the extent to which deterioration and decay may be allowed over a given period of time.

DURABILITY

The durability factors concerned with the effects of fire, explosions and structural movements will be mentioned under the appropriate performance requirement to follow. The problems of daily wear and tear by the users, and attack from vermin, vandals and burglars, also form an important part of the design criteria for the building. All involve the careful selection of appropriate materials and functional detailing as well as thoughtful overall planning which will lessen the likelihood of their occurrence.

3.3 Intended life span

Ideally a chosen design and construction method should be able to resist all detrimental effects and provide prolonged durability. As economic considerations make this impossible, the selection of materials and assembly technique for a design should be made to ensure that their rate of deterioration will not impair the functional performance of a building, including appearance, during its *intended life span*. This period is difficult to quantify and, therefore, selection processes are generally influenced by an inter-relationship between *initial costs* of materials and likely future *maintenance costs*. When deciding, it should be remembered that maintenance costs not only include labour and materials for renovation and for cleaning, but also sometimes financial losses resulting from the temporary curtailment of trade of business by the building owner or tenants while remedial work is being executed. Furthermore, where access for maintenance is difficult because of design detailing, consideration must be given to the cost of hiring scaffolding, special plant or even the provision of permanent gantries, mobile or otherwise. The latter can form a dominant feature of the external or internal appearance of a building.

The degree to which subsequent maintenance of a building is possible, desirable, or *necessary* provides the 'fine tuning' which asists in achieving an optimum design in terms of initial running and future costs. A designer must be expert in the choice of materials or carry out necessary research into the physical and chemical properties of them, before devising construction techniques which will take full advantage of their potentialities. Continual rising costs is forcing serious consideration of designing buildings to give a *guaranteed, but limited life span* (limit state design). In this way, the life of a building is predetermined by quantifiable characteristics, including predictability of certain detrimental influences. For example, it would be unnecessarily costly to incorporate an earthquake-resisting structure in a building with an intended useful life of twenty years if the earthquake 'cycle' where the building is to be erected does not fall within this period. If available, the money saved by not using elaborate structural solutions can be effectively used on materials and constructional details with guaranteed performance requirements during the 20 years' life of the building.

This approach can be extended to the study of statistical analysis about freak wind turbulances and amounts of rainfall in otherwise predictable climatic regions. However, financial savings in capital costs of construction methods (materials and technology) can only be made after thorough study and research have revealed the precise degree of durability achievable and therefore the life span of a building with minimal maintenance.

Further specific reading

Mitchell's Building Series

Environment and Services: Chapter 2 *Moisture*

Materials: Chapter 1 *Properties generally* (Deterioration)
Components: Chapter 5 *Doors* 5.2(b) Durability
Chapter 6 *Windows* 6.2(h) Durability and maintenance
Chapter 7 *Glazing* 7.2(h) Durability and maintenance
Chapter 11 *Demountable partitions* 11.2(g) Durability and maintenance
Chapter 12 *Suspended ceilings* 12.2(f) Durability and maintenance
Chapter 13 *Raised floors* 13.2(f) Durability and maintenance
Chapter 14 *Roofings* 14.2(g) Durability and maintenance

Finishes: All Chapters listed under *Appearance*, page 9

Mitchell's Professional Library

Movement Control in the Fabric of Buildings, Philip Rainger

Building Research Establishment Digests

BRE Digest 18:	*Design of timber floors to prevent decay*
BRE Digest 45:	*Design and appearance* Part 1
BRE Digest 46:	*Design and appearance* Part 2
BRE Digest 69:	*Durability and application of plastics*
BRE Digest 89:	*Sulphate attack on brickwork*
BRE Digest 98:	*Durability of metals in natural waters*
BRE Digest 139:	*Control of lichens, moulds and similar growths*
BRE Digest 144:	*Asphalt and built-up felt roofings: durability*
BRE Digest 176:	*Failure patterns and implications*
BRE Digest 177:	*Decay and conservation of stone masonry*
BRE Digest 200:	*Repairing brickwork*
BRE Digest 217:	*Wall cladding defects and their diagnosis*
BRE Digest 238:	*Reducing the risk of pest infestation: design recommendations*
BRE Digest 250:	*Concrete in sulphate-bearing soils and ground waters*
BRE Digest 251:	*Assessment of damage in low-rise buildings*
BRE Digest 263:	*The durability of steel in concrete* Part 1
BRE Digest 264:	*The durability of steel in concrete* Part 2
BRE Digest 265:	*The durability of steel in concrete* Part 3
BRE Digest 268:	*Common defects in low-rise traditional housing*
BRE Digest 269:	*The selection of natural building stone*
BRE Digest 280:	*Cleaning external surfaces of building*
BRE Digest 286:	*Natural finishes for exterior timber*
BRE Digest 296:	*Timbers: their natural durability and resistance to preservative treatment*
BRE Digest 304:	*Preventing decay in external joinery*
BRE Digest 330:	*Alkali aggregate reactions in concrete*

4 Dimensional Suitability

The dimensional suitability of a construction method involves the consideration of two areas:

- the manner in which *movement* of materials causes dimensional variations in a building, or parts of a building during its life; and,
- *appropriate sizes* for the parts of a building which suit the materials available to fulfil specific design functions, cost ratios, manufacturing processes and assembly techniques.

4.1 Movement

A building never remains inert because dimensional changes will occur in the materials from which it is made as a result of changes in environmental conditions and/or loading. Variations in moisture content and temperature produce movements in a building which tend to occur in relation to the stronger 'fixed points' in the building – between foundations and first floor; between top floor and roof; between partitions and main structure; or between panels and supporting frames (figure 8).

4.2 Irreversible and reversible movement

The moisture content of porous building materials can cause *irreversible* movement or *reversible* movement. The former is generally associated with establishing a 'normal' or atmospheric moisture level in the materials of a component which have recently been manufactured. For example, clay bricks leaving a kiln will be very dry and will immediately begin to absorb moisture from the air which causes expansion. Conversely, calcium silicate bricks will be more saturated than normal bricks because they are cured by autoclave processes and will immediately shrink after manufacture as their moisture content moves towards an equilibrium with that of the atmosphere. For this reason, newly manufactured bricks should not be immediately used for building walls as cracking will inevitably occur (figure 9). Reversible moisture movement occurs in materials which are in use and

generally involve expansion on wetting and shrinkage on drying. These movements have both immediate and long term effects on the fabric of a building and thoughtful detailing is essential if damage is to be avoided. Care must be taken to ensure movements are reduced to acceptable amounts by limiting the uninterrupted heights and lengths of components and

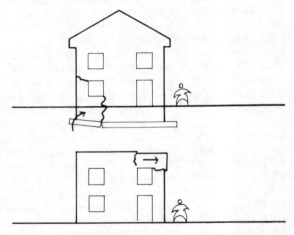

Movements due to effect of moisture and/or temperature change

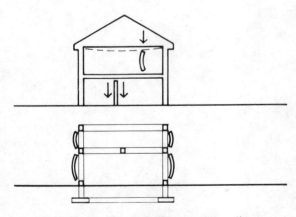

Movements due to effects of loading

Figure 8   *Typical movements likely to occur in a building*

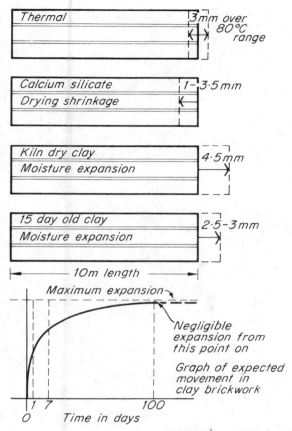

Figure 9 *Moisture and thermal movements in calcium silicat and clay brick walls (Based on Brick Development Association data)*

Severe damage to walls can be caused by attempting to restrain beams and slabs – particularly where temperature ranges are likely to be great. Fortunately, dramatic failures of this nature are not too common. Daily temperature ranges (diurnal) are, however, a frequent cause of damage to a building which can occur immediately in the form of buckling metal cills, cracking glass, etc, or over a period by causing gaps in a weather impermeable construction which allows the free penetration of moisture. The same care needed in constructional detailing and the provision of movement joints, which is required to limit moisture movements, is also necessary when considering thermal movements. In fact, often both problems are closely interrelated. In general terms, irreversible moisture movement in porous building materials is greater than reversible movement, and reversible moisture movement is usually less than movements due to temperature changes.

4.3 Softening and freezing

Detrimental effects resulting from temperature changes are also caused as a result of *softening* and of *freezing*. The majority of materials used for a building will not become softened by normal climatic temperature. However, those containing bituminous or coal tar pitch (eg asphalt and bituminous felts used for roofs, floor finishes, damp proof courses, etc) are liable to become more and more plastic as temperatures rise. This can result in indentation and perforation under load, or in elongation causing their displacement. A bituminous damp proof course can soften sufficient for the load of a wall above to squeeze it outwards from its bedding and even upset the stability of the wall. This problem is most likely to occur on exposed south facing walls.

Freezing causes a rather special form of thermal movement and in this respect is not so dissimilar from chemical attack referred to in chapter 3 *Durability*. In some cases water may penetrate into a structure and freeze along a junction between two materials. In forming ice lenses, the water expands by about 10% and causes considerable damage. Water freezing in air pockets or other fissures in the mortar of brickwork can cause spalling of the joint and

elements. This can be achieved through a precise knowledge of the characteristics of the materials involved and the incorporation of *movement joints* at centres beyond which excessive movement is likely to occur (figure 10). These movement joints should be positioned so as to take account of their visual effect on a building. It is important not to confuse changes in the size of materials due to absorption of moisture with the problems associated with moisture movement *through* materials. The latter will be dealt with in chapter 6 *Weather Exclusion*.

Most building materials also expand to a greater or lesser extent with rises in temperature, and if they are restrained could induce considerable stress resulting in cracking, bowing, buckling or other forms of deformation.

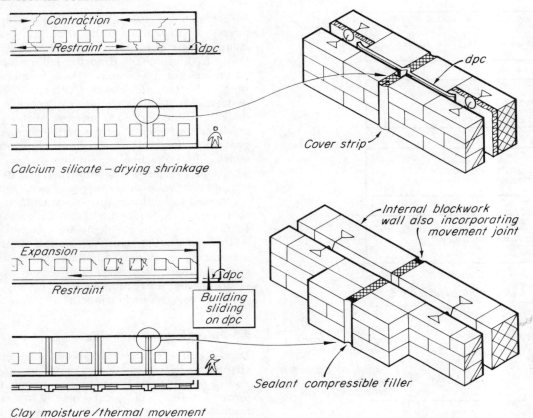

Figure 10   Moisture and thermal movements in brick walls

the adjoining arrises of brickwork. If the water freezes within the body of a material a similar disintegration process will occur. A clear understanding is required about the ratio between amount of water absorbed, and the volume and distribution of pore space available if this phenomenon is to be avoided. The presence of moisture in certain materials, even in very minute quantities can cause chemical changes resulting in movement. The corrosion processes of iron and steel form a layer of rust which, being porous, becomes liable to expansion. This action can cause spalling of the concrete cover when the metal bars of reinforced concrete beams become rusted.

4.4 Shrinkage

By way of contrast, the chemical action resulting in the setting of Portland cement causes a volumetric shrinkage. Constructional detailing should take this into account by allowing adequate gaps or tolerances between *in situ* concrete components and other materials. This is particularly true when detailing the junction between an *in situ*-concrete frame and a brick infill panel which is liable to expansion due to moisture absorption and climatic temperature increases (figure 11).

4.5 Loading

As most materials used for buildings are *elastic* to some degree a certain amount of *plastic flow* or *creep* will occur over a period of many years depending upon the type and amount of load or force to which they are subjected. Regardless of this long-term movement, however, structural elements such as beams and columns of any material, are likely to be subject to initial *deflec-*

16

tion. This will occur even as their superimposed loads accumulate during construction and continue until full loading conditions are achieved. Construction methods should take into account the possibility of these movements by ensuring adequate tolerances between structural and non-structural parts of a building to avoid the effects of crushing and cracking. For example, the details for an internal partition of a framed building should ensure that the performance requirements (particularly strength and stability) are not negated by any deflection inherent in the nature of the materials from which beams and columns are formed (figure 12). If the tendency towards deflection or creep becomes too great in the external fabric of a building, secondary problems could develop resulting from moisture penetration through cracks, etc. Similar problems could arise if a supporting or loading condition of a building is altered and the new conditions cannot be accommodated by the existing construction methods.

4.6 Appropriate sizes

Even the simplest form of building requires thousands of individual components for its construction (figure 13). With traditional building practice, these products were completely unrelated in dimension and necessitated the use of skilled craftsmen to scribe, cut, fill, lap and fit them together on the site. Such techniques were notoriously wasteful of both materials and labour. In order to make a building financially viable today, it is necessary that products are manufactured to sizes which co-ordinate with each other so that they can be assembled on site without the need for alteration. This means that the manufacturers' role becomes much more important as they must ensure that the overall dimensions between various products co-ordinate and jointing methods between each allow connection with other related products. Products which can be easily assembled together using simple jointing methods are, theoretically, likely to be less reliant on exacting site skills to fulfil their intended functions. However, in practice the ease and success of the joint again relies on adequate interpretation by the manufacturer of all performance requirements, labour skills, and actual site conditions.

4.7 Dimensional co-ordination

Apart from providing savings in materials and labour, the use of *dimensionally co-ordinated* products for a building eases the processes of selec-

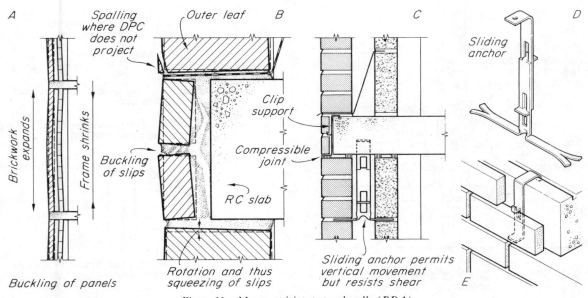

Figure 11   Movement joints to panel walls (BDA)

17

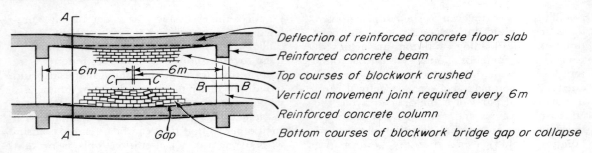

ELEVATION: stability of blockwork internal wall affected by deflecting floor slab

- Deflection of reinforced concrete floor slab
- Reinforced concrete beam
- Top courses of blockwork crushed
- Vertical movement joint required every 6m
- Reinforced concrete column
- Bottom courses of blockwork bridge gap or collapse

Gap

6m   6m

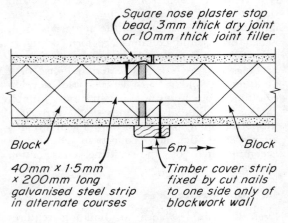

Square nose plaster stop bead, 3mm thick dry joint or 10mm thick joint filler

Block   Block

40mm x 1·5mm x 200mm long galvanised steel strip in alternate courses

Timber cover strip fixed by cut nails to one side only of blockwork wall

6m

PLAN CC: vertical movement joint

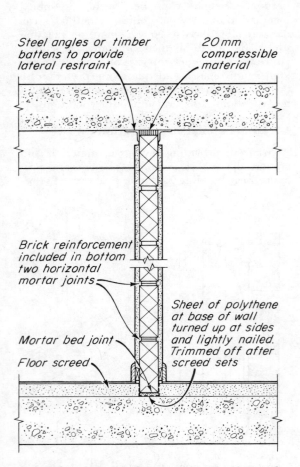

Steel angles or timber battens to provide lateral restraint

20mm compressible material

Brick reinforcement included in bottom two horizontal mortar joints

Sheet of polythene at base of wall turned up at sides and lightly nailed. Trimmed off after screed sets

Mortar bed joint

Floor screed

SECTION AA: detail to overcome effects of deflecting floor slab

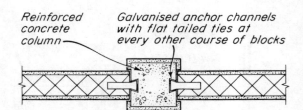

Reinforced concrete column

Galvanised anchor channels with flat tailed ties at every other course of blocks

PLAN BB: detail to allow movement of blockwork at column

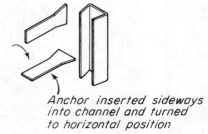

Anchor inserted sideways into channel and turned to horizontal position

Figure 12   Deflection of reinforced concrete beams and the stability of non-loadbearing blockwork internal wall

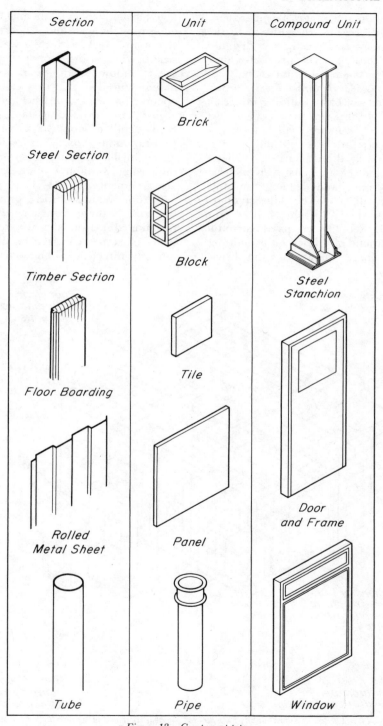

Figure 13   Component types

tion by allowing a greater range of similar items to be available (figure 14). In addition to the availability of home-based products, a designer can select from those abroad providing they follow the same system of dimensional co-ordination. See 15.6 *Manufacturing Team*.

Many of the sizes of basic building materials used today derive through history and are directly related to human scale (anthropometrics). The clay brick, for example, was dimensionally standardized during medieval times relative to its weight, so as to enable easy manipulation by one hand, leaving the other hand free to operate a trowel. This standard size, now translated to 215 mm × 102.5 mm × 65 mm (BS 3921 : 1985), has given rise to the conventional aesthetic of a brick built building. When used in large quantities, the bricks

together with their joints and bonding method, determine the scale and proportion of the overall shape and thickness of the walls as well as the sizes of window and door openings, etc. Provided the window and door components are manufactured so that they can be fitted into an opening formed by the brick dimensions without adjustment, the building shell can be constructed easily and economically. Similarly, internal partition units, floor joists, cupboards, floor and wall finishes should also be available to dimensions suitable to those of brickwork. See also comments under 17.11(b) *Joinery*.

Certain manufacturing processes and some materials cannot conveniently conform with common dimensional standards. For example, walls of stone, concrete or plastics will have different dimensional characteristics to walls of

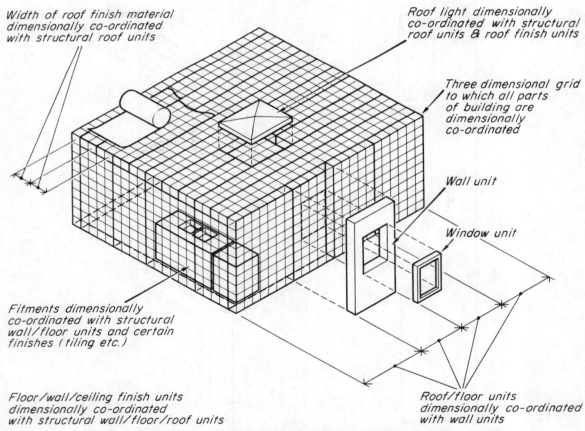

Width of roof finish material dimensionally co-ordinated with structural roof units

Roof light dimensionally co-ordinated with structural roof units & roof finish units

Three dimensional grid to which all parts of building are dimensionally co-ordinated

Wall unit

Window unit

Fitments dimensionally co-ordinated with structural wall/floor units and certain finishes (tiling etc.)

Floor/wall/ceiling finish units dimensionally co-ordinated with structural wall/floor/roof units

Roof/floor units dimensionally co-ordinated with wall units

Figure 14   *The use of dimensionally co-ordinated components which fit together to form a building*

brick. It is necessary, therefore, that manufacturers should be given a guide on the likely
range of dimensions for which particular components will be required, and the variations for
which they should allow within this range.
Accordingly, a range of overall dimensions of
components related to building use, anthropometric requirements and manufacturing criteria
has been recommended. Figure 15 illustrates the
recommendations for the vertical dimensions
used in the construction of a house. There are

similar recommendation for the horizontal dimensions. In order not to overstretch the resources
of manufacturers and further rationalize the
available range of components, BS 6750 : 1986
Specification for modular co-ordination in building
states that components should be manufactured
in basic incremental sizes of 300 mm as first
preference, or 100 mm as second preference with
50 mm and 25 mm being allowed up to 300 mm
(figure 16). For a more detailed explanation see
MBS: Components Chapter 1, *Component Design*.

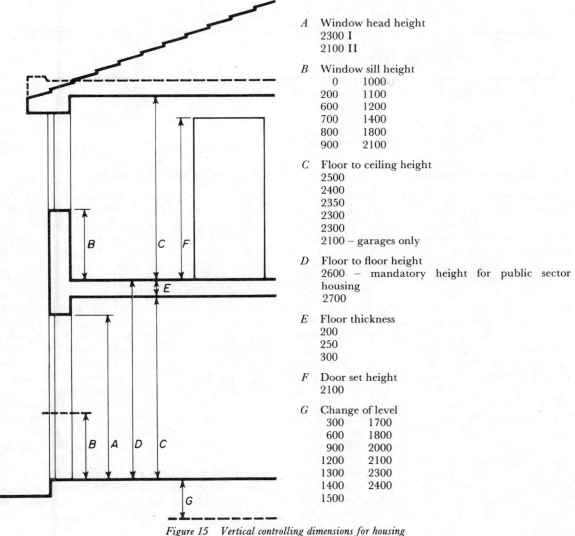

A   Window head height
    2300 I
    2100 II

B   Window sill height
    0   1000
    200   1100
    600   1200
    700   1400
    800   1800
    900   2100

C   Floor to ceiling height
    2500
    2400
    2350
    2300
    2300
    2100 – garages only

D   Floor to floor height
    2600 – mandatory height for public sector
    housing
    2700

E   Floor thickness
    200
    250
    300

F   Door set height
    2100

G   Change of level
    300   1700
    600   1800
    900   2000
    1200   2100
    1300   2300
    1400   2400
    1500

Figure 15   Vertical controlling dimensions for housing

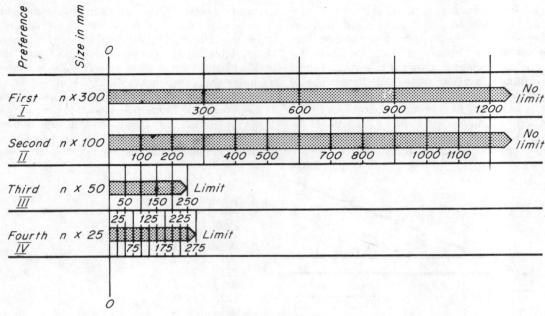

Figure 16   Preferred linear sizes for components used in a building

4.8 Modular co-ordination

Attempts have been made to persuade manufacturers to agree to manufacture components in standardized *modular* increments of 100 mm, which is the dimension most common in building products and at home and abroad. This system is known as *modular co-ordination* and, when adopted, means that a building is designed within a three-dimensional framework of 100 mm cubes. A successful design relies on the certainty that a vast range of products will be available for the construction – a range which can be selected from almost any country adopting the dimensional system. Products requiring to have dimensions greater than the basic module are manufactured so as to be a *multiple of the basic module* (figure 17); or if they are required to be smaller, are manufactured so that when placed together their combined dimensions suit either the basic module or a multiple thereof. Although the idea of modular co-ordination has been in existence for a number of years, many manufacturers are reluctant to change their existing manufacturing equipment to suit a 100 mm modular increment.

4.9 Jointing and tolerances

Whether modular or otherwise, components cannot be manufactured to precise dimensions to suit a regular three-dimensional grid. Allowance must be made for *jointing* the component, as well as for inaccuracies in the products due to *material properties* (shrinkage, expansion, twisting, bowing, etc) and *manufacturing processes* (changes in mould shapes, effects on materials, etc). Furthermore, allowance must be made for *positioning* the components on site. This is particularly important for very large and heavy components which are manoeuvred into position by cranes, etc, and final location could vary by as much as 50 mm. Allowances of this nature are known as *tolerances*: if manufacturers supply products without these tolerances, accumulative errors in trying to place each component on a grid line would not only result in a building of incorrect overall dimensions, but also a great deal of frustration in the fixing of secondary items such as windows, or furniture and finishes within the building (figure 18). An important document covering the aspects of tolerances is BS 5606: 1978 *Code of practice for accuracy in building.*

22

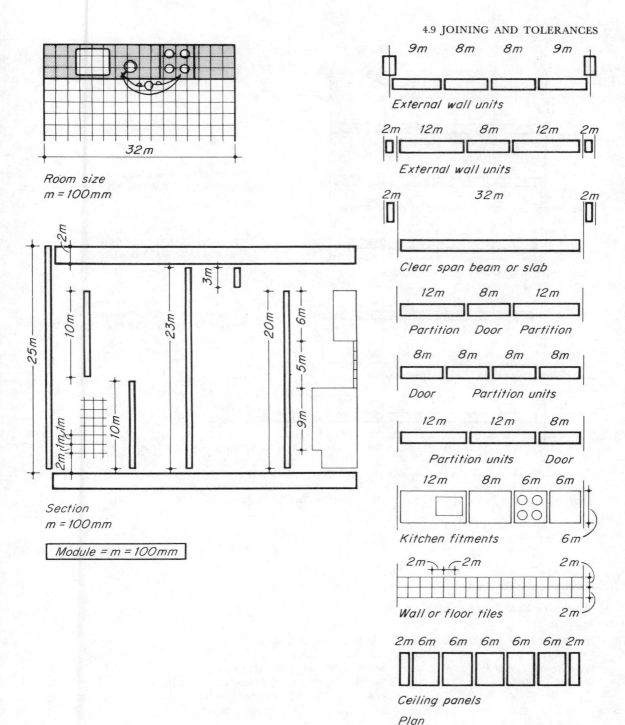

Room size
m = 100mm

32m

External wall units
9m   8m   8m   9m

External wall units
2m   12m   8m   12m   2m

Clear span beam or slab
2m   32m   2m

Partition   Door   Partition
12m   8m   12m

Door   Partition units
8m   8m   8m   8m

Partition units   Door
12m   12m   8m

Kitchen fitments
12m   8m   6m   6m   6m

Wall or floor tiles
2m   2m   2m   2m

Ceiling panels
2m 6m   6m   6m   6m   6m 2m

Plan
m = 100mm

Section
m = 100mm

Module = m = 100mm

25m   2m(1m)1m   10m   10m   3m   23m   20m   5m 6m   9m

Figure 17   The use of modular co-ordinated sizes in building (m = 100 mm)

23

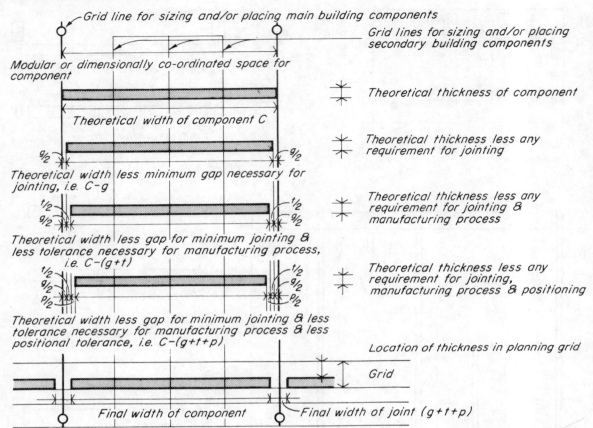

Figure 18   Tolerances allowed when designing components. Note: this illustrates a simplified method of arriving at the 'final dimension' of a component and is based on BS 3626 : 1965 which is now withdrawn. More accurate (and more complicated) methods of calculations are now available in BS 6954 : 1988 Tolerances for building *Parts 1, 2 and 3*

Further specific reading

Mitchell's Building Series

Materials: Chapter 1 *Properties generally* (Deformations)
Structure and Fabric Part 1: Chapter 2 *The production of buildings*
                                      Chapter 2 *Structural behaviour*
Structure and Fabric Part 2: Chapter 4 *Walls and piers* (Movement control)
                                       Chapter 5 *Multi-storey structure* (Movement control)
                                       Chapter 6 *Floor structures* (Movement control)
                                       Chapter 9 *Roof structures* (Movement control)
Components: Chapter 1 *Component design*
                    Chapter 2 *Economic criteria*
                    Chapter 3 *Industrialized system building*

Mitchell's Professional Library

Movement Control in the Fabric of Buildings, Philip Rainger

Building Research Establishment Digests

BRE Digest 35:   *Shrinkage of natural aggregates in concrete*
BRE Digest 75:   *Cracking in buildings*
BRE Digest 137:   *Principle of joint design*
BRE Digest 157:   *Calcium silicate brickwork*
BRE Digest 163:   *Drying out buildings*
BRE Digest 199:   *Getting good fit*
BRE Digest 223:   *Wall cladding: designing to minimize defects due to inaccuracies and movements*
BRE Digest 227:   *Estimation of thermal and moisture movements and stresses* Part 1
BRE Digest 228:   *Estimation of thermal and moisture movements and stresses* Part 2
BRE Digest 229:   *Estimation of thermal and moisture movements and stresses* Part 3
BRE Digest 234:   *Accuracy in setting-out*

5 Strength and Stability CI/SfB (J)

The *strength* of a building refers to its capacity to carry loads without failure of the construction method; *stability* refers to the ability of a building to resist collapse, distortion, localized damage and movement.

5.1 Dead, live and wind loads

A building is required to resist loads imposed by gravity as well as other externally and internally applied forces: loads and forces from roofs, floors, and walls must be transferred by load carrying mechanisms to the supporting ground.

The loads and forces acting on a building are diagrammatically illustrated in figure 19 and comprise of the following:

(a) the weight of all the materials from which it is made (bricks, mortar, concrete, timber, plaster, glass, nails, screws, etc). These weights are more or less constant during the life of a building and are called *dead loads*: they can be calculated from tables for weights of materials, etc;

(b) the weight of people using a building, and their furniture, goods, storage, etc. These weights are called *live loads* or *imposed loads* and, as they will vary, an average maximum load can be assumed from tables giving values applicable to the particular use of a building;

(c) various forces may be applied to a building during its life such as those resulting from wind, physical impact by people, machines, or explosion, and ground movements caused by changes in soil characteristics, earthquakes, mining subsidence, etc. The calculation of these forces is much more problematic and relies on adequate research and experience. Maximum *wind loads* (gusts) for various locations in the

BRITISH STANDARD 648
Weights of Building Materials

Example:
Clay brickwork 51–59 kg/sq m per 25mm
Natural aggregate concrete 2307 kg/cu m
Soft wood 480–672 kg/cu m
Gypsum plaster 22 kg/sq m per 12·7mm
Clay roofing tiles 63 kg/sq m

Soil resistance   *Soil resistance*

Figure 19(a)   Building loads: dead loads

BS 6399 : Part 1: 1984
Code of Practice for dead
and imposed loads

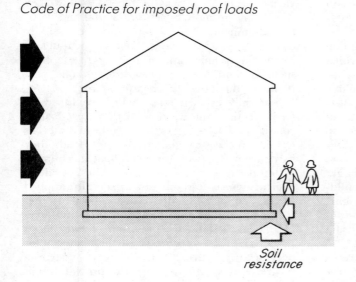

5.1 DEAD, LIVE AND WIND LOADS

Example:

Houses	DL	1·5	CL	1·4 kN/sq m
Churches		3·0		2·7 kN/sq m
Gymnasia		5·0		3·6 kN/sq m
Offices		5·0		4·5 kN/sq m
Shops		4·0		3·6 kN/sq m

DL = distributed load
CL = concentrated load

Soil
resistance

Soil
resistance

Figure 19(b)   Building loads: imposed loads

BS 6399 : Part 3: 1988
Code of Practice for imposed roof loads

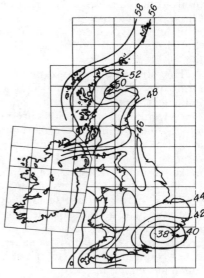

Soil
resistance

U.K. basic wind speeds in m/s
*Maximum gust speed likely to
be exceeded on the average once
in ten years at 10m above open
ground level (Meteorological Office)*

Figure 19(c)   Building loads: wind loads

country have been tabulated, and should be consulted before finalizing the structural requirements for a building.

These loads and forces must be resisted by the supporting soil so that a building remains in equilibrium. A building can be visualised, therefore, as being 'squeezed' between the downward applied loading and the upward reactions of the supporting soil.

5.2 Structural organization

There are basically four methods of structural organization which can be employed in a building to ensure loads, forces, and soil reactions act together in providing equilibrium to a building. The choice of which is most suitable for a particular building is initially dictated by strength and characteristics of the soil providing support, and analysis of the precise nature of all the structural influences. The latter is closely related to the function of the building and involves consideration of such factors as whether long or short spans are required, the height of the building, and the weight of materials necessary to fulfil other performance requirements, etc. Often the juxtaposition of existing buildings provide positive guide lines, if not limitations on the selection of structural form for a new building.

The structural organization of a building forms one of the most important aspects which influences appearance and other functions. Since technological solutions are now available which make almost every structural organization possible, an increasing burden of responsibility is being placed on designers to make rational decisions. The rigorous limitations imposed by simpler construction methods (materials and technology) no longer exist, and it is now feasible to develop a building with volumetric spaces (plan and height) greater than ever before; smaller, to complicated configuration; or to the same size using much less material.

Nevertheless, although perhaps interpreted with greater understanding, certain basic structural principles still remain. Expressed simply, the four basic methods involved in the use of construction methods resist the combined building loads by compression, tension, or a com-

bination of the two. These methods are defined by various terms but here will be called: Continuous structures; Framed structures; Panel structures; and, Membrane structures.

5.3 Continuous structures

These are continuous supporting walls which transfer the combined loads and forces through its construction mainly by direct *compression* (figure 20). Materials commonly used for this purpose are stone, horizontal timber logs, brick, block, or concrete. In this respect, therefore, this may be the oldest form of structural organization.

Walls constructed from fairly small units such as bricks, blocks, or stones rely on their strength by being laid in horizontal courses so that their vertical joints are staggered or *bonded* across the face of the wall (figure 21). In this way the compression loads which may initially affect individual or a series of bricks, blocks, or stones can be successfully distributed through a greater volume of the wall. The units are held together by an adhesive mixture known as *mortar*, thereby completing the structural (and environmental) enclosure. This mortar also serves the function of taking up any dimensional variations in the bricks or blocks so that they can be laid in more or less horizontal and vertical alignment. Most mortars today usually consist of water-activated binding mediums of cement and lime, and a fine aggregate filler such as sand, in the proportion of one part binder to three parts aggregate.

Laterally applied forces which could create *tension* in the wall are resisted by its pre-loaded condition. This is to say, the weight of materials forming the wall together with any loads they carry from floors or roof, combine in counteracting the tendency for horizontal movement or overturning as a result of horizontally applied forces, eg wind. Alternatively, where preloading is insufficient, the action of lateral forces can be resisted by the provision of *buttresses* at predetermined centres to resist the tendency for overturning (figure 20). In this way a system of buttresses or piers can be used to provide stability to a long length of wall. Walls which are serrated or curved in plan will also be stronger than straight walls because they are more able to resist laterally-applied forces.

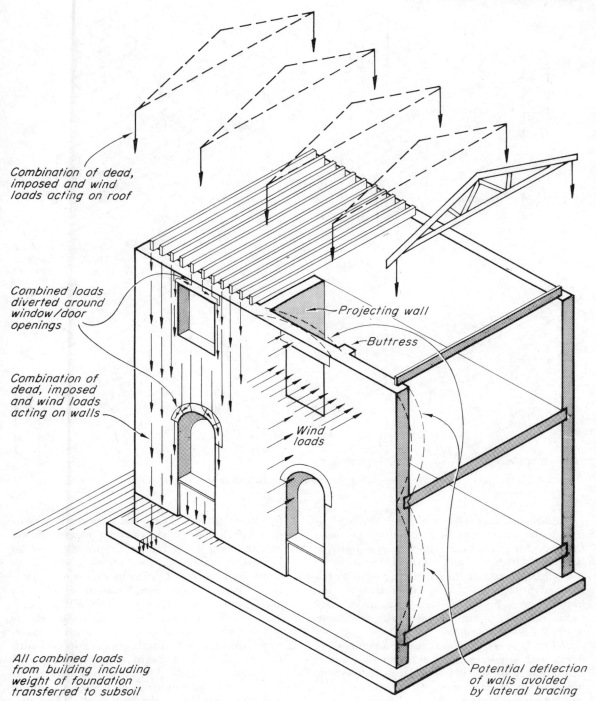

Combination of dead, imposed and wind loads acting on roof

Combined loads diverted around window/door openings

Combination of dead, imposed and wind loads acting on walls

Projecting wall

Buttress

Wind loads

Potential deflection of walls avoided by lateral bracing

All combined loads from building including weight of foundation transferred to subsoil

Figure 20   The transfer of loads in continuous structures

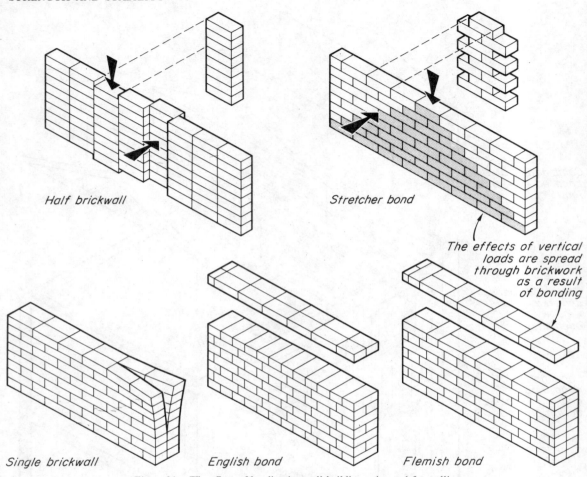

Half brickwall

Stretcher bond

The effects of vertical loads are spread through brickwork as a result of bonding

Single brickwall

English bond

Flemish bond

Figure 21   The effects of bonding in small building units used for walling

Additional stability can also be provided by the floor(s) and roof of a building provided there is adequate connection at the junction between horizontal and vertical elements. A wall which is laterally braced in this manner has the advantage of using considerably less material to support the same load as a thick unbraced wall (figure 22). See also comments under 5.7 *Slenderness ratio*.

When it is required to provide openings (eg doors and windows, etc) in a building using continuously supporting walls, it is necessary to use a beam or *lintol* made from a material or combination of materials capable of resisting both *compression and tension* forces resulting from

the loads above. These loads must be transferred to the sides of the opening or *jambs*. Stone can be used for a lintol but provides only a limited resistance to tensile forces relative to its depth and, therefore, permits only small openings. Timber, steel, a combination of steel bars and concrete (reinforced concrete) or steel angles and bricks can be suitable materials for lintol construction, although care must be taken to ensure that their durability corresponds with the intended life of the building (see chapter 7 *Construction Method*). Arches formed over openings use materials in direct compression only, in a similar manner to the wall itself. Therefore, arched openings are often considered by design

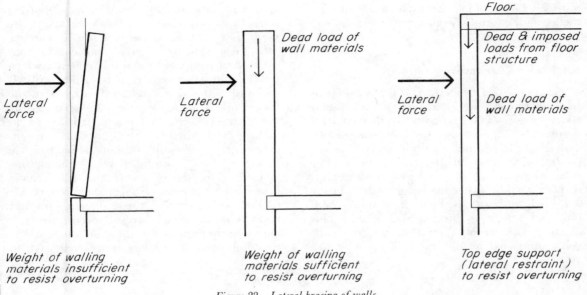

Floor

Lateral force →

Dead load of wall materials

Lateral force →

Dead & imposed loads from floor structure

Dead load of wall materials

Weight of walling materials insufficient to resist overturning

Weight of walling materials sufficient to resist overturning

Top edge support (lateral restraint) to resist overturning

Figure 22   Lateral bracing of walls

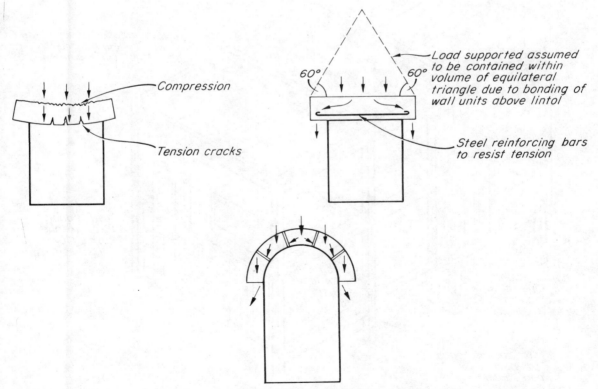

Compression

Tension cracks

60°   60°

Load supported assumed to be contained within volume of equilateral triangle due to bonding of wall units above lintol

Steel reinforcing bars to resist tension

Figure 23   Reinforced concrete lintols and block arches

31

purists to be more compatable with the aesthetic of this form of wall construction (figure 23).

5.4 Framed structures

These consist of a framework of timber, steel or reinforced concrete consisting of a regular system of horizontal beams and vertical columns (figure 24). The beams resist both compressive and tensile forces and transmit loads from the floors, roof and walls to the columns. The columns are required to resist mainly compressive forces and transfer the beam loads (and the self weight of beam and column) to the foundation and finally to the supporting soil. This obviously results in more concentrated loads being supported by the soil than for a similar weight of building using continuous supporting walls, unless special forms of foundations are used. The infil *panels* between the framework used to provide the external wall can be constructed of any suitable durable material which fulfils performance requirements satisfactorily. If the wall material is positioned away from the framework so as to be externally or internally free of the columns and beams, it is known as *cladding*. Both panel and cladding walls are generally non-loadbearing although in practice they must carry their own weight (unless suspended from above); resist the wind forces acting on their external face; perhaps provide support for internal fixtures, shelves, etc; and resist localized impact forces. However, depending upon the form of construction adopted, the resulting loads are usually transferred back to

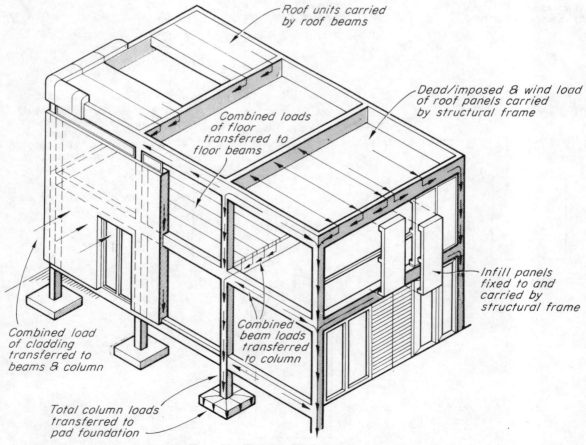

Figure 24 Transfer of loads in home structures

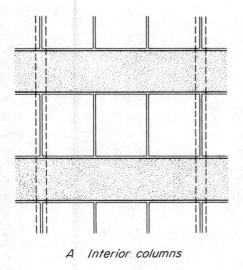

A   Interior columns

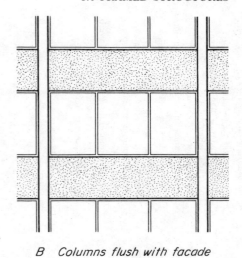

B   Columns flush with facade

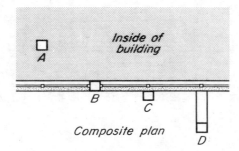

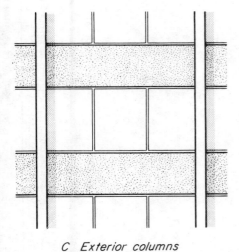

C Exterior columns

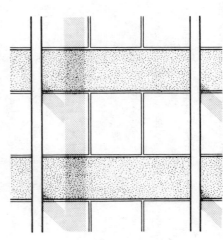

D Exterior columns

Figure 25   Position of structural frame (column) and the effect on appearance of a building. (Based on Multi-storey Buildings in Steel: Hart, Henn and Sontag, Granada)

	TIMBER	STEEL	REINFORCED CONCRETE
Availability	Mostly imported as dimensionally co-ordinated sections	Most iron ore imported	UK supplies of aggregates and cement Iron is mostly imported
Conversion	Factory converted sections and components	Iron works produce bulk sections and components produced in factory under controlled conditions	Factories and iron works convert into basic use materials (aggregates, cement, bar reinforcement)
	Preformed components manufactured to fine tolerances under factory conditions	Components manufactured to fine tolerances under factory conditions	Precast concrete components manufactured to fine tolerances under factory conditions
Site operations	Erected by skilled and semi-skilled operatives	Site erected by skilled operatives	*In situ* structures erected by semi-skilled and unskilled operatives
	Components relatively lightweight and to fine tolerances	Very accurate within small tolerances Heavy sections used need crane	Various shapes possible dependent upon potential of formwork material (up to 40% of total cost) May need crane although *in situ* materials can be pumped to higher levels Precast components erected by skilled operatives
Site progress	Components relatively lightweight and can be quickly erected	Heavy components but can be erected quickly	Slow progress when necessary to wait for hardening before commencing next sequence or trade
	Foundation work less Progress dependent upon weather conditions unless protection provided	Mistakes difficult to correct Progress dependent upon weather conditions unless protection provided	Multi-storey structures can be formed near ground level and raised into position
	Once erected floor/roof components can be placed and building sealed against weather	Once erected, floor/roof components can be placed and building sealed against weather	Slow sealing against weather – design can help by using repeat formwork
Fire protection	Not considered to be good fire risk although designs can take account of 'sacrificial' sections to provide insulation (see text)	Not considered to be good fire risk and generally must be insulated for all but very small structures Possible to expose sections providing design permits shielding isolation or cooling (see text)	Steel reinforcement insulated by concrete cover
	Poor spread of flame characteristics, but can be improved by chemical treatment Damaged timber irreparable	Spread of flame characteristic depends on type of insulation provided Damaged steel irreparable	High inherent degree of fire protection and nil spread of flame
			Fire resistance periods can be improved by selecting aggregate to reduce spalling Damaged reinforced concrete can be repaired
Adaptability	Easily adapted during construction	Difficult to adapt on site owing to pre-cut lengths or difficulty in cutting	Can be adapted
	Extensions easy to provide	Extension can be achieved providing access to original sections available through fire protection	Difficult to provide extension of existing structure (steel reinforcement must be exposed) and separate new structure required)
Maintenance	Finished with preservative stain, or varnish, or paint system Intumescent paint can be decorative and provides fire resistance Designs must provide protection against insect and fungal attack	Needs regular maintenance if not encased or weathering steel Intumescent paint can be decorative and provides fire protection	Self finish quality depends on site skills and formwork quality

Figure 26 Comparison between timber, steel and reinforced concrete as structural framing

the supporting columns and beams by their fixing method. A structural framework and panels or cladding walling is an example of *composite construction*, where the use of different materials to provide independent functions requires careful constructional detailing and workmanship to ensure an entirely successful enclosure.

The external appearance of a framed structure will vary according to the location of the beams and columns relative to the external wall. Figure 25 indicates the basic permutations. When not located within an enclosing panel wall, additional precautions may be necessary to protect the structural frame against possible detrimental effects resulting from an outbreak of fire (see 9.3 *Spread of fire*); and also from the effects of weathering when the frame is external

to the wall. Although this may present no special problems for reinforced concrete – other than a slight increase in cross-sectional area – the use of timber and exposed steel frames require special consideration. Figure 26 provides a brief check list of the structural materials used for a framed building. (See also the comments under *Durability* regarding the use of *in situ* concrete framework with clay brick infil panel wall construction page 16).

5.5 Panel structures

These include preformed loadbearing panel construction for the walls, floors and roof which carry and transfer loads without the use of columns and, sometimes, beams (see figure 27). This is similar to continuous supporting wall

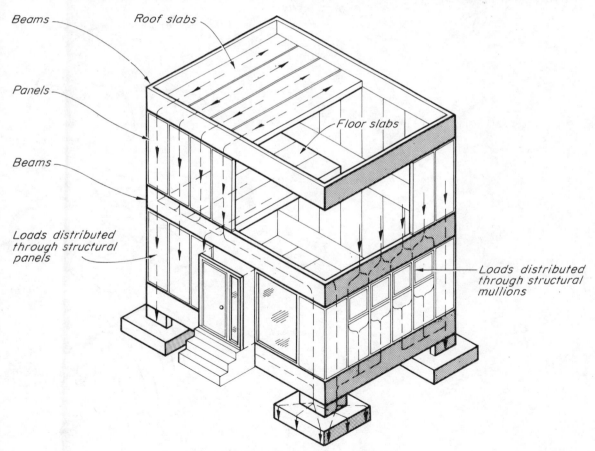

Figure 27   Transfer of loads in panel structures

35

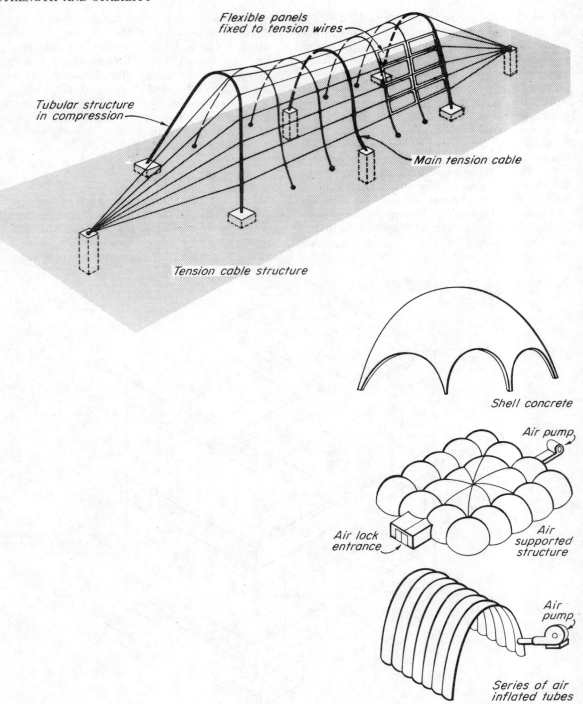

Flexible panels
fixed to tension wires

Tubular structure
in compression

Main tension cable

Tension cable structure

Shell concrete

Air pump

Air lock
entrance

Air
supported
structure

Air
pump

Series of air
inflated tubes

Figure 28   The use of tension cables and membrane structures

construction but each panel is designed to resist its own imposed loads, as well as other performance requirements. They are generally more slender than most other forms of construction and are dimensionally co-ordinated so as to be interchangeable within their specific functional requirement. The main structural material of a panel is generally of steel or timber and this can be faced with a suitable material (plywood, flat or profiled metal sheet) and incorporate thermal insulation to form a *sandwich construction*. (NOTE: certain forms of sandwich construction are also used for non-loadbearing panel cladding for a framed building). Panels can incorporate window and door openings. The combined loads which this

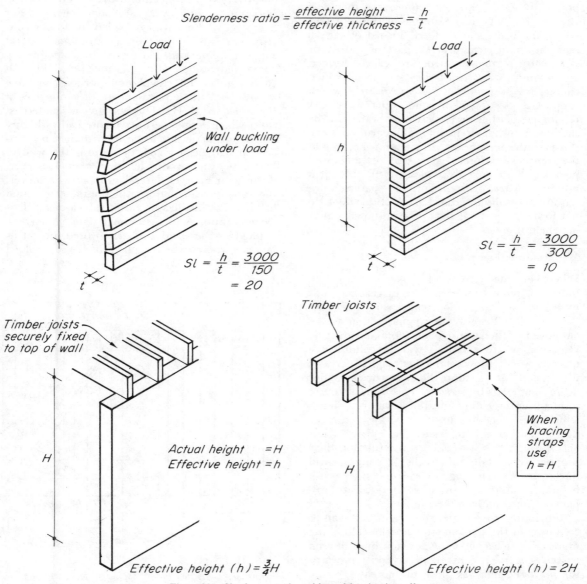

$$\text{Slenderness ratio} = \frac{\text{effective height}}{\text{effective thickness}} = \frac{h}{t}$$

Load

Wall buckling under load

h

t

$$Sl = \frac{h}{t} = \frac{3000}{150} = 20$$

Load

h

t

$$Sl = \frac{h}{t} = \frac{3000}{300} = 10$$

Timber joists securely fixed to top of wall

H

Actual height $= H$
Effective height $= h$

Effective height $(h) = \frac{3}{4}H$

Timber joists

H

When bracing straps use $h = H$

Effective height $(h) = 2H$

Figure 29   Slenderness ratio and lateral bracing in walls

37

form of construction collects can be transferred to the supporting soil by continuous distribution or by concentrating them in a similar manner adopted for framed buildings.

5.6 Membrane structure

Thin non-structural membranes forming walls and roof (often combined in one place) are supported by tension and/or compression members (figure 28). A typical example of this is a tent where the walls and roof are formed of canvas and the main structural support of timber or steel. Most permanent structures can be formed by columns (*compression members*) from which cables are suspended (*tension members*) which support a plastics membrane. Alternatively, a reinforced plastics or canvas membrane can be supported by air as in inflatable structures. In this case the membrane is in tension because of the compression forces exerted by the air under pressure. Both these examples are suitable for a building where certain of the performance requirements discussed in Part A of this book do not form an essential part of a proposed building enclosure.

5.7 Slenderness ratio

Whenever the structural organization of a building involves the resistance of vertical loads by a wall or column in compression, adequate *thickness* of sufficiently *strong* material must be employed in their construction to avoid *crushing*. However, whereas a short wall or column can ultimately fail by crushing, as height increases ultimate failure is more likely to occur under decreasing loads by *buckling*. This form of failure results from lack of stiffness in a wall or column which causes bending to occur because, in practice, it is impossible to ensure that vertical loads act through the vertical centre line of their support. Very tall, thin walls or columns will buckle before crushing, short squat ones crush before buckling, and those of intermediate proportions may fail by either method.

Obviously, the greater the height of a wall or column and the tendency towards buckling, the more critical becomes the relationship between *thickness and height*. This relationship is known as the *slenderness ratio*; as this ratio increases, so the load carrying capacity of the wall or column

decreases (figure 29). Because the stiffness of a wall or column can be increased by lateral bracing as described in 5.3 *Continuous Structures*, for calculation purposes the dimensions used for the *effective* height and thickness can vary from the *actual* height and thickness in order to obtain a realistic slenderness ratio. The amount of variation depends on the degree and effectiveness of connection provided between the horizontal and vertical components. For example, where a wall is loaded by a floor construction which provides continuous lateral support (reinforced concrete slab or adequately connected timber joists), the height of the wall can be taken for calculation purposes to have *effective height* (h) equivalent to three-quarters of the *actual height* (H). This will give a slenderness ratio which either permits theoretically slightly less strong materials to be used than if no concession had been given, or permits the wall to be thinner and occupy less plan area. However, if the floor gave no lateral support whatsoever, the rules of calculation will double the actual height of the wall and vastly increase the slenderness ratio. A further *correction factor* may be applied to the slenderness ratio used for calculation purposes when applied loads are resolved

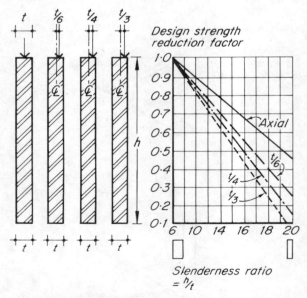

Figure 30   Correction factors applied to walls according to amount of eccentricity of loads (BDA)

eccentrically to the centre of the wall (figure 30). The design of columns is also subject to similar requirements regarding slenderness ratios and correction factors for eccentric loadings (figure 31).

5.8 Diagonal bracing

For frame buildings the provision of effective lateral bracing will vary according to how well the materials employed will permit a rigid joint to be made between horizontal and vertical components. Very rigid joints can easily be made between the beams and columns formed in reinforced concrete, but it is more difficult when they are formed in steel, and very hard when of timber. When jointing techniques cannot provide sufficient lateral restraint, the structural frame can be made more rigid by inserting *diagonal bracing* in various locations around a building, or by using shear panels (figure 32).

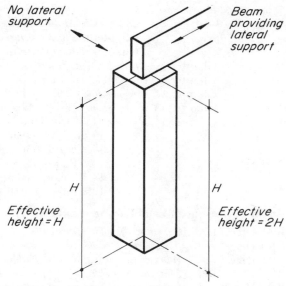

Figure 31   Slenderness ratio and lateral bracing in columns (BDA)

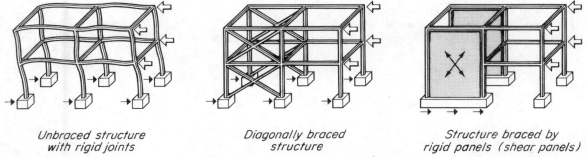

Unbraced structure with rigid joints

Diagonally braced structure

Structure braced by rigid panels (shear panels)

Figure 32   The use of diagonal bracing in frame structures. (Based on Multi-storey Buildings in Steel: Hart, Henn and Sontag, Granada)

Further specific reading

Mitchell's Building Series

Materials: Chapter 1 *Properties generally* (Strength)
Structure and Fabric Part 1: Chapter 1 *The nature of buildings and building
Chapter 3 *Structural behaviour*
Chapter 4 *Foundation* (Functional requirements)
Chapter 5 *Walls and piers* (Functional requirements)
Chapter 6 *Frame structures* (Functional requirements)
Chapter 7 *Roof structures* (Functional requirements)
Chapter 8 *Floor structures* (Functional requirements)
Chapter 9 *Fireplaces, flues and chimneys* (Functional requirements)
Chapter 10 *Stairs* (Functional requirements)

STRENGTH AND STABILITY

Building Research Establishment Digests

BRE Digest 12:   *Structural design in architecture*
BRE Digest 119:   *Assessment of wind loads*
BRE Digest 141:   *Wind environment around tall buildings*
BRE Digest 160:   *Mortars for bricklaying*
BRE Digest 235:   *Fixings for non-loadbearing precast concrete cladding panels*
BRE Digest 246:   *Strength of brickwork and blockwork walls: design for vertical load*
BRE Digest 281:   *Safety of large masonry walls*
BRE Digest 283:   *The assessment of wind speed over topography*
BRE Digest 284:   *Wind loads on canopy roofs*
BRE Digest 287:   *Specifying structural timber*
BRE Digest 332:   *Loads on roofs from snow drifting against vertical obstructions and in valleys*

6 Weather Exclusion

Weather exclusion is concerned with methods of ensuring that wind and water (rain and snow) do not adversely affect the fabric of a building or its internal environment.

6.1 Wind and water penetration

Wind can cause direct physical damage by collapse or removal of parts of a building. In addition, it can cause dampness by driving moisture into or through a building fabric, and also excessive heat losses from the interior of a building by uncontrolled air changes.

Water penetration can result in rapid deterioration as discussed in chapter 2 *Durability*, and cause the fabric of a building to become moist enough to support life, including bacteria, moulds, mildew, other fungi, plants and insects. Saturated materials also permit the quick transferral of heat (water is a good conductor) and this, together with the other factors mentioned, will cause an uncomfortable, unhealthy and uneconomical building.

The sources of water likely to penetrate a building not only include those from rain and snow, but also those from moisture contained in soil or other material in immediate contact with the building fabric. For water to penetrate there must be openings or passages in the building fabric through which it can pass, and a force to move it through these openings or passages. Without these two factors, the building fabric would remain in a watertight condition. Most buildings, however, have window and door openings, are made from lapped or jointed parts, or from porous materials ready to absorb moisture: wind currents and eddies are also normally present.

Condensation may also create moisture problems which are considered in chapter 8 *Thermal Comfort*. Damage by flooding is beyond the scope of this book.

6.2 Exposure zones

Construction methods of earlier periods were generally capable of permitting a certain amount of wind and water to penetrate through to the interior of a building. Shapes for buildings were devised for particular climatic exposures which best provide an initial defence, and the constructional detailing endeavoured to provide a final barrier.

By way of progress, modern construction methods are expected to give almost total exclusion against wind and water penetration. With ever changing 'fashions' for building shapes – sometimes borrowed from areas vastly different in climatic influences – care must be taken to ensure that constructional methods are suitable for the exposure conditions dictated by the *specific* location and disposition of a building. That is to say, before considering form and construction method, research must be carried out to reveal the degree to which a proposed building will be exposed to driving rain. In the United Kingdom, initial assessment can be obtained by reference to the *Driving Rain Index* (DRI) for the particular location as published on maps by the Building Research Establishment (figure 33). Values are obtained by taking the mean annual wind speed in metres per second (m/s) and multiplying by the mean annual rainfall in millimetres. The product is divided by one thousand and the result is used to produce contour lines linking areas of similar annual driving rain index in m^2/s throughout the country. Briefly:

Sheltered exposure zone refers to districts where the DRI is 3 or less,
Moderate exposure zone refers to districts where the DRI is between 3 and 7,
Severe exposure zone refers to districts where the DRI is 7 or more.

The value for a particular location within 8 km of the sea, or a large estuary, must be modified to the next zone above (Sheltered to Moderate, and Moderate to Severe) to take account of unusual exposure conditions. Furthermore, modifications may also be necessary to allow for local topography, special features which shelter the site or make it more exposed, roughness of

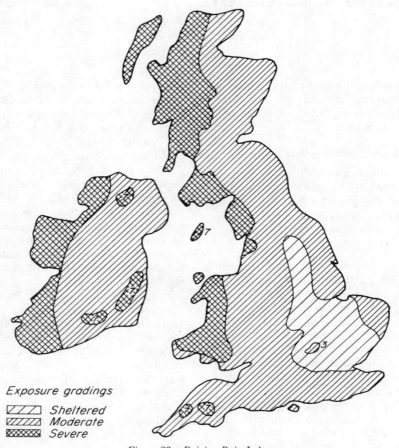

Exposure gradings

Sheltered
Moderate
Severe

Figure 33   Driving Rain Index

terrain, height of proposed building, and altitudes of the site above sea level.

The proportion of driving rain from various directions within one particular location can be obtained by reference to *Driving rain rose diagrams* (figure 34). In conjunction with the DRI, the use of these 'roses' permits the approximate amount of driving rain to be calculated which can be expected to affect *each face* of a building. This of course does not mean that a designer need adjust construction detailing for different elevations, but it will reveal which is the most vulnerable aspect of a building and give guidance about the location of openings and other design features more vulnerable to climatic influences. One of the most important lessons to be learnt from driving rain indexes and roses is that design and construction details, of necess-

ity, may vary from one exposure zone to another: details suitable in a 'sheltered exposure zone' would probably leak if simply transferred to a 'severe exposure zone' without modification.

6.3 Macro- and micro-climates

However, there can exist a danger in using only meteorological climatic data of this nature for the final selection of appropriate material and constructional detailing. This data reveals the general or *macro-climate* conditions liable to affect a building in a particular location. There is also a *micro-climate* surrounding the immediate outer surface of a building, ie not more than 1 m from surface of a building. This is created by specific environmental conditions arising from the precise form, location, juxtapositions, and surface geometry of a building. A designer is expected to know when design ideas are liable to cause the micro-climate to significantly vary from the macro-climate and make adjustments in materials and/or constructional detailing accordingly. Where there is no past experience, it may be necessary to make models of a building and its surroundings, test it under simulated environmental conditions, and record the precise effects. For example, detailed analysis has revealed that tall buildings can receive more rainwater on their walls than on their roofs – especially on elevations facing the wind. Under

these circumstances, rain is often driven *vertically up* the face of the building, making it necessary to use constructional details which are different to those considered suitable for lower buildings in the same exposure zone (figure 35).

6.4 Movement of water

As indicated earlier, there are a very wide range of materials and combinations of materials, as well as numerous construction methods from which to choose when designing a building. Providing they are carefully matched to exposure conditions, all can function with equal effi-

Figure 34 Driving rain rose diagram

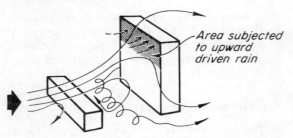

Area subjected to upward driven rain

Figure 35   Alteration of general climate conditions as a result of building disposition and shape

ciency in controlling the *movement of water* from the exterior to the interior of a building. The precise way this is achieved will vary, and depend upon the properties of the materials employed and the manner in which they are arranged to form the weather barrier. Figure 36 illustrates the three basic arrangements for the external fabric of a building where:

- walls initially act as a *permeable barrier* which permits subsequent evaporation of absorbed water
- walls and roofs act as *semi-impermeable* which permits a certain amount of water to penetrate until reaching a final barrier
- walls and roofs act as an *impermeable barrier* which diverts water on contact.

When rain falls or is driven on *porous* building materials such as most brick types, some stones, or blocks, it is absorbed and then subsequently removed by natural evaporation. This process occurs when water adheres to the pores of the

material and, if the adhesive force between the water molecules and the wall material is greater than the cohesive force between the molecules themselves, the water is drawn in by *capillary action*. A strong wind increases the rapidity of absorption and only evaporation resulting from changes in the climatic conditions (rain ceases, temperature rise, and warmer air currents) will prevent the moisture penetrating through the thickness of the material. Earlier construction forms ensured that this thickness was sufficient to prevent water penetration by capillary action. Current economic trends and the need for energy conservation (saturated material loses about ten times more heat through it than when dry) have now firmly established construction methods using materials of thinner cross sections. These are employed to provide an external initial weather check which is combined with materials used to provide *internal* thermal insulation. The external and internal components are separated by a water barrier (or air cavity) to interrupt the continuous flow of water from outside to inside. In this way, potentially wet areas are isolated from those which must remain permanently dry.

Constructional detailing must attempt to reduce the movement of water as much as possible. It is particularly important to ensure that the water barrier is incorporated in such a way that moisture is not trapped and kept in a position for a period of time liable to cause damage.

For similar reasons, when rain falls or is driven on theoretically entirely *impervious* build-

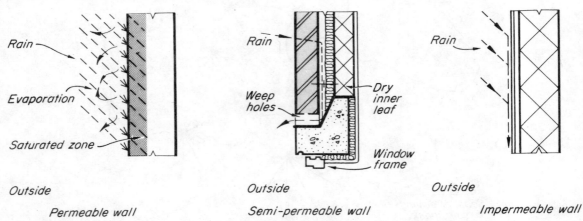

Figure 36   Movement of water for walls of different construction methods

44

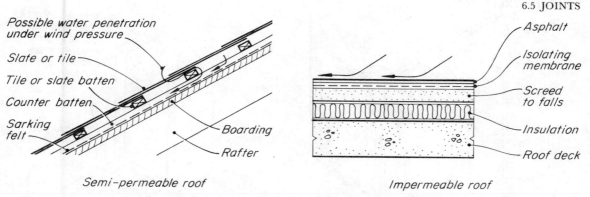

Possible water penetration under wind pressure

Slate or tile

Tile or slate batten

Counter batten

Sarking felt

Boarding

Rafter

Semi-permeable roof

Asphalt

Isolating membrane

Screed to falls

Insulation

Roof deck

Impermeable roof

Figure 37   Movement of water for roofs of different construction methods

ing materials such as dense concrete, glass, metal, bituminous products, or plastics, precautions must be taken to ensure quick and efficient run off. The quantity of water must never be underestimated – on a glass wall of a building it can be as much as five litres per ten metres square of façade.

A typical flat roof construction to provide a weather resisting barrier would consist of sealed lengths of multi-layer bituminous felt or a continuous homogenous layer of asphalt (figure 37). A comparable pitch roof construction to resist the penetration of water under the influence of gravity would incorporate outer surface finish of lapped tiles or slates backed by an impervious water barrier. Further comments about this form of construction have been included in chapter 17 *Construction Methods*.

6.5 Joints

The need for joints arises because of the necessity to link, lap or bond materials together when providing the continuous and efficient weather enclosure for a building. Unless the many inter-related factors which influence their position and type are particularly carefully considered, they can form the 'weak link' in the enclosure.

The first rule is to ensure that as much water as possible is kept away from this vulnerable point where two or more materials are brought together – each perhaps having different characteristics, and connected in some way with yet other types of materials. The designs for roofs and walls can provide shelter to their joints, or channel free-flowing water in pre-determined directions to permit either collection or dis-

charge to less damaging areas. The effects of wind-driven rainwater must always be taken into account in this respect, and boldly profiled upstands and overhangs at joint positions are desirable when it is difficult to form a continuous 'membrane type' seal between two building components.

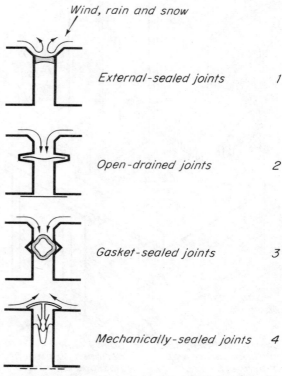

Wind, rain and snow

External-sealed joints   1

Open-drained joints   2

Gasket-sealed joints   3

Mechanically-sealed joints   4

Figure 38   Types of jointing used to resist the penetration of water

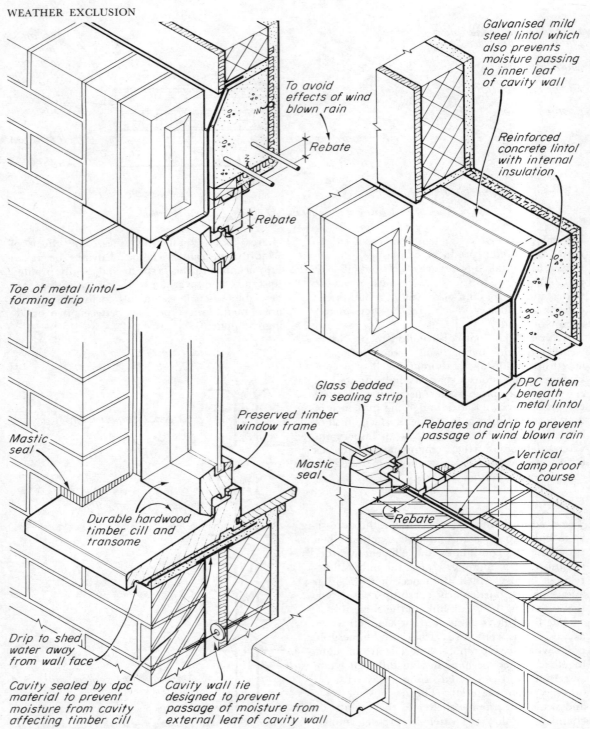

To avoid effects of wind blown rain

Rebate

Rebate

Galvanised mild steel lintol which also prevents moisture passing to inner leaf of cavity wall

Reinforced concrete lintol with internal insulation

Toe of metal lintol forming drip

DPC taken beneath metal lintol

Mastic seal

Preserved timber window frame

Glass bedded in sealing strip

Mastic seal

Rebates and drip to prevent passage of wind blown rain

Vertical damp proof course

Durable hardwood timber cill and transome

Rebate

Drip to shed water away from wall face

Cavity sealed by dpc material to prevent moisture from cavity affecting timber cill

Cavity wall tie designed to prevent passage of moisture from external leaf of cavity wall

Figure 39   Construction method used to resist water and wind penetration through a window opening in a brick/block cavity wall

The actual method of forming a joint will initially depend upon the physical and chemical properties of the materials involved. The comments made in chapter 5 *Dimensional Suitability* regarding *movements* and appropriate sizes are particularly relevant. The joint can be expected to behave in a similar manner to the surrounding surfaces by stopping water penetration at the outermost places; or to allow water to be caught in its recesses, then collected before directing it to the outside again. Within these two extremes, there are a vast range of jointing possibilities (figure 38).

Some of the main joints necessary to provide a weather-resisting enclosure for a simple design of a timber-framed window in a brick/block cavity wall are illustrated in figure 39. These include the use of mortar for the brickwork outer leaf of the wall, an impervious water barrier membrane (dpc) between brickwork and window frame, and a seal between window frame and glass. The profiles between the fixed and opening parts of the window frame are also specially designed to reduce the movement of wind borne water to the interior of the building. Draught-excluding devices can also be fitted in this gap to eliminate the flow of air and the possibility of heat loss. The relationship between the type of brick and the type of mortar used in the outer leaf of brickwork is also important, as indicated by figure 40.

One of the most important aspects of joints in a building involves their effect on appearance.

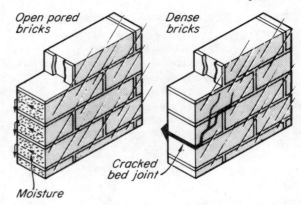

Figure 40   *Ways water can penetrate brickwork*

The particular type of brick bond (stretcher, flemish, dutch, quetta, etc) and the width, profile and colour of mortar joint can have as much varying impact on the appearance of brickwork as the colour and shape of the bricks themselves. Similarly, the precise location and profile of the joints in preformed panel and *in situ* reinforced concrete walls will assist in determining not only the overall scale and proportion of a building, but also the pattern and rhythm of features on the façade. The surface texture created by the need to channel water away from widely spaced joints of preformed wall units also helps in creating the particular character and expression of a building.

Further specific reading

Mitchell's Building Series

Building Research Establishment Digests

BRE Digest 54: *Damp-proofing solid floors*
BRE Digest 77: *Damp-proof courses*
BRE Digest 127: *An index of exposure to driving rain*
BRE Digest 137: *Principles of joint design*
BRE Digest 217: *Wall cladding defects and their diagnosis*
BRE Digest 261: *Painting woodwork*
BRE Digest 262: *Selection of windows by performance*
BRE Digest 304: *Preventing decay in external joinery*
BRE Digest 312: *Flat roof design: the technical options*

7 Sound Control

The control of sound in a building must be considered from two aspects:

- the elimination or reduction of *unwanted sound* generated by sources within or outside a building (sound attenuation)
- the creation of good *listening conditions* within a building where clarity of speech and music is required which must not be marred by sound reverberation and echoes.

7.1 Unwanted sound

Figure 41 indicates the main ways by which sound can be transmitted into and through a building. This involves movement through air or other elastic media formed from solids, liquids or gases (*airborne sound*) or movement through a solid structure resulting from an impact force (*impact sound*). Both are transmitted by direct paths from source to recipient or by indirect paths along adjoining elements. This latter method is known as *flanking transmission*. High levels of unwanted sound, or noise, can lead to the breakdown in the mental health or even damage to the hearing of a recipient. Unwanted lower levels of sound are a nuisance and become a source of constant irritation resulting in a loss of concentration.

7.2 Noise outside a building

Apart from industrial operations, *external noise* nuisance is most often caused by motor traffic and it is necessary for the designer to be familiar with the *noise climate* liable to affect the performance of a building. This is the range of sound levels achieved for 80% of the time – the remaining 20% being divided equally between sound levels occurring above and below the main range. The upper limit of the noise climate is called the 'ten percent level' (L_{10}) and has become the unit used for specifying extreme exposure conditions to traffic noise which, when existing for 18 hours per day, permits Government compensation or grants to be paid for control of sound levels in houses adjoining motorways, etc. Similar compensation is avail-

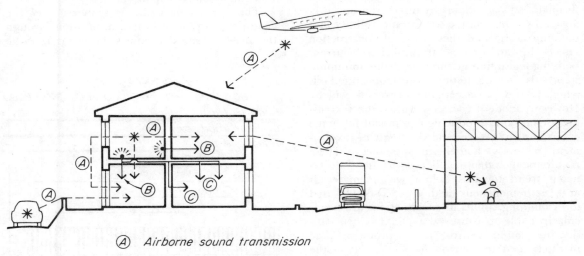

Ⓐ *Airborne sound transmission*

Ⓑ *Impact sound transmission*

Ⓒ *Flanking sound transmission*

Figure 41   Unwanted sound (J E Moore)

able for exposure to aircraft noise: this is assessed in terms of *Noise and Number Index* (NNI) and takes into account the number of movements during the day and the loudness of each. Local authorities have powers and duties to control noise nuisance under the Control of Pollution Act 1974 which provide means of creating *Noise Abatement Zones* for the long-term control of noise from fixed sources such as may exist in areas of mixed residential and industrial development. This Act also provides the power to control noise on construction and demolition sites which, although usually shortlived, may inflict severe discomfort on normally peaceful neighbourhoods. Nevertheless, prevention is better than cure, and various documents exist which attempt to control the initial output of noise such as BS 5228, *Noise Control on construction and open sites*; *The Motor Vehicles (Construction and Use) Regulation* 1969; and the Health and Safety Executive Act 1974, Code of Practice for *Reducing the exposure of employed persons to noise*.

7.3 Noise inside a building

As indicated by figure 41, the movement of either airborne or impact noises *inside* a building is a complex process involving transmission through walls and floors by direct and/or flanking paths. The relative weight and rigidity of a building fabric and the nature of construction affect the amount of transmission. Current building legislation attempts to define minimum standards for domestic construction methods which provide an acceptable degree of control. However, modern society requires the increasing use of sound producing equipment for home entertainment devices and the frequent involvement of noisy household appliances. These requirements often conflict with the simultaneous trend towards lighter weight materials and less homogenous methods of assembly used for a building. Joints in constructions are particularly liable to cause 'weak links' when considering sound control as a problem. The methods used to control the movement of the sound within a building are similar to those adopted to control sound from external sources, although the forms of construction needed for the latter are often assisted in this respect by the

performance requirements relating to weather exclusion, ie greater thicknesses and weights.

7.4 Frequency, intensity and loudness

When analysing the control of noise it is useful to clarify the two basic factors – frequency and intensity – which initially influence the kind of sound received by the human ear.

Sound is normally created in the air when a surface is vibrated and sets up waves of alternating compression and rarefaction: the centres of compression being known as the wavelength of the sound which for human hearing, vary from about 20 mm to 15 m. The number of complete movements or cycles from side to side made by the particles in the air (or any other elastic medium) during the passage of sound waves determines the *frequency* of the sound and this is measured by the complete number of cycles made in a second or Hertz (Hz). The greater the number of cycles per second or Hertz caused by the vibrations, the higher the pitch of the sound: people are most affected by frequencies between 500 Hz and 6000 Hz (figure 42). Similar waves to those described can also be caused by air turbulence resulting from explosive expansion of air or a combination of vibration and explosive expansion.

The *intensity* of sound is a measure of the acoustic energy used in its transmission through the air. Intensity determines the strength of the

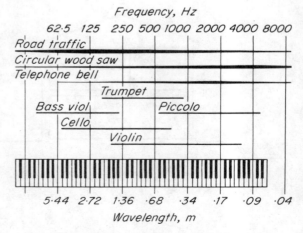

Figure 42   Sound frequencies (J E Moore)

50

sound, and is expressed in *decibels* (dB) conforming to a logarithmic scale (regular proportionate increments rather than equal increments) which happens to approximate closely the way sound is heard. It also gives a manageable scale for a wide range of sounds (figure 43). The actual decibels produced by a particular external airborne sound will be reduced according to the amount and characteristics of the intervening space between source and recipient. There is a theoretical reduction of 6 dB each time the distance from the source is doubled.

Four-engine jet aircraft at 100 m	120 dB
Riveting of steel plate at 10 m	105 dB
Pneumatic drill at 10 m	90 dB
Circular wood saw at 10 m	80 dB
Heavy road traffic at 10 m	75 dB
Telephone bell at 10 m	65 dB
Male speech, average, at 10 m	50 dB
Whisper at 10 m	25 dB
Threshold of hearing, 1000 Hz	0 dB

Figure 43   Typical sound intensity levels (J E Moore)

In practice, this amount may be modified by such factors as whether the source is of single point, continuous line, or area origin; its height; the amount reflected during its transmission; the effectiveness of screening devices provided by trees, other buildings, embankments, etc; and meteorological conditions. In addition, the ability to hear a given sound within a building will depend to a considerable extent on the background noise generally existing within the interior. General room sounds created by radio, television, and conversation often make traffic noises far less noticeable. Furthermore, the number of decibels created by a particular source does not necessarily provide an indication of how *loud* it sounds because the human ear is more sensitive to high frequency sounds than low. Therefore, the *subjective* loudness of a noise is measured by a weighted scale known as *dBA*, which gives an overall intensity with a bias towards high frequency sounds. Other weighted scales are also available (dBB, dBC).

7.5 Defensive measures

From the foregoing it will be appreciated that the first and obvious defence against the intru-

sion of unwanted airborne sound lies in placing as much distance as possible between the source and the recipient. For example, activities accommodated by the function of a building requiring a quiet environment (sleeping, studying, lecturing, etc) can be placed remote from the external noisy distraction of motorways, sports stadiums and industrial application. Further reduction can then be provided by the fabric of a building, although the precise nature of material and construction technique most suitable to reduce intrusive sound can be fairly complicated to assess. Nevertheless, data is available which specifies desirable sound levels within a building according to functions (figure 44), and construction methods can be selected which provide the necessary sound reduction to achieve these goals relative to the anticipated external noise environment. Similar consider-

	dBA
Banks	55
Churches	35
Cinemas	35
Classrooms	35
*Concert halls	30
Conference Rooms	30
Court rooms	35
Council chambers	35
Department stores	55
Hospitals, wards	35
Hotels, bedrooms	35
Houses, living	45
Houses, sleeping	35
Lecture rooms	35
Libraries, loan	45
Libraries, reference	40
*Music rooms	30
Offices, private	40
Offices, public	50
*Radio studios	30
Restaurants	50
*Recording studios	30
Shops	55
Telephoning, good	50
Telephoning, fair	55
*Television studios	35
*Theatres	30

Figure 44   Acceptable intrusive noise levels in respect of broad-band random frequency noise, e.g. road traffic. These are optimum levels and may be exceeded by 5 dBA except in the critical situation marked by an asterisk (J E Moore)

ations apply to the reduction of unwanted sounds which may occur inside a building.

The amount of sound control or *sound attenuation*[1] provided by certain construction methods have been established over a 100 to 3150 Hz range of frequencies (roughly corresponding to the lowest/highest normally experienced in a building). For this reason they are useful when it is necessary to provide sound attenuation for broad-band frequency noises, but may be misleading when comparing dissimilar but adjoining methods of construction, or when noise is concentrated predominantly at selective frequencies, eg related to dBA scale.

For practical purposes, however, the airborne sound attenuation of a construction is controlled by four factors:

1 Mass or weight per unit area – attenuation increases by approximately 5 dB for a doubling of weight or doubling of frequency.
2 Discontinuity or the elimination of direct sound paths where mass is insufficient by isolating those surfaces which receive the sound from those which surround the listener – the effect of a cavity between depends on its dimension in relation to the wavelength of sound to be controlled. Generally, a minimum practical gap of 50 mm is suitable for high frequencies but a wider gap is necessary for low frequencies. Discontinuity must not incorporate any bridging along which sound may travel.
3 Stiffness or elimination of vibration by sound waves – when incident sound waves have frequencies similar to those created by vibration of the construction, the sound attenuation will be less and unrelated to that theoretically provided by its mass.
4 Uniformity or elimination of direct air paths through the construction – this applies not only to openings, holes and cracks, but also to materials which are not in themselves airtight. In the case of doors and windows, no matter how heavy the surrounding wall may be, the net sound

[1] NB: The term 'sound insulation' should be avoided so as to avoid confusion with 'thermal insulation'. Whereas the former relies on *greater* density (weight) for effective control, the latter relies on *less* density.

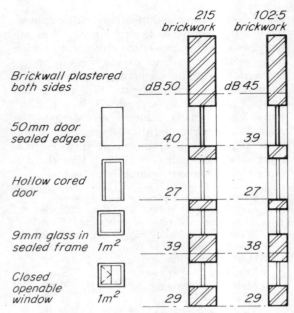

Figure 45   Effect of openings in walls have on sound attenuation performance (BDA)

attenuation will be limited to a maximum of about 7 dB *above* that of the door (figure 45).

The control of the effects of *impact* noise in a building requires different consideration to those given above. For example, the noise heard below a floor subject to impact noise will bear little relation to the airborne noise it causes in the room above. For the room below, weight has no advantage and the only defence is to prevent the transmission of the impact sound to the structure by using a soft floor finish or a finish which is isolated from the structure.

7.6   Listening conditions

Sound within a room consists of two components – the *direct*, which travels in a straight line through the air from the source to the recipient, and *reverberent*, which is the sum of all the sound reflections from the room surfaces (figure 46(a)). As discussed under *Unwanted sound*, the direct noise decreases at the rate of 6 dB for each doubling of distance from the source, whereas, the reverberent sound is theoretically constant throughout the room. This

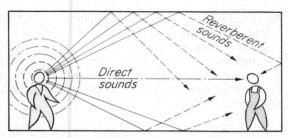

Figure 46(a)   Direct and reflected (reverberant) sound

means that in a room containing a single source there is a zone where the direct sound predominates; immediately beyond this another zone will occur where neither direct nor reverberent sound predominates; and then finally a zone where the reverberent component predominates. The sizes of these zones depend upon the dimensions of the room and the quantity/quality of absorbent surfaces present.

7.7 Blurring and echoes

Reverberant sound can be extremely useful in large rooms such as auditoria, where controlled reflections can send the sound levels to positions normally too distant to receive adequate sound from the direct path. However, focusing reflections in this way is a relatively skilled process, and if misjudged, reverberant sound and direct sound will be heard at slightly different intervals, thereby creating blurring (reflection arriving 1/30 to 1/15 seconds after direct sound) or

echoes (more than 1/15 second discrepancy). Further understanding of the precise science used to achieve audibility and clarity of performance in auditoria is beyond the scope of this book, and reference should be made to one of the many excellent monographs available on this subject for further clarification.

7.8 Sound absorption

It is important to differentiate between sound attenuation as described earlier, and *sound absorption*. The former is concerned with the passage of sound energy through a barrier; the latter with the sound reflection capabilities of the surfaces within a room or space. Sound produced by speech or music may be reflected several times from walls, floors or ceiling before the residual energy in the sound waves is negligible – the continuance of sound during this period is known as reverberation, and the interval between the production of the sound and its decay to the point of inaudibility is known as the reverberation time. The sound absorbing properties (absorption coefficient) of a surface are measured by the amount of sound reduction which occurs after waves strike the surface. Coefficients vary from 0 to 1, ie perfect reflection (hard surface) to total absorption (open window). See figure 46(b).

The sound levels in a room build up to a level which is determined by the absorption characteristics of the room. This can cause dis-

Item	Unit	Absorptive coefficient		
		125 Hz	500 Hz	2000 Hz
Air	m³	0	0	0.007
Audience (padded seats)	person	0.17	0.43	0.47
Seats (padded)	seat	0.08	0.16	0.19
Boarding or battens on solid wall	m²	0.3	0.1	0.1
Brickwork	m²	0.02	0.02	0.04
Woodblock, cork, lino, rubber floor	m²	0.02	0.05	0.1
Floor tiles (hard)	m²	0.03	0.03	0.05
Plaster	m²	0.02	0.02	0.04
Window (5 mm)	m²	0.2	0.1	0.05
Curtains (heavy)	m²	0.1	0.4	0.5
Fibreboard with space behind	m²	0.3	0.3	0.3
Ply panel over air space with absorbent	m²	0.4	0.15	0.1
Suspended plaster-board ceiling	m²	0.2	0.1	0.04

Figure 46(b)   Sound absorption coefficients (J E Moore)

comfort to the occupants, and increase the likelihood of noise being transferred to adjoining rooms where the original attenuation standards were sufficient. The provision of sound absorbent materials on the surfaces of the room containing the sound source will eliminate these possibilities. If a room has mostly hard surfaces, no soft furnishings and few occupants to absorb sound effectively, it is usually not difficult to make a fourfold increase in absorption and obtain a reduction of 6 dB. However, care must be taken when selecting absorbent materials that the coefficient matches the frequency of the offending sound(s).

Further specific reading

Mitchell's Building Series

Environment and Services: Chapter 6 *Sound*
Materials: Chapter 1 *Properties generally* (Acoustic properties)
Structure and Fabric Part 1: Chapter 5 *Walls and piers* (Functional requirements)
Chapter 7 *Roof structures* (Functional requirements)
Chapter 8 *Floor structures* (Functional requirements)
Chapter 10 *Stairs* (Functional requirements)
Components: Chapter 5 *Doors* 5.2(d) Thermal control and sound control
Chapter 6 *Windows* 6.2(g) Sound control
Chapter 7 *Glazing* 7.2(e) Sound control
Chapter 11 *Demountable partitions* 11.2(e) Sound control
Chapter 12 *Suspended ceilings* 12.2(e) Sound control
Chapter 13 *Raised floors* 13.2(e) Sound control
Chapter 14 *Roofings* 14.2(c) Sub-structure

Building Research Establishment Digests

BRE Digest 140:   *Double glazing and double windows*
BRE Digest 162:   *Traffic noise and overheating in offices*
BRE Digest 185:   *Prediction of traffic noise* Part 1
BRE Digest 186:   *Prediction of traffic noise* Part 2
BRE Digest 187:   *Sound insulation of lightweight dwellings*
BRE Digest 192:   *The acoustics of rooms for speech*
BRE Digest 193:   *Loudspeaker systems for speech*
BRE Digest 203:   *Noise abatement zones* Part 1
BRE Digest 204:   *Noise abatement zones* Part 2
BRE Digest 226:   *Thermal, visual and acoustic requirements in buildings*
BRE Digest 293:   *Improving the sound insulation of separating walls and floors*
BRE Digest 333:   *Sound insulation of separating walls and floors* Part 1: *walls*
BRE Digest 334:   *Sound insulation of separating walls and floors* Part 2: *floors*
BRE Digest 337:   *Sound insulation: basic principles*
BRE Digest 338:   *Insulation against external noise*

8 Thermal Comfort

A building must provide a satisfactory thermal environment for its occupants as well as for the mechanical systems it accommodates. Energy within the human body produces uninterrupted heat at varying rates in order to maintain an ideal temperature of 37°C in the internal organs. However, the heat is lost by the body through radiation, convection, and evaporation from the skin and the lungs, and there is a continuous process of adjustment to ensure a thermal balance between heat produced and heat lost. It is particularly important that the brain temperature is maintained at a constant rate. The factors in the local environment which govern heat loss include not only air temperature but also air movement, relative humidity, and the radiant temperatures from surrounding surfaces.

The fabric of clothing and of a building perform similar functions by maintaining temperature control through *passive means* which regulate natural flows of heat, air, and moisture vapour. As a building involves a volume many times larger than that contained by clothing, and provides environmental conditions suitable for many occupants in different spaces, it must also provide *active means* for thermal comfort not unlike that achieved by the human body itself (figure 47).

8.1 Passive means

Simply stated, the creation of thermal comfort within a building by passive means involves the reduction of the rate of *heat losses* from the inside to the outside in colder climates, and the reduction of *heat gains* from the outside to the inside in warmer climates. In both cases the transfer of heat is regulated by the external fabric of a building which can provide varying degrees of thermal insulation – the type and amount necessary to achieve ideal conditions varying according to resource availability, climate of locality, and the degree of exposure (figure 48).

In the United Kingdom, the thermal performance of a building fabric is directly affected

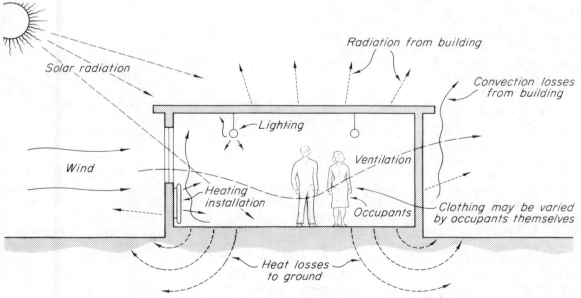

Figure 47   Heat balance of building in relation to external and internal influences

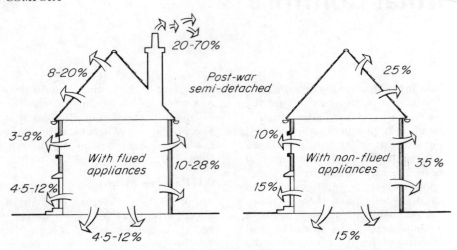

8-20%

20-70%

Post-war
semi-detached

25%

3-8%

With flued
appliances

10-28%

10%

With non-flued
appliances

35%

4·5-12%

15%

4·5-12%

15%

Figure 48   Percentage heat losses from domestic buildings

by such detail criteria as seasonal changes and extremes of temperature; temperature differences between day and night (diurnal range); sky conditions (amount of sunlight and overshadowing); incoming and outgoing heat radiation; rainfall and its distribution; the effects of water absorption/repulsion on materials and their forms (weathering); air movements; and other special features influenced by location and orientation. A designer must interpret these requirements in conjunction with fashions in appearance for a building, together with the current legal requirements which endeavour to conserve the use of fuel for space heating by regulating the amount of heat loss from a building to the external environment.

8.2 Heat transfer

Heat may be lost from a building fabric to the external environment by *convection* (movement of heat by hot liquid or gas); *conduction* (transfer of heat through solids); and/or by *radiation* (transfer of heat from a surface across a vacuum, gas or liquid filled space, or a transparent solid). Natural ventilation causes convection losses and takes the form of warm air rising in a building, and eventually escaping to be replaced by colder air. Whilst some degree of ventilation is desirable as one of the functions of a building (see chapter 10), excessive heat losses should be minimised. Solid building materials

used for a building fabric loose heat from warm to cold face by conduction. The ability of a material to conduct heat is known as its *conductivity* (λ) value, and is measured by the amount of heat flow (Watts) per square metre of surface area for a temperature difference of 1°C per *metre thickness*, ie $Wm/m^2°C$ or $W/m°C$. When comparing the insulation properties of a material to that of others, it is generally more convenient to use its *resistivity value* (r) because it takes no account of size or thickness. Resistivity, therefore, is expressed as the reciprocal of the conductivity, ie m°C/W.

In order to calculate the actual *thermal resistance* (R) offered by a particular material of *known* thickness, the formula used is actual thickness (m) × m°C/W, ie $R = m^2 °C/W$. The higher the R value, the better the thermal resistance and the insulating performance of the material (figure 49). Also, the less dense a material is, the greater insulation value it will provide.

It is unfortunate that generally these relationships of density are counter to the noise control and structural qualities required from a material. A solid monolithic wall may be up to three times the thickness required for structural purposes in order to provide an acceptable level of thermal insulation. If a dense structural material is filled with pockets of air, its thermal insulation characteristics can be vastly improved while at the same time maintaining

The thermal resistance of a structural component is expressed as:

$$R = \frac{L}{k}$$

where
L = Thickness of element in metres
k = Thermal conductivity W/m°C

Material	Conductivity W/m°C	Resistance m² °C/W
Brickwork 105 mm	0.84	0.125
220 mm	0.84	0.262
335 mm	0.84	0.399
Plaster 15 mm hard	0.5	0.03
15 mm lightweight	0.16	0.09
10 mm plaster board		0.06
Cavity, unventilated	—	0.18
with foil face		0.3
behind tile hanging		0.12
Tile hanging	0.84	0.038
25 mm expanded polystyrene	0.033	0.76
13 mm expanded polystyrene	0.033	0.39
Glass fibre 50 mm	0.035	1.43
75 mm	0.035	2.14
100 mm	0.035	2.86
Aerated concrete 100 mm	0.22	20.45
150 mm	0.22	0.68
Softwood 100 mm	0.13	0.77
Weatherboarding 20 mm	0.14	0.14
External surface		0.055
Internal surface		0.123

Figure 49   Thermal conductivity and thermal resistances of some materisl (CIBSE Guide)

much of its strength characteristics. However, if these air pockets are allowed to become filled with water from exposure to rainfall or ground moisture, their insulation value will be cancelled altogether. This is because water is a much better conductor of heat than air: a saturated material could permit about ten times more heat to be transferred through it when compared with its 'dry' state.

Any heated material will radiate heat from its surface; bright metallic surfaces generally radiate least and dark surfaces the most. In this respect, radiation and absorption characteristics of materials correspond. Some construction methods incorporate air cavities which reduce the amount of heat transfer by conduction (and

the passage of moisture) from inner warm materials to outer cold materials. Some heat, however, will be transferred across cavities by radiation and the absorption (and reflection) properties of the adjacent surfaces across the cavity will be significant to the thermal insulation value of the construction. (The performance of the cavity as an insulator will be impaired if convection currents take place.) Radiation losses generally depend on the *emissivity* – the rate of radiant heat omission – from the surface and values depend upon roughness of surface, the rate of air movement across it, its orientation or position, and with the temperature of the air and other bodies facing it.

8.3 Thermal insulation values

It is unrealistic to rely on empirical rules to establish forms of constructions which provide satisfactory resistance to heat transfer *as well as* weather resistance, strength and stability and many of the other performance criteria. Moreover, current construction methods are generally less bulky than those used years ago as much reduced thicknesses of materials are employed in combination with each other; each layer, leaf or skin perhaps fulfilling only one very specific function. The thermal comfort properties of these materials used in combination can only be assessed by calculating the amount of heat transfer from internal to external air (in cold climates). The thermal resistance properties of each layer must be taken into account, along with other factors relating to their surface texture and juxtapositions within the construction.

The internal to external thermal transmission rate of all the layers of a construction is known as a *U value* and is more accurately defined as the number of Watts transmitted through one square metre of construction for each single °C temperature difference between the air on each side of the construction, ie W/m² C. It is calculated by taking the reciprocal of the sum of all the thermal resistance values (R values of all materials used and any air cavities) as well as the internal and external surface resistances (figure 50).

The calculated U value for a particular construction is unlikely to provide an entirely accurate thermal transmittance rate because the heat flow conditions through the construction

The U value (or thermal transmittance) of a building element is obtained by combining the thermal resistances of the component parts and taking the reciprocal, expressed as:

$$U = \frac{1}{R}$$

where R is the sum of all the separate resistances for different materials in the element:

$$U = \frac{1}{Rsi + R1 + R2 + R3 \ldots Ra + Rso}$$

will vary with the amount of solar radiation, moisture conditions and the effects of prevailing winds. The consistancy of the construction method is also an important factor, as illustrated in figure 51. Here most of the construction is of layered form but is 'bridged' at intervals by a material or an air space providing less thermal resistance. This results in variations in the

where

U = Thermal transmittance W/m² °C
Rsi = Inside surface resistance m² °C/W
R1, R2, R3 = Thermal resistance of structural components m² °C/W
Ra = Airspace (cavity) resistance m² °C/W
Rso = Outside surface resistance m² °C/W

Example:

U value calculation for typical wall

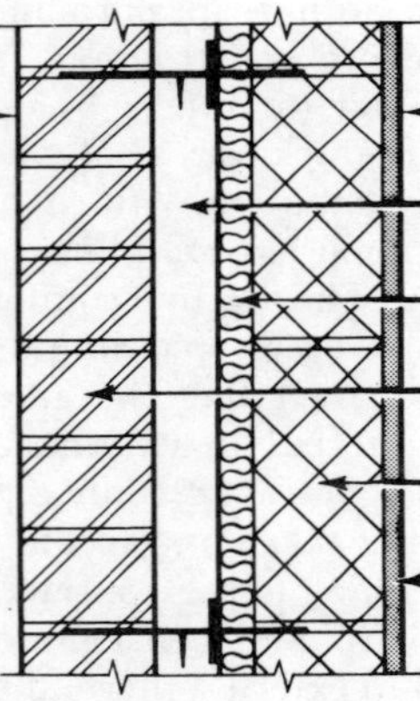

		L	÷ k value	= R value
		Thickness metres	Conductivity W/m °C	Thermal resistance m² °C/W
External surface → Internal surface	Sum of external and internal surface resistances			= 0·180
Cavity	Air cavity resistance			= 0·180
Insulation		0·025	÷ 0·036	= 0·694
Facing brick		0·100	÷ 0·900	= 0·111
Thermalite or similar		0·100	÷ 0·184	= 0·543
Dense plaster		0·013	÷ 0·500	= 0·026
Total thermal resistance R =				1·734

Figure 50 'U' value calculation (RIBA)

thermal transmittance values along the construction and the occurence of *cold bridging*. The slightly cooler internal surface at the point of the *cold bridge* provides a dew-point temperature at which the warm internal air cools causing droplets of water (condensation) as well as dust to be deposited locally, ie *pattern staining*.

Nevertheless, calculated U values are a reasonably accurate guide to the *thermal insulation value* of a construction and provide the main guide as to whether or not the heat losses from a building are within the acceptable limits imposed by building legislation. It is important to

To calculate U value of element:

$$U = \frac{1}{R} = \frac{1}{1·734} = 0·576 \ W/m^2 \ ^\circ C$$

remember that higher U values mean poorer thermal insulation properties, and that U values refer to the *total constructional thickness* of a building element or component whether consisting of a single layer of material or a combination of materials, with or without separating cavities.

The Building Regulations 1985: Approved Document L: Conservation of fuel and power provides four independent procedures of increasing complexity

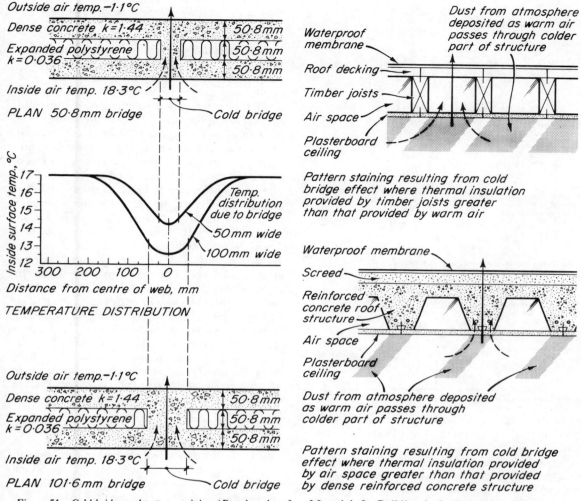

Figure 51   Cold bridge and pattern staining (Based on data from Materials for Building by Lyall Addleson, Iliffe 1972)

for meeting the thermal control requirements in dwellings and in buildings other than dwellings ie those having a floor area greater than 30 m² which are residential buildings (hotels, etc), shops, offices, or used for recreational, educational, business, industrial or storage purposes.

Procedure 1 specifies thermal insulation material and construction thicknesses for external elements, whereas *Procedure 2* requires certain U-values to be met (figure 52). The need to conserve energy in buildings has resulted in careful consideration of the effect windows and rooflights have on the *overall heat loss* from a building and for these Procedures maximum areas of single glazed windows and rooflights are specified. Window and rooflight areas which are double glazed may have up to twice the single glazed area; those which are triple glazed, or are double glazed and incorporate a low emissivity coating, may have up to three times the permitted single glazed area. When calculating these percentages it may be necessary to include the surface area occupied by lintels, jambs and sills as these often fall below the thermal insulation standards of the body of the wall. Furthermore, an external door with 2 sq m or more of glazed area should be counted as part of the percentage allowed for windows and rooflights.

Procedure 3 allows 'trade off' between windows and rooflights and, independently, between

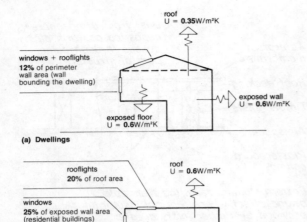

roof
U = **0.35**W/m²K

windows + rooflights
12% of perimeter
wall area (wall
bounding the dwelling)

exposed wall
U = **0.6**W/m²K

exposed floor
U = **0.6**W/m²K

(a) Dwellings

rooflights
20% of roof area

roof
U = **0.6**W/m²K

windows
25% of exposed wall area
(residential buildings)
35% of exposed wall area
(offices, shops – not display
windows and assembly
buildings)

exposed wall
U = **0.6**W/m²K

exposed floor
U = **0.6**W/m²K

(b) Residential, offices, shops and assembly buildings

roof
U = **0.7**W/m²K

rooflights
20% of roof area

windows
15% of exposed wall area

exposed wall
U = **0.7**W/m²K

exposed floor
U = **0.7**W/m²K

(c) Industrial, storage and other buildings

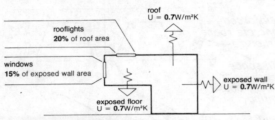

◀ *Figure 52   Calculated rate of heat loss*
Source: *The Building Regulations 1985* AD L2/3

exposed walls, roofs and floors. In the case of dwellings where the thermal insulation is better than the minimum allowed in the *Regulations*, larger glazed areas are permitted although no account is taken of heat gains. In calculating the rate of heat loss that would occur, the U-value of windows and rooflights is taken as 5.7 W/m²K if single glazed, 2.8 for double glazing, and 2.0 for triple glazing or for double glazing incorporating a low emissivity coating.

Procedure 4 (not to be used for dwellings) also allows 'trade-off' between glazed and solid areas, but account can be taken of useful heat gains as well, including solar gains and those resulting from artificial light, industrial processes, etc. Acceptance of this method of calculation relies on proof that sufficient heating controls can be provided in respect of the contribution of useful heat gains.

8.4 Condensation and interstitial condensation

In practice, it is also necessary to prevent the occurrence of *condensation* at the surface or within material(s) used for a building fabric (figure 53). This takes place when atmospheric temperature (dry bulb) falls below the dew-point temperature – a property that depends upon the water vapour content of the air and, therefore, upon the vapour pressure. The amount of water vapour contained in the atmosphere will fall according to temperature and relative humidity (ratio of vapour pressure present to completely saturated air).

Surface condensation is likely to occur when air containing a given amount of water vapour is cooled by coming into contact with a cold plane. High relative humidity – more than 80% – will cause mould growths on the surface of organic materials or droplets of water on non-organic surfaces. This risk can be reduced by keeping internal surfaces at a higher temperature than the dew-point of the internal air which involves carefully balancing the building fabric insulation *and* internal heating and ventilating requirements. Simply to increase the insulation value of the building fabric may only move the dew-point of the air *into* the body of the material, and cause *interstitial condensation* resulting in a loss of insulation and deterioration.

One solution to the damage caused by surface and interstitial condensation is to use a construction method which makes the condensation or dew-point coincide with a *cavity* which separates external weathering layers from internal insulating layers – see also 6.4 *Movement of water*. This cavity should be ventilated, and the moisture formed by condensation (together with any which penetrates the external weathering layers) must be adequately collected and diverted to the outside. When a cavity cannot be incorporated in the correct position relative to the point of condensation, then a *vapour barrier* must be used. This consists of a thin impermeable sheet (eg plastics or reinforced aluminium foil) and should be placed on the *warm side* of the insulating material to prevent transference of moist air from the warm interior of a room into the fabric of the building (figures 153 and 175). Care must be taken to ensure that the sheets are adequately lapped and folded at their edges to prevent moisture penetration. A vapour barrier must also be incorporated for composite construction methods employing dif-

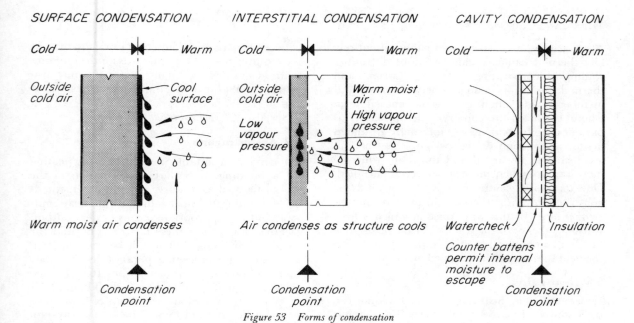

SURFACE CONDENSATION

Cold ———————►◄——— Warm

Outside cold air / Cool surface

Warm moist air condenses

Condensation point

INTERSTITIAL CONDENSATION

Cold ———————►◄——— Warm

Outside cold air

Low vapour pressure

Warm moist air

High vapour pressure

Air condenses as structure cools

Condensation point

CAVITY CONDENSATION

Cold ———————►◄——— Warm

Watercheck / Insulation

Counter battens permit internal moisture to escape

Condensation point

Figure 53   Forms of condensation

ferent layers of materials providing varying thermal and vapour resistance properties. The problems associated with the practical application of vapour barriers are discussed under 17.9(e) *Roof Construction: Thermal insulation and ventilation.*

Modern forms of construction and living patterns have greatly increased the risk of condensation, not only upon internal surfaces, but also within wall, floor, and roof constructions. Traditional construction allowed moisture vapour to pass more easily through its fabric and through air gaps around doors and windows. There was often less moisture vapour to be vented owing to lower temperature requirements and greater ventilation rates at moisture sources (air bricks and chimney flues, etc). With modern consruction methods, it is essential to estimate the degree of condensation by means of scientific calculation (see 16 *Communication*, figure 103).[1]

[1] See also *The Building Regulations 1985*: Approved Documents F for details of natural ventilation for rooms and roofs in domestic buildings.

8.5 Thermal capacity

The selection processes for a building fabric to provide satisfactory thermal comfort conditions must include consideration of the form of space

heating to be employed. When the heat source remains virtually constant, it may be considered expedient to adopt a building fabric which will store heat or have a *high thermal capacity*. For example, masonry walling material will gradually build up a reservoir of heat while it is being warmed – when the heat source ceases for a short period (ie overnight) or when the external temperature drops below normal, the stored heat will be slowly given back. At any event, the internal wall surfaces will feel relatively warm and comfortable, and the possibility of condensation occurring will be less likely – except during the period when the heat source is initially resumed. In hot climates, very thick and heavy construction buffers the effect of very high external daytime temperatures on the internal climate of the building.

When intermittent heat sources are used, or other marked fluctuations from steady temperatures occur, dense construction will be slow to warm up and lower surface temperatures may be present for sometime. This will give rise to discomfort for the occupants and condensation. In rooms used occasionally and, therefore, requiring only intermittent heating for comfort, lining the interior surfaces with material of *low thermal capacity* incorporating a vapour barrier, will reduce the amount of heat required and produce a quick thermal response.

61

Alternatively, a construction method of entirely low thermal capacity can be adopted (timber walling incorporating vapour barrier and thermal insulation in spaces between structural members), and often this is the chosen form where the need to conserve fuel resources is paramount. In both cases, internally insulated dense construction or lightweight construction, less heat will be stored and the room will cool more rapidly when the heat source is curtailed. One acceptable compromise is to adopt a dense construction with thermal insulation on the *external* face. In this way, once a wall has been warmed, there will be a long period before it is finally dissipated completely, during which time the heat source may be resumed again.

8.6 Heat gains

Unwanted solar heat gain into a building can be a source of considerable thermal discomfort and interrupt the working of normal methods of space heating. However, heat gains of this nature can be modified by careful adjustment of the amount of glazed areas (either fixed or openable); the type of glass used for windows; building and configuration; thermal character of the building fabric; surface colours and texture; and the degree of absorption permitted by exposed materials. The usual methods adopted are: shading or screening devices; increasing the thermal capacity of the building fabric; and adopting a reflective outer surface. Nevertheless, with special references to colder climates like the United Kingdom, the beneficial physiological and biological effects upon humans, animals, and plants should be considered before deciding to eliminate solar heat gains altogether.

Apart from solar sources, however, considerable heat gains can be encountered when the function of a building requires the use of large amounts of energy. Office lighting systems and mechanical ventilation plant, exhibition display cases and lamps, and even the heat from people crowded together in supermarkets, cinemas, swimming pools and disco halls can create a considerable amount of heat. In large buildings the heat generated in this way may sometimes result in cooling or ventilation plant being required rather than heating plant, even during the cold winter season. The designer of such a building must ensure the correct *thermal balance*

between heat requirement, heat inputs/gains and the required insulation standards for the building fabric. Considerable financial benefits may also be obtained if the heat gains can be transformed and stored for later use.

See also *8.3 Thermal insulation values*.

8.7 Active means

To arrive at a suitable method of space heating for a building, it requires examination of the fuel to be used as a source of heat; the methods of distributing the heat source to the heat emitters; and the appropriate means of heat output. Detailed analysis of these factors is beyond the scope of this book, but it will be realised from the immediate preceding paragraphs that they exert a profound influence on the thermal comfort standards achieved by a building.

With few exceptions, any form of heat emission can be powered by any fuel. The choice of fuels currently available include fossil fuels (wood, coal, gas and oil); electricity; and, to a lesser extent, renewable resources (solar, animal wastes, geothermal, tidal, wind, wave, and ambient energy from light fittings, mechanical plant, and even human beings). Fossil fuels are being consumed at an increasingly rapid rate to keep up with comfort standards for today, and there is likely (pending new discoveries) to be only sufficient gas for the next 40 years, oil for about 30 years, and coal for at least 300 years. Wood is a renewable resource, but as materials for building purposes are also being depleted, it is becoming an extremely valuable commodity which is far too important to burn. Electricity is generated mainly from fossil fuels, and for this reason *ideally* should not be used as a direct heating source because its production by this means is both inefficient and expensive.

The increasing use of nuclear reactors for the production of electricity, using natural occurring resources, certainly makes a more viable heating source. Developments in nuclear fusion (there is still ample supply of uranium) could make electricity the major future source for thermal comfort, provided conversion to a useful form can be accomplished in safety and without danger of radiation or pollution to people and the countryside. Fear of these happening is one of the reasons for increasing interest in the use of certain renewable resources and

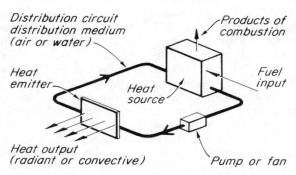

Distribution circuit
distribution medium
(air or water)

Products of
combustion

Heat
emitter

Fuel
input

Heat
source

Heat output
(radiant or convective)

Pump or fan

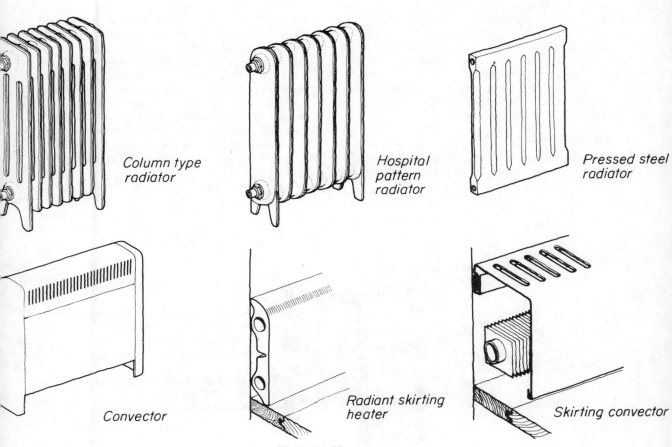

Column type
radiator

Hospital
pattern
radiator

Pressed steel
radiator

Convector

Radiant skirting
heater

Skirting convector

Figure 54   Heat emitters

ambient energy for generating electricity (windmills, waves, etc).

One of the most important decisions, as far as user comfort is concerned, is the choice of the form of *emitter* (radiator, convectors, ducted warm air, etc). See figure 54. The selection is influenced by the response of a building fabric to changes in external air temperature; to temperature variations caused by the sun; and to the heat gains from changing occupancy, cooking, or the use of other heat-producing equipment. If discomfort is to be avoided, either from underheating or overheating, the response of the heating system must be equal to or shorter than the response of the building fabric. This latter response is determined by the mass of the building fabric, the degree and position of insulation, the reflectivity of external surfaces where exposed to the sun, and the area and orientation of windows. Furthermore, controls have a vital role to play in achieving economy of operation: the generally available controls are not yet able to provide immediate response to changes in temperature, and a lapse of time resulting in discomfort can occur.

8.8 Solar energy

Notwithstanding the comments made under *Heat gains*, the re-use of solar energy is becoming increasingly important as an alternative means of both space and water heating, and various forms of construction have been devised which 'capture' this free resource. For example, a glazed conservatory on the south side of a building will create an accumulation of heat which can be absorbed into an interior wall of high thermal capacity (figure 55). The heat stored in this manner will be emitted into the interior spaces at night or during other periods of no sunshine. The amount of heat absorbed by an internal wall of this nature can be regulated by louvres opening in the conservatory so that excessive heat gains will not occur and cause discomfort. This system is known as a *passive* solar energy resource.

Active solar systems are also now available where solar energy is used to heat water, and this helps to provide warm water for washing or space heating. Systems usually consist of an exposed glass collector behind which are located pipes containing water – the water gradually becomes heated and circulates to be stored in a vessel located in a convenient position in a building. Alternatively, the heated water can also pass through a heat exchanger and be used to supplement other forms of heat source, such as may be used for ducted warm air central heating.

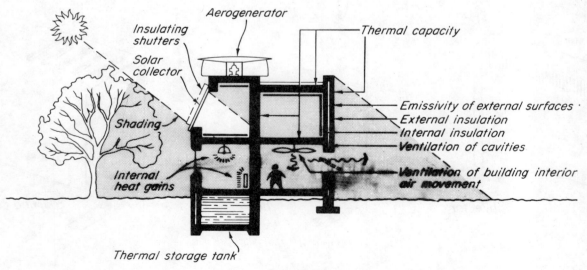

Figure 55   The use of solar energy

Further specific reading

Mitchell's Building Series

Environment and Services: Chapter 5 *Heat*
                                Chapter 7 *Thermal insulation*
Materials: Chapter 1 *Properties generally* (Thermal properties)
Structure and Fabric Part 1: Chapter 5 *Walls and piers* (Functional requirements)
                                          Chapter 7 *Roof structures* (Functional requirements)
                                          Chapter 8 *Floor structures* (Functional requirements)
                                          Chapter 9 *Fireplaces, flues and chimneys* (Functional requirements)
Components: Chapter 5 *Doors* 5.2(d) Thermal control and sound control
                   Chapter 6 *Windows* 5.2(c) Ventilation
                                   5.2(e) Thermal control
                   Chapter 7 *Glazing* 7.2(c) Thermal
                   Chapter 14 *Roofings* 14.2(d) Thermal control
                                   14.2(e) Vapour control

Mitchell's Professional Library

Practical Thermal Design in Buildings, Peter Burberry

Building Research Establishment Digests

BRE Digest 108:   *Standard U-values*
BRE Digest 110:   *Condensation*
BRE Digest 140:   *Double glazing and double windows*
BRE Digest 145:   *Heat losses through ground floors*
BRE Digest 162:   *Traffic noise and overheating in offices*
BRE Digest 180:   *Condensation in roofs*
BRE Digest 190:   *Heat losses from dwellings*
BRE Digest 226:   *Thermal, visual and acoustic requirements in buildings*
BRE Digest 232:   *Energy conservation in artificial light*
BRE Digest 236:   *Cavity insulation*
BRE Digest 254:   *Reliability and performance of solar collector systems*
BRE Digest 270:   *Condensation in insulated domestic roofs*
BRE Digest 277:   *Built-in cavity wall insulation for housing*
BRE Digest 297:   *Surface condensation and mould growth in traditional-built dwellings*
BRE Digest 302:   *Building overseas in warm climates*
BRE Digest 306:   *Domestic draughtproofing: ventilation considerations*
BRE Digest 319:   *Domistic draughtproofing: materials, costs and benefits*
BRE Digest 324:   *Flat roof design: thermal insulation*
BRE Digest 336:   *Swimming pool roofs: minimising the risk of condensation using warm-deck roofing*
BRE Digest 339:   *Condensing boilers*

It is ironical that whilst fire is used in the manufacture of most materials and can provide the thermal conditions required in a building, it can also be highly destructive to a building and its occupants. Death or injury by fire is particularly horrifying and incidents naturally receive special concern in the minds of most building occupiers. Account must also be taken of the financial loss of a building and its contents when considering the full effects of devastation by fire. Losses can be in excess of £350 m per annum when taking account of the current inflatory trends in the cost of material and labour.

The design and construction method employed for a building, therefore, must safeguard occupiers from death or injury and also minimise the amount of destruction. These goals can be achieved through an understanding of the nature of fire and its affects on materials and construction used in a building; methods of containing a fire and limiting its spread; methods of ensuring the occupants of a building being attacked by fire can escape to safety; and methods of controlling a fire once it has started.

9.1 Combustibility

Fire is a chemical action resulting in heat, light, and flame (a glowing mass of gas), accompanied by the emission of sound. To start, a fire needs a combustible substance, oxygen, and a source of heat such as that resulting from a flame, friction, sparks, glowing embers, or concentrated solar rays. Once a fire has started, heat is usually produced at a rate that is in excess of that at which it is dissipated to its surrounding and, therefore, temperature rises with time. The fire will burn out once the fuel is removed, become smothered if oxygen is not available, or die if the heat is removed by, for example, water. Before this happens however, a fire is likely to ignite nearby material resulting in it spreading through a building and perhaps, by processes which include radiation, to adjoining buildings.

In the context of fire protection, materials used in a building fall into two broad categories: *combustible* and *non-combustible*. British Standard tests define precisely the meaning of these terms (BS 476: Part 4), and others establish the ease of ignition which provides a basis for separating particularly hazardous materials (BS 476: Part 5). It will generally be found that inorganic materials, such as stone, brick, concrete or steel are non-combustible; whilst organic materials such as timber, or the by-products of timber (fibreboard, plywood, particle board, etc), and petro-chemical products including plastics are combustible.

9.2 Fire resistance

In situations where it is imperative that a building must not contribute fuel to a fire, only noncombustible materials should be employed. However, even if practical, the sole use of noncombustible materials will not necessarily avoid the spread of fire which has been generated by the burning contents of a building. Avoidance of this requires the parts of the building – materials and construction used for elements (walls, floors, etc) – to have *fire resistance*. This is the term used to describe the ability of an element of building construction to fulfil its assigned function in the event of a fire without permitting the transfer of the fire from one area to another. Tests according to BS 476: Part 8 establish a time period during which an element can be expected to perform this function, and requires a sample to be subjected to a simulated building fire which research has indicated rises in temperature according to duration – see time/temperature curve, figure 56. The tested element is then rated according to the time ($\frac{1}{2}$, 1, 2, 3, 4, 5 or 6 hours) it is able to fulfil three performance criteria. These criteria approximate those necessary to ensure the reasonable safety of occupants and contents in a building including time to discover the fire and make a safe escape. The performance criteria (figure 57) applied in the tests are:

(a) *Stability*
A *loadbearing* construction must support its full load during the fire and continue this

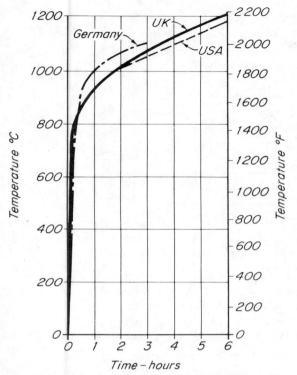

Figure 56   Comparison of the time/temperature curves used in different countries

support 24 hours after the heating period. Failure for floors, flat roofs and beams is said to occur when deflection reaches more than 1/30th of span. *Non-loadbearing* construction fails on collapse.

(b) *Integrity*
Failure occurs when cracks or other openings form through which flame or hot gases can pass which would cause combustion on the side of the element remote from the fire.

(c) *Insulation*
Failure occurs when the temperature on the side of the element remote from the fire is increased *generally* by more than 140°C, or, *at any point* by more than 180°C above the initial temperature.

The period of fire resistance suitable for particular elements of a building depends on the functions to be accommodated, and the volume, height and floor areas of the spaces involved (figure 58).

Combustibility cannot be equated with fire resistance: one is a characteristic of a material, the other relates to the performance of an element as a whole. For example, an external wall of corrugated asbestos sheeting (non-combustible) on a mild steel frame (non-combustible) has no notional period of fire resistance because a construction of this nature would rapidly permit fire to spread by heat transfer, ie would not meet the 'insulation' requirement of fire resistance. Conversely, an external wall of timber cladding (combustible) on timber studs (combustible) with a plasterboard internal lining (combustible because of paper sheathing) will provide a full fire resistance for a period of half an hour. However, there are cases where legislation requires both fire resistance and non-combustibility, eg external walls of a building located in close proximity to a boundary.

9.3 Spread of fire

The *spread of fire* within a building and from one building to another can most effectively be restricted by the identification and isolation of potential hazards. Therefore, the first defence involves siting. Having established that poten-

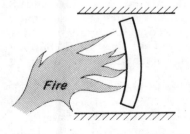

Collapse or excessive deflection

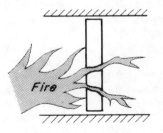

Maintain integrity

Average temperature rise not more than 140°C or 180°C at any one point

Insulation

Figure 57   The meaning of fire resistance

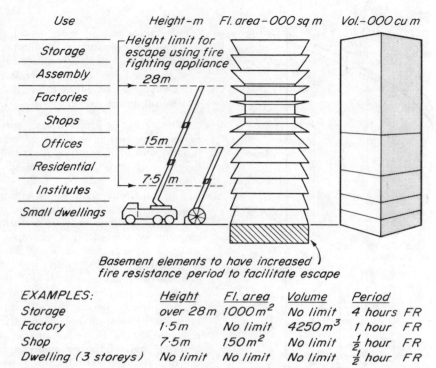

Use	Height – m	Fl. area – 000 sq m	Vol. – 000 cu m
Storage	Height limit for escape using fire fighting appliance		
Assembly	28 m		
Factories			
Shops			
Offices	15 m		
Residential			
Institutes	7·5 m		
Small dwellings			

Basement elements to have increased fire resistance period to facilitate escape

EXAMPLES:

	Height	Fl. area	Volume	Period
Storage	over 28 m	1000 m²	No limit	4 hours FR
Factory	1·5 m	No limit	4250 m³	1 hour FR
Shop	7·5 m	150 m²	No limit	$\frac{1}{2}$ hour FR
Dwelling (3 storeys)	No limit	No limit	No limit	$\frac{1}{2}$ hour FR

Figure 58   Factors which determine period of fire resistance for elements of structure forming part of a building

tial fire risks exist by the nature of the functions accommodated, it is necessary to select the appropriate siting for a building relative to safety of nearby properties. An extreme example might be that a building accommodating particularly dangerous fire hazards (eg, manufacture and/or storage of highly flammable chemicals). This should be located in a remote part of the countryside, away from properties likely to be damaged by heat radiation caused as a result of the fire.

Building legislation exists (*The Building Regulations 1985*: Approved Document B *Fire Spread*) which attempts to limit the spread of fire by stipulating fire resisting periods and construction methods appropriate to the function of a building as discussed earlier. In addition, it limits the use of certain construction methods according to their distance from other properties and/or boundaries. For example, the amount of timber cladding (combustible) allowed on an external wall of a house is directly related to its distance from a boundary: none is permitted if a wall is within 1 m of the boundary; or the whole area of a wall which is over 6 m from the boundary. Windows, doors, and other openings in external walls are also carefully controlled since, unless special forms are used, they do little in stopping the spread of fire from inside or into a building. Regulations restrict the positioning, size, and amount of these openings according to function (fire risk of a building as well as location of the external wall relative to other properties and/or boundaries).

Whereas the general approach on walls and floors is to 'contain' the fire, there is no legislative requirement for the external surfaces of a roof to have fire resistance. Instead, reference is made to test procedures in BS 476: Part 3, which grades the suitability of certain roof coverings according to their ability to resist radiant heat and burning brands from a fire emanating from adjoining properties.

The spread of fire within a building can similarly be restricted by the special segregation of particular fire hazards. Also, fire resisting *compartments* or 'cells' can be used to limit the spread of a fire through a building – the precise location, enclosed volume, and period of fire

68

resistance required again depending on the nature of activities accommodated. In addition the height of a building and the ease with which its occupants can escape, and/or firefighting can be successfully carried out, are also deciding factors (figure 59). Structural organization (see 5.2 *Structural organization*) is also important in this respect. Whereas continuous masonry wall construction with reinforced concrete floors use materials inherently capable of providing high standard of fire resistance, the 'infill' type of constructions needed for framed buildings, for example, must be carefully selected to provide adequate fire resistance as well as fulfilling other performance requirements which includes compatibility and continuity with the support system.

One aspect of the fire resisting requirements is that the structural organisation of a building should not collapse or deform before the occupants can escape safely. The eventuality of collapse will, of course, be of a major importance when considering the spread of fire (and also fire fighting – see later). For the reasons already stated, a framed structural organisation may need special consideration relative to the materials employed. Providing sufficient insulating cover of concrete is provided to the steel bars of reinforced concrete columns and beams, adequate defence against the untimely collapse or deformation is reasonably easy to achieve. However, steel columns and beams (figure 60) can present bigger problems since, depending on the precise physical composition, their ultimate strength is about 50% when subjected to a temperature of 550°C. This can be achieved within 15 minutes in a BS 476 building fire, as already indicated in figure 56. Unless special design techniques are employed which keep temperatures below this level for at least the

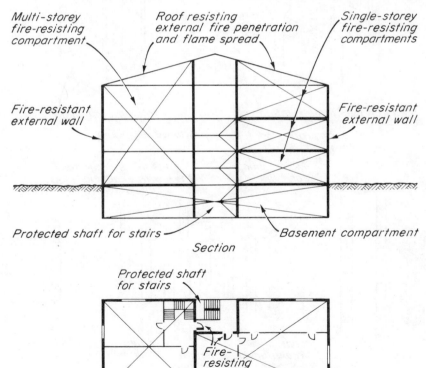

Figure 59   The use of 'compartments' to limit spread of fire in a building (AJ)

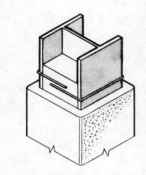

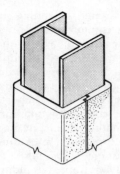

A

Solid concrete encasement   Hollow vermiculite encasement

Fire protection conceals steelwork

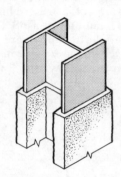

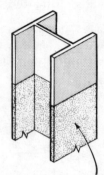

B

Spray vermiculite insulation   Intumescent paint

Fire protection follows profile of steelwork

Direction
of fire ➤

Mains supply
for filling

Water storage tank
with expansion
capacity

Vent

Multi-storey zone

C

Loops allow
circulation along length
and breadth of building

Water
filled
column

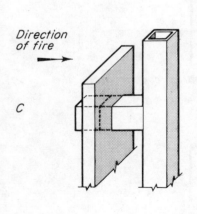

Heat shielded sections   Water filled sections   Concrete filled hollow sections

Exposed steelwork

Figure 60   Methods of providing fire resistance to steel columns (CONSTRADO)

required fire resisting period of a building, structural steel sections must be insulated with a solid or a hollow casing. This protection, of course, must also be compatible with the prevailing environmental and functional conditions. External casings should be weather and impact resistant, durable, and of acceptable appearance to maintain their effective life in a building. Similarly, internal fire insulating casings may be required to withstand knocks from trolleys, provide fixings for fitments and space for service pipes or cables, as well as be aesthetically suitable.

Interestingly, timber columns and beams can provide greater fire resistance than unprotected steel. Figure 61 indicates the use of 'sacrificial timber' which provides effective insulation to structural sections subjected to a fire.

Apart from normal fire resisting requirements (including doors, etc), care must be taken in the selection of finishes which will not contribute too much to the spread or growth of fire. To insist on the use of non-combustible materials in all circumstances would be too onerous and restrictive. For this reason, the *Surface Spread of Flame* test set out in BS 476: Part 7 was devel-

oped to classify the relative risk of various combustible materials, and these can be used in certain positions – even if for a limited area. Building legislation has introduced a further surface spread of flame classification which covers not only non-combustible materials but also materials with a surface finish giving low fire propagation properties (BS 476: Part 6) which, therefore contribute little to the growth of a fire.

9.4 Means of escape

The occupants of a building must be provided with clearly defined and safe escape routes in the event of the occurrence of a fire. These routes must be kept clear of obstruction, be easy to manoeuvre and, very importantly, should be free of the effects of flames, heat and smoke (figure 62). For this reason escape routes need special consideration regarding safety from fire; access points must be shielded by lobbies and fire resisting (and smoke resisting) doors; walls, and floors should have sufficient fire resistance to allow escape and subsequent access for firemen to fight the fire. Sometimes it may be

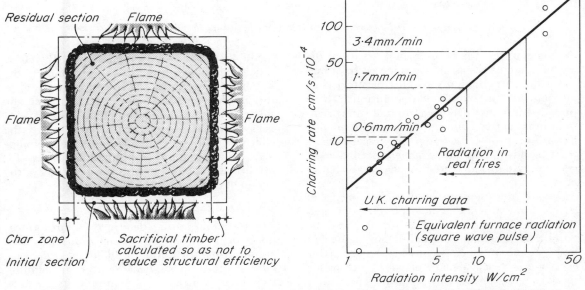

Figure 61   The effect of fire on structural timber sections. Once the structural size has been established, additional thickness can provide insulation. The thickness required for a specific period of fire protection can be established according to the charring rate of the particular type of timber, ie density. (Based on Fire Research Note 896, November 1971)

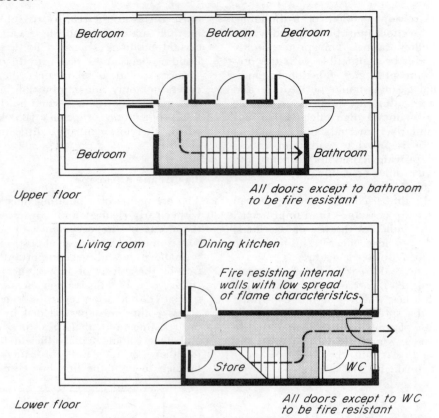

Upper floor

All doors except to bathroom to be fire resistant

Lower floor

All doors except to WC to be fire resistant

Figure 62   The use of fire resisting walls and doors to ensure safe escape from a building which is burning

necessary to provide positive air pressure in the escape route to ensure smoke is forced back into the body of the building making exit easier, or alternatively, an extract ventilation unit can be provided to remove smoke. Internal surface finishes must not contribute to the fire or provide hazard to people escaping, and escape areas must be adequately lit, in some cases by emergency only lights activated by a separate generator.

Major tragedies can occur unless means of escape from a building are correctly designed and sited. In the event of a fire it should be possible to evacuate a building in reasonable time ($2\frac{1}{2}$ minutes being considered normal for everyone to reach a place of safety, except for special premises such as hospitals). The width, the size of treads and risers, and heights of handrails for staircases used for escape purposes must also conform to similar safety require-

ments. Escape routes' planning, therefore, should be related to the use of the building, the number of occupants, the risks involved, and to the heights of floors above ground and the shapes and dimensions of floors. The risk of persons being trapped or overcome by the effects of a fire is greatest in multi-storey buildings and access routes and standing positions for fire brigade appliances are critical factors in the design of escape routes.

9.5 Fire fighting

The techniques so far described for fire protection can be classified as *passive measures* as they are an in-built feature of a building. *Active measures* of fire protection incorporate fire fighting techniques which can be in-built (sprinklers, dry/wet risers, alarm systems, fusable links to doors, shutters, firemen's lifts, etc) or be brought to a building in distress (firemen and appliances). The siting of a building must allow easy access for firemen and their appliances in an emergency. Hard standing areas must be included in the landscaping and located at prescribed distances from a building to facilitate fire fighting activities.

9.6 Fire insurance

The influences associated with the insurance of a building against damage or loss through the occurrence of a fire are important aspects of initial design considerations. Fire insurers often look for higher standards of fire protection than are stipulated in the clauses of fire legislation documents or recommendations given direct by the fire brigade. Failure to consider this point at an early stage of design, or during the selection processes for appropriate construction methods, may result in high premiums for fire insurance, or the adoption of costly modifications.

Further specific reading

Mitchell's Building Series

Environment and Services: Chapter 18 *Firefighting equipment*
Materials: Chapter 1 *Properties generally* (Deterioration)
Structure and Fabric Part 1: Chapter 5 *Walls and piers* (Functional requirements)
Chapter 6 *Frames structures* (Functional requirements)
Chapter 7 *Roof structures* (Functional requirements)
Chapter 8 *Floor structures* (Functional requirements)
Chapter 9 *Fireplaces, flues and chimneys* (Functional requirements)
Chapter 10 *Stairs* (Functional requirements)
Structure and Fabric Part 2: Chapter 10 *Fire protection*
Components: Chapter 5 *Doors* 5.2(e) Fire precautions
Chapter 6 *Windows* 5.2(f) Fire precautions
Chapter 7 *Glazing* 7.2(d) Fire precautions
Chapter 9 *Ironmongery* 9.2(d) Performance requirements
Chapter 10 *Balustrades and barriers* 10.2 Performance requirements
Chapter 11 *Demountable partitions* 11.2(d) Fire precautions
Chapter 12 *Suspended ceilings* 12.2(d) Fire precautions
Chapter 13 *Raised floors* 13.2(d) Fire precautions
Chapter 14 *Roofings* 14.2(d) Fire precautions

Building Research Establishment Digests

BRE Digest 158: *Honeycomb fire dampers*
BRE Digest 208: *Increasing the fire resistance of timber floors*
BRE Digest 225: *Fire terminology*
BRE Digest 230: *Fire performance of walls and linings*
BRE Digest 233: *Fire hazards from insulating materials*
BRE Digest 260: *Smoke control in buildings: design principles*
BRE Digest 285: *Fires in furniture*
BRE Digest 288: *Dust explosions*
BRE Digest 294: *Fire risk from combustible cavity insulation*
BRE Digest 300: *Toxic effects of fires*
BRE Digest 317: *Fire resistant steel structures: free standing blockwork-filled columns and stanchions*
BRE Digest 320: *Fire doors*

10 Lighting and Ventilating

Together with the provision of thermal comfort and sound control, lighting and ventilating provide the initial environmental aspects of a building which ensure the physiological and psychological well-being of its occupants. However, the advancement of technology has led to some degree of complacency on the part of designers concerning the influence which the provision of acceptable lighting and ventilating standards has on the overall appearance and construction method of a building. Apart from purely functional requirements, they provide the principal means of creating aesthetic atmosphere and character. Nevertheless, whereas these functions once derived purely through an inter-relationship between architectural form and construction method, it is becoming increasingly possible to design a building where fashion can be *made* to function by the use of artificial devices.

In all but the simplest form of building, often it may be considered normal to make 'corrections' in the lighting levels not achieved from the designed building by the use of electric light systems. Similarly, air conditioning apparatus can be made to compensate for the lack of sufficient natural ventilating openings (windows, chimney openings, etc). Indeed, artificial systems may even be oversized to make the thermal environment acceptable because of ill-considered decisions about siting, orientation, or the amount of glazing in a building envelope.

10.1 Standards

Although the size, position and amount of window openings is necessarily controlled by current needs to conserve the use of energy and to ensure safety from the spread of fire or the intrusion of unwanted sound, a sensible balance must be achieved between these aims and acceptable lighting/ventilating standards. This can be accomplished by detailed analysis of each requirement, careful design decisions and adoption of suitable construction methods. Obviously, artificial lighting is required for certain activities to assist concentration without

eye strain, and also after natural light periods. A building with a deep plan form required by virtue of optimum function, ie large office floor spaces, warrant continuous artificial light sources. In this case, a suitably designed system can convert the otherwise wasted heat, generated by the lamp source, into useful back-up or supplementary space heating for a building. The residue energy can be similarly used from artificial ventilating systems required by a large building, or a building with the external envelope entirely sealed against noise or extremes of climate.

10.2 Natural lighting

Amongst its many other vital functions, the sun provides the sources of *natural light* which create the first psychological connection between the inside and the outside of a building. The influence of the sun in creating shaded areas affecting the vegetation and human enjoyment of spaces is a critical factor when considering the dimensions and shape of a building and its distance from others. The use of 'natural' coloured daylight to illuminate interior spaces can create interesting effects caused by variations in intensity during the day influencing the shading of planes, and the hue and depth of coloured surfaces.

Daylight is admitted into a building through 'holes' in external fabric (windows, rooflights, etc), which, in adverse climates, generally incorporate glass or an alternative transparent material to control the effects of heat loss and/or inclement weather on the interior spaces. The amount of light received inside a building is usually, of course, only a small fraction of that received outside because of modifications imposed by the size and position of openings and will also constantly vary owing to the influences imposed on the 'whole sky' illumination level by clouds, buildings and/or other reflecting planes. Therefore, it is impracticable to express interior daylighting in terms of the illumination actually obtainable inside a building at any one time, for within a few minutes that figure is liable to change with corresponding changes in the luminance of the sky.

For practical purposes, use is made of the *Daylight Factor* – a percentage ratio of the instantaneous illumination level inside to that simultaneously occurring outside in an unobstructed position. Typical daylight factors are indicated in figure 63: daylight reading of a reference point in a room can be made up of three components – *sky component*, or the light received directly from the sky; *externally reflected component* which is the light received after reflection from the ground, building or other external surface,

and *internally reflected component* which is the light received after being reflected from the surfaces inside a building. Therefore, the design of a building must take into account these three factors if the 'correct' amount of daylight is an essential factor in its function and, as already said, design and construction method are closely interrelated. Figure 64 indicates various arrangements which can achieve acceptable daylight factors within a building for specific visual functions. The arrangement of the windows and other openings in the walls provide the main architectural character of a

Building type	Location	Daylight factor %
Dwellings	Living rooms (over $\frac{1}{2}$ depth, but for minimum area 8 m²)	1
	Bedrooms (over $\frac{3}{4}$ depth, but for minimum area 6 m²)	0.5
	Kitchens (over $\frac{1}{2}$ depth, but for minimum area 5 m²)	2
Offices and banks	General offices, counters, accounting book areas, public areas	2
	Typing tables, business machines, manually operated computers	4
Drawing offices	General	2
	Drawing boards	6
Assembly and concert halls	Foyers, auditoria, stairs (on treads)	1
	Corridors (on floors)	0.5
Churches	Body of church	1
	Chancel, choir, pulpit	1.5
	Altars, communion tables (depending on lighting emphasis required)	3–6
	Vestries	2
Libraries	Shelves (on vertical surfaces of book spines), reading tables (additional lighting on book stacks)	1
Art galleries and museums	General	1
	On pictures (but special provision for conservation where required)	6 (max)
Schools and colleges	Assembly and teaching areas	2
	Art rooms	4
	Laboratories (benches)	3
	Staff rooms, common-rooms	1
Hospitals	Wards	1
	Reception rooms, waiting rooms	2
	Pharmacies	3
Sports halls	General	2
Swimming pools	Pool surfaces	2
	Surrounding floor areas	1

Note: The minimum daylight factors recommended do not necessarily apply to the whole area of the interior. Unless otherwise stated, the values are for a horizontal reference plane at table or desk height (0.850 m above floor level).

Figure 63   Typical recommended minimum daylight factors for rooms with side lighting only

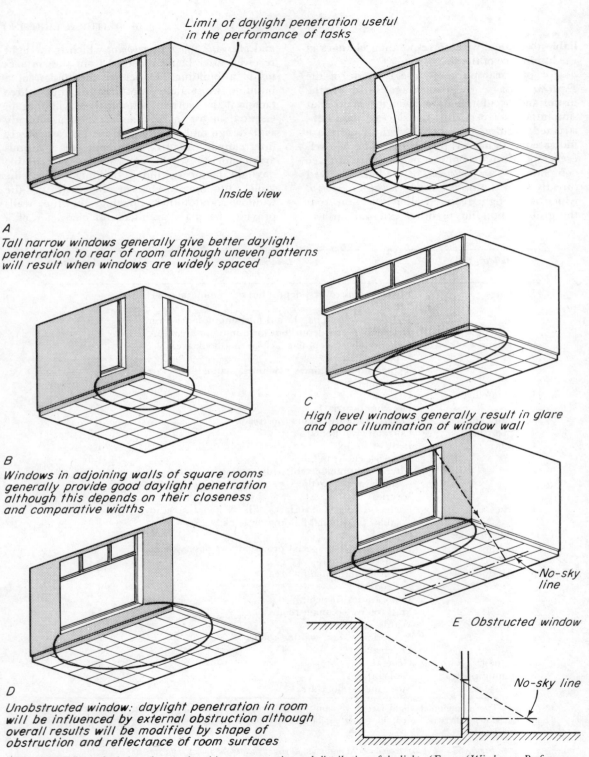

Limit of daylight penetration useful in the performance of tasks

Inside view

A
Tall narrow windows generally give better daylight penetration to rear of room although uneven patterns will result when windows are widely spaced

B
Windows in adjoining walls of square rooms generally provide good daylight penetration although this depends on their closeness and comparative widths

C
High level windows generally result in glare and poor illumination of window wall

No-sky line

D
Unobstructed window: daylight penetration in room will be influenced by external obstruction although overall results will be modified by shape of obstruction and reflectance of room surfaces

E Obstructed window

No-sky line

Figure 64   *Effects of window shape and position on penetration and distribution of daylight. (From: 'Windows: Performance, Design and Installation' by Beckett and Godfrey CLS/RIBA 1974)*

building, and their external appearance is generally referred to as *fenestration*.

No account is taken of the influence of direct sunlight, but various methods of calculating daylight factors have been devised for overcast sky conditions. Modifications to values can be made for glazing materials other than clear glass, dirt on the glass, and reductions caused by the window framing.

If windows or skylights are within the normal field of vision inside a building, they are likely to be distractingly bright compared with other things occupants may wish to study. To reduce this apparent brightness, or *glare*, the openings should generally be placed away from interior focal points, or by reducing the brightness of the light source (tinted glass, louvres or curtaining) whilst at the same time increasing the brightness of the interior spaces by better light distribution techniques such as the use of lighter colours for surfaces, or in extreme conditions, by supplementary artificial lighting.

Although exerting a very pleasing influence, brightening interior colours and providing both psychological and physical warmth, direct sunlight in a building can cause intensive glare, overheating (see chapter 8 *Thermal Comfort*) and fading of surface colours. For this reason, sunlight used to illuminate a building is also often diffused, or reflected to reduce its intensity. Shading and reflecting devices include trees, vines, overhangs, awnings, louvres, blinds, shades and curtains. Overhead shading devices (*brise soleil*) block or filter direct sunlight allowing only reflected light from the sky and ground to enter a building whilst louvres and blinds are capable of converting direct sunlight into a softer, reflected light.

10.3 Artificial lighting

The chief drawback of daylighting is its inconsistency, especially its total unavailability after dusk and before sunrise. Artificial lighting can be instantly and constantly available, is easy to manipulate and can be controlled by the occupants of a building. However, daylighting and artificial lighting should be regarded as complementary; artificial lighting being used mainly for *night-time illumination*, and as *day-time supplement* where daylighting alone is insufficient.

An acceptable balance of brightness within a building can be accomplished by an integration between the design of natural daylight sources and *artificial supplementary lighting* (figure 65), to provide the combined level of light appropriate to a specific visual task. During daylight hours, natural light should appear dominant wherever possible. However, quite apart from artificial light sources supplementing lighting levels, its use in a building could lead to more flexible internal planning arrangements and to the incorporation of fewer or smaller size windows. Thus day-time supplementary artificial lighting schemes directly affect the appearance of a building and its economy of construction. Against this must be levied the probability of greater energy usage, although reference has already been given to the effects which artificial

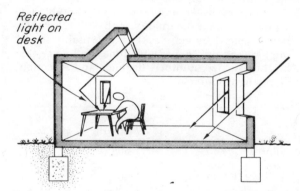

Reflected light on desk

Space designed to provide natural lighting to area remote from external wall

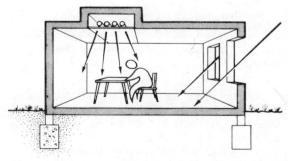

Space designed to supplement natural lighting with artificial lighting to area remote from external wall

Figure 65 Artificial supplementary lighting

lighting installations have upon the heating load for a building, and the possible economic advantages obtained by the re-cycling of heat generated by lamps, etc.

The objective of lighting design is to achieve an appropriate brightness or *luminance* for a visual task to be performed. When establishing desired luminance levels account must be taken of the appearance (position, colour, shape, and texture) of all wall, ceiling and floor surfaces, as well as the selection of suitable light fittings to not only light the task to be performed, but also provide appropriate amounts of reflected light. *Luminance* should not be confused with *illuminance*. The latter is the measure of light falling on a surface (lumens per sq m or lux), whereas the former refers to light reflected from it – or emitted by it (candela per sq m, or alternatively apostilb-illuminance × reflection factor).

Figure 66 lists illumination levels suitable for a range of tasks: the quality of these levels could be influenced by glare and an acceptably limiting index is also shown which may influence the brightness, size, number, distance and direction of any light sources.

	Illumination level lux	Limiting glare index
Offices	500	19
Drawing offices		
General	500	16
Boards	750	16
Auditorium and foyers	100	—
Shops	500	19
Living rooms		—
General	50	—
Reading	150	
Sewing	300	

Figure 66   Illumination levels and limiting glare index for various functions

As seen from figure 65, various basic decisions have to be made concerning lighting objectives and whether the system involves daylight, electric light or a combined system. With electric or combined systems, further decisions must be taken concerning the way light is distributed by a particular fitting, and upon their positions relative to each other as well as in relation to the surface to be illuminated. As with daylighting, light coloured highly-reflective room surfaces help to provide more illumination from the same amount of energy source.

For artificial lighting there is also a problem concerning the way the internal colours of a building may be changed depending upon the way they are affected by the light source. Designers must always ensure that particular light fittings not only provide the correct level of illumination in the required direction, but also that the light (energy) source allows the desired colour rendition of the objects to be illuminated. In this respect the reflectance value of the surrounding surfaces and the contrast created between them also play an important role by reducing the effects of glare. Particular attention should be paid when two or more sources are visible together, such as daylights and supplementary artificial light; or tungsten (incandescent) and fluorescent fittings (see also 12.5 *Accidents*).

The ease with which the maintenance and cleaning of artificial lights can be carried out will depend upon design of fittings and how they are incorporated into a building. Generally, fittings, lamps and auxiliary gear should be readily accessible, and it is an advantage if fittings can be easily removed for replacement and servicing. Access for servicing, whether by means of reaching, ladders, demountable towers, winches, catwalks, or external access by roof, will depend upon the fixing height, space allocation, and the general structural/constructional design details adopted for a building. However, maintenance of light systems must not be considered in isolation since it generally forms part of similar needs of other services, equipment, and perhaps even of window cleaning procedures (see 12.3 *Accidents*).

A building of special importance is often floodlit either for prestige or for security purposes. Frequently, there is also need for emergency or safety lighting – buildings where public assembly takes place, eg theatres, cinemas, department stores, etc. This is generally supplied from an independent energy source and could involve the use of gas, batteries or an automatic-start diesel generator. The provision and this type of lighting must again form an essential part of the design of a building and its construction method.

10.4 Natural ventilation

The simplest ventilation system in a building uses external air as its source, the wind as its motive force, and openings in the external enclosure for fresh air intake on the windward side and stale air extract on the leeward side (figure 67). In a tightly constructed building, air infiltration is slow, but in a building with loose fitting components, doors and windows, air movements will be excessive and cause draughts and high heat losses. The idea, therefore, is to create a naturally ventilated building using correctly fitting components of sizes and configuration which provide the optimum amount of air changes for the occupants according to their activities, and also permits a minimum amount of heat loss. Figure 68 indicates the desirable minimum fresh-air requirements for persons taking part in various activities. These figures are also known as *ventilation rates* or *air change rates*.

In practice, these rates may be very difficult to achieve by natural methods of ventilation because air flow will be governed by areas of openings, and the degree to which their use can be controlled by obstruction within a building restricting air movements, and by the pressure differences causing the flow. Also, the recommended quantities of air indicated in figure 68 will need some adjustment relative to the likely presence of offensive fumes or smells (including

tobacco smoke), as well as for the moisture content of the ventilating air (relative humidity). It is usually considered that relative humidities of between 30% to 70% are acceptable as healthy, and increased ventilation rates will be required to reduce higher levels.

For natural ventilation, windows are used to control the volume, velocity and direction of airflow, and they are designed so as to provide openings capable of adjustment. Without wind forces, airflow through a building is simply produced by the migration of air from a high pressure zone to a low pressure zone by convection currents created through the difference in density between warmer air and cooler air. Unless employing sealed duct devices, fuel burning plant such as fireplaces, boilers or furnaces draw oxygen from the external air via interior spaces, and produce a *stack effect* which will also assist ventilation (figure 69). This technique often avoided the build-up of air in an interior space which contains a large amount of water vapour and thus reduced the possibility of condensation.

However, the need to conserve energy and reduce room heat losses has given rise to a technology which endeavours to reduce air movements in an interior space to a minimum compatible with the functions to be carried out. The spaces created by joints between components are now reduced, finer tolerances are possible, and where gaps are inevitable, rubber

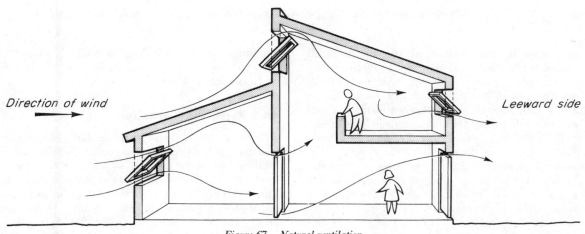

Direction of wind

Leeward side

Figure 67   Natural ventilation

Types of buildings	Recommended minimum rates of fresh-air supply to buildings
Assembly halls	28 m³ per hour per person
Canteens	28 m³ per hour per person
Factories and workshops*	
Work rooms	22.6 m³ per hour per person
Lavatories and WCs	2 air changes per hour
Hospitals	
Operating theatres and X-ray rooms	10 air changes per hour
Wards	3 air changes per hour
Houses and flats	
Bathroom and WCs	2 air changes per hour
Halls and passage	1 air change per hour
Kitchens	56 m³ per hour
Living rooms & bedrooms:	
8.5 m³ per person	20.5 m³ per hour per person
11.5 m³ per person	18.5 m³ per hour per person
14 m³ per person	12 m³ per hour per person
Pantries and larders	2 air changes per hour
Places of entertainment	28 m³ per hour per person
Restaurants	28 m³ per hour per person
Schools	
Occupied rooms (classrooms, laboratories, practical rooms, etc):	
2.8 m³ per person	42 m³ per hour per person
55.6 m³ per person	28 m³ per hour per person
8.5 m³ per person	20.5 m³ per hour per person
11.2 m³ per person	18.5 m³ per hour per person
14 m³ per person	12 m³ per hour per person
Cloakrooms	3 air changes per hour
Corridors, lavatories and WCs	2 air changes per hour

* The conditions in factories are regulated by the Factories Acts, and regulations made thereunder. From BSCP 3, Chapter I (c) *Ventilation*

Figure 68   Recommended minimum rates of fresh-air supply to buildings for human habitation

or synthetic seals are used to ensure small amounts of air circulation from the exterior to the interior of a building. This often means that fresh air ducts or grilles now have to be provided to allow sufficient air for both combustion of fuel and the well-being of the occupants in a building.

Building legislation (*The Building Regulations 1985*: Approved documents F) requires open spaces outside ventilating openings in domestic buildings to ensure an adequate volume of air for ventilation. Nevertheless, when a building is subjected to high velocity winds which are likely to cause excessive ventilation or draughts, as well as high heat losses, care should be taken to locate openings in the building exclusively on the leeward side, or protect them by shielding devices such as fences or trees, etc.

10.5 Artificial ventilation

Where a reliable and positive flow of air for ventilation is required, fans can be installed in a building to *extract* stale air which is immediately replaced by fresh air from the outside flowing through gaps around windows/door frames, or through purposely designed grilles. Fans in more elaborate ventilation schemes are connected to systems of ductwork for better air dis-

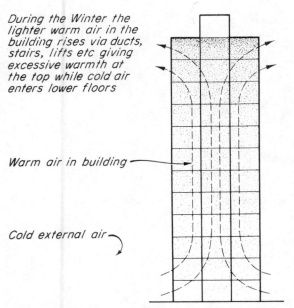

During the Winter the lighter warm air in the building rises via ducts, stairs, lifts etc giving excessive warmth at the top while cold air enters lower floors

Warm air in building

Cold external air

Figure 69   Stack effect in naturally ventilated tall building

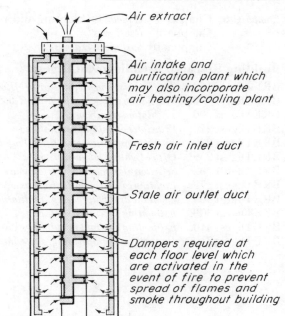

Air extract

Air intake and purification plant which may also incorporate air heating/cooling plant

Fresh air inlet duct

Stale air outlet duct

Dampers required at each floor level which are activated in the event of fire to prevent spread of flames and smoke throughout building

Figure 70   Method of obtaining clean air uniformly through building using artificial ventilation system

tribution throughout a building. Separate duct systems can be installed to pull away stale air from those used to distribute clean air. Often such systems are coupled with heating and cooling plant in such a way that the clean air is distributed at the selected temperature for thermal comfort (air conditioning) (figure 70).

Artificial ventilation systems are usually employed for internally located rooms; crowded rooms where natural ventilation is insufficient; special rooms needing closely controlled humidity and/or freedom from any dust (eg computer rooms and operating theatres); and where polluted air is required to be either removed or prevented from entering internal spaces.

Certain high buildings will need artificial air movement control to ensure a balanced thermal environment because of the otherwise exaggerated 'stack effect' causing the highest parts being too hot as a result of rapidly rising warm air.

The selection, design and integration of artificial ventilation and, more particularly, air conditioning systems into a building requires specialist knowledge, techniques and skills. It is essential that the implications of a chosen scheme are realized at a very early stage in the design of a building. The effects of plant; exposed or concealed ductwork; additional fire protection to prevent the spread of fire through ductwork, etc; suspended ceilings; and additional sound control must all be carefully considered as they may have a profound effect on not only the appearance of the building, but also other technical aspects, including construction method. The integration of plant and ductwork within the dimensional discipline of the structure of a building requires particular attention.

Further specific reading

Mitchell's Building Series

Environment and Services: Chapter 3 *Air movement* (Ventilation)
Chapter 4 *Daylighting*
Chapter 7 *Thermal insulation* (Ventilation systems)
Chapter 8 *Electric lighting*

Components: Chapter 4   *Windows*   (Lighting and ventilation)
            Chapter 6   *Rooflights*
            Chapter 10 *Suspended ceilings*

Building Research Establishment Digests

BRE Digest 162:   *Traffic noise and overheating in offices*
BRE Digest 170:   *Ventilation of internal bathrooms and WCs in dwellings*
BRE Digest 206:   *Ventilation requirements*
BRE Digest 210:   *Principles of natural ventilation*
BRE Digest 232:   *Energy conservation in artificial lighting*
BRE Digest 256:   *Office lighting for good visual task conditions*
BRE Digest 262:   *Selection of windows by performance*
BRE Digest 272:   *Lighting controls and daylight use*
BRE Digest 306:   *Domestic draughtproofing: ventilation considerations*
BRE Digest 309:   *Estimating daylight in buildings* Part 1
BRE Digest 310:   *Estimating daylight in buildings* Part 2
BRE Digest 319:   *Domestic draughtproofing: materials costs and benefits*

Some of the essential activities taking place within and around a building are liable to encourage the growth of bacteria, insects, and vermin, which could result in pollution, disease, and foul smells. It is, therefore, necessary to control carefully the conditions most favourable to the development of these unwanted phenomena, ie the provisions for *drinking water, food preparation and washing*; and the generation of *waste products, refuse and dirt*. As far as the building designer is concerned, this involves a careful analysis of suitable water supply and storage systems, and the effective methods of waste and refuse removal. Decisions in these areas must be closely related to the selection of materials least subject to contamination, and the provision in design for efficient cleaning and freedom from deterioration.

11.1 Drinking water, food preparation and washing

In the UK, the supply of water to a building is subject to statutory undertakings mostly controlled by the provisions of the Water Acts 1945 and 1973. Clean and potable water is, therefore, generally in ample supply even during times of drought, and as the same source is used for all purposes in a building, installation are of simple design (figure 71). The principal mechanical requirements for the system are that all *pipework* used should be non-corrodible, capable of being tightly jointed, and offer resistance to deformation and mechanical damage; the *layout* of pipework should provide the minimum resistance to water flow and should be protected from freezing; and, *forming*, *supporting*, and *connecting* techniques to appliance should reduce the possibility of noise generation and transmission. The water supply system or *plumbing* allows the water not intended for human consumption to be stored in a replenishable cistern incorporating its own feed system to sanitary appliances such as wcs, bidets, basins and baths; wash down points; and the hot water system supplied via another storage vessel and cylinder. The provision of

separate water storage facilities, isolated from the mains, reduces the risk of mis-use and contamination of the water within the rising main. Mis-use could adversely affect the supply source, as well as the water to be used for direct human consumption (drinking and food preparation) within a building. The storage cistern also ensures a continued supply of water in the event of the mains supply being cut off for a short period due to damage or maintenance work. By installing the storage tank at high level in a building sufficient supply pressure can be ensured. However, if the pressure of the water in the mains supply is insufficient or likely to fluctuate dramatically owing to demand, the mains water must be pumped to the required level for a building to be adequately serviced. As already indicated, it is usual to provide a cold water storage cistern as this assists in reducing the size of the pipe used for the mains supply. To avoid the possibility of pollution, there must be a clear horizontal gap between the outlet of the supply pipe and the top surface of the stored water — the gap being maintained by siting an overflow pipe at a distance *below* the supply pipe. Regulation of the flow of water is achieved by a float valve, usually in the form of a hollow ball (see figure 71).

A similar gap is provided between the outlet of a tap on sanitary appliances and the maximum permitted height of water (controlled by an overflow) which the appliances are likely to contain; eg the drinking water supply to a kitchen sink is isolated from the hot water supply in this manner. However, as a result of research based on systems in America and some European countries, some UK local authorities now allow taps to most sanitary appliances to be fed directly from the mains without involving water storage cisterns. Figure 72 indicates such a plumbing system, and pollution is prevented by using an anti-vacuum valve which maintains a positive pressure in the system to prevent any possibility of water backflow (back syphonage). Sealed hot water heating systems are currently used in the UK, but the diagram also indicates

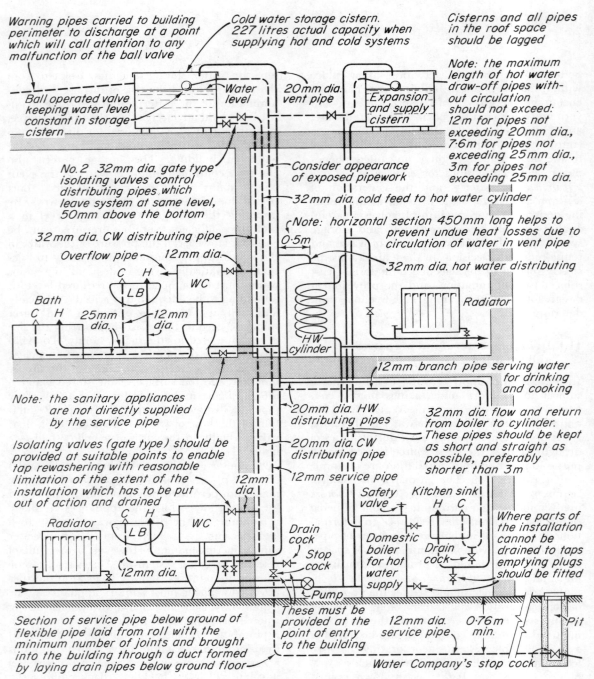

Warning pipes carried to building perimeter to discharge at a point which will call attention to any malfunction of the ball valve

Cold water storage cistern. 227 litres actual capacity when supplying hot and cold systems

Cisterns and all pipes in the roof space should be lagged

Ball operated valve keeping water level constant in storage cistern

Water level

20mm dia. vent pipe

Expansion and supply cistern

Note: the maximum length of hot water draw-off pipes without circulation should not exceed: 12m for pipes not exceeding 20mm dia., 7·6m for pipes not exceeding 25mm dia., 3m for pipes not exceeding 25mm dia.

No. 2 32mm dia. gate type isolating valves control distributing pipes which leave system at same level, 50mm above the bottom

Consider appearance of exposed pipework

32mm dia. cold feed to hot water cylinder

32mm dia. CW distributing pipe

Overflow pipe

12mm dia.

0·5m

Note: horizontal section 450mm long helps to prevent undue heat losses due to circulation of water in vent pipe

32mm dia. hot water distributing

Bath
C H

25mm dia.

12mm dia.

C H
LB
WC

HW cylinder

Radiator

Note: the sanitary appliances are not directly supplied by the service pipe

12mm branch pipe serving water for drinking and cooking

20mm dia. HW distributing pipes

32mm dia. flow and return from boiler to cylinder. These pipes should be kept as short and straight as possible, preferably shorter than 3m

Isolating valves (gate type) should be provided at suitable points to enable tap rewashering with reasonable limitation of the extent of the installation which has to be put out of action and drained

20mm dia. CW distributing pipe

12mm service pipe

12mm dia.

Kitchen sink
H C

Where parts of the installation cannot be drained to taps emptying plugs should be fitted

Radiator

C H
LB
WC

12mm dia.

Drain cock

Stop cock

Pump

Safety valve

Domestic boiler for hot water supply

Drain cock

Section of service pipe below ground of flexible pipe laid from roll with the minimum number of joints and brought into the building through a duct formed by laying drain pipes below ground floor

These must be provided at the point of entry to the building

12mm dia. service pipe

0·76m min.

Pit

Water Company's stop cock

Figure 71   Typical domestic water supply system

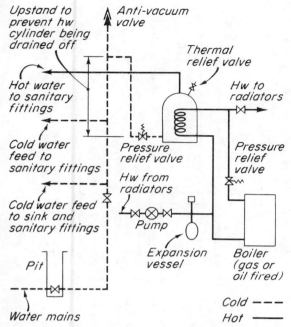

Upstand to prevent hw cylinder being drained off

Anti-vacuum valve

Thermal relief valve

Hot water to sanitary fittings

Hw to radiators

Cold water feed to sanitary fittings

Pressure relief valve

Pressure relief valve

Hw from radiators

Cold water feed to sink and sanitary fittings

Pump

Expansion vessel

Boiler (gas or oil fired)

Pit

Cold − − −

Water mains

Hot ⸻

Figure 72 Proposed plumbing arrangements based on current EEC practice

a sealed-system for the hot water supply. The latter is similar to that which is widely employed in other EEC countries and has now been introduced into the UK.

In addition to the mechanical and technical criteria, it is very important that detailed consideration is given to the *appearance* of a water supply system within a building. Pipework, ancillary devices, and appliances must be carefully integrated as an essential part of the design of a building. No plumbing system should 'occur' in a building as an afterthought; if adequately considered during the early stages of design the system can enhance the appearance of purely practical areas, or be successfully concealed in other areas by incorporation with other services within structural elements, eg wall, floor and roof construction.

11.2 Waste products, refuse and dirt

Depending upon its precise function, usually a building must also provide installations which facilitate the disposal of water-borne organic waste matter including human excrement

(sewage); permit the collection of rainwater otherwise liable to cause some form of deterioration; and, allow other organic and non-organic waste (refuse) to be removed.

Sanitary fittings (wcs, bidets, urinals, baths, showers, basins and sinks) which receive the sewage are designed on the basis of specific anthropometric data for efficient function, and also to eliminate the possibility of bacteria, as well as foul smells, building up. They are manufactured from non-porous, smooth, durable, and easily cleaned materials, and incorporate a water-filled trap to prevent gases escaping into the building from the pipework beyond. Further precautions against smells and other forms of pollution are provided by building designs which promote good ventilation in rooms used for bathrooms, laboratories, operating theatres, abatoirs, etc, where sanitary appliances are situated; ensuring adequate space around a building and its correct location within a particular environment.

11.3 Drainage systems

The sanitary fittings discharge sewage along a network of gas and watertight pipes which together form the *drainage system* of a building. This system conveys both solids and liquids to a treatment plant in the local authority sewage works, or, if this plant is not available, to a *cesspool*, or *septic tank* and filter bed in close proximity. Pipes which convey liquids only are generally referred to as *waste pipes*, and those which convey solid matter and liquids as *soil pipes*.

Efficient and economic design considerations usually result in the waste and soil pipe system in a building taking one of two forms. The simplest occurs when sanitary appliances can be fairly closely grouped around vertical stacks (soil and waste pipes) which then conveys sewage by the most direct route to underground drains (figure 73). Such installations are found in many types of buildings, including blocks of flats, where the activities to be accommodated allow vertically repetitive planning arrangements. It is now usual to combine the soil and the waste pipe into a *single pipe system* and the overall dimensions of this pipe is dictated by the amount and frequency of solid material it must transfer to the drains. Great care must be taken

85

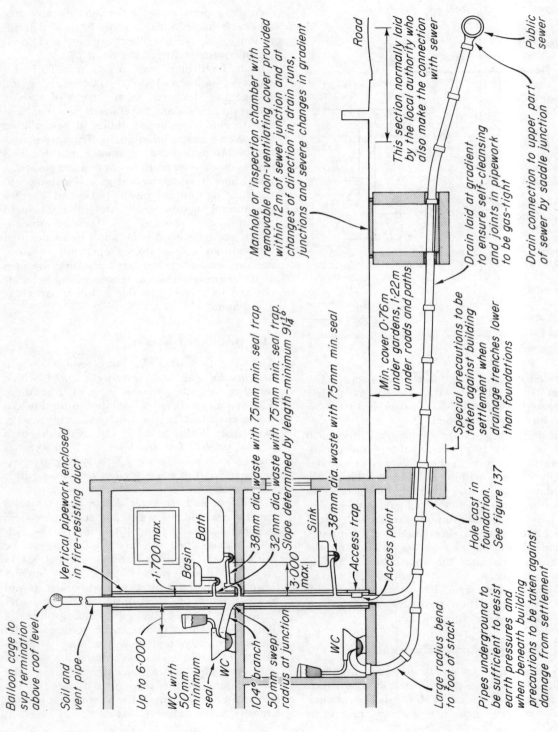

Road

Public sewer

This section normally laid by the local authority who also make the connection with sewer

Manhole or inspection chamber with removable non-ventilating cover provided within 12m of sewer junction and at changes of direction in drain runs, junctions and severe changes in gradient

Drain laid at gradient to ensure self-cleansing and joints in pipework to be gas-tight

Drain connection to upper part of sewer by saddle junction

Min. cover 0·76m under gardens, 1·22m under roads and paths

Special precautions to be taken against building settlement when drainage trenches lower than foundations

Balloon cage to svp termination above roof level

Vertical pipework enclosed in fire-resisting duct

Soil and vent pipe

Up to 6·000

1·700 max.

Basin

Bath

38mm dia. waste with 75mm min. seal trap

32mm dia. waste with 75mm min. seal trap. Slope determined by length–minimum 9¼°

3·000 max.

Sink

38mm dia. waste with 75mm min. seal

Access trap

Access point

Hole cast in foundation. See figure 137

WC with 50mm minimum seal

WC

104° branch 50mm swept radius at junction

WC

Large radius bend to foot of stack

Pipes underground to be sufficient to resist earth pressures and when beneath building precautions to be taken against damage from settlement

Figure 73 Typical sanitary installation and drainage for small building

when designing the system to ensure that the water seals or traps in the sanitary fittings are not prevented from re-forming after use. Seals may not reform because of *induced syphonage*, which occurs when the seal is sucked away by the force of water discharging from a branch pipe on the floor(s) and passing down the vertical pipe. The rush of water down the pipe absorbs air from the branch pipe of the fittings below and this results in the external air pressure on the seal also forcing the water down the vertical pipe. This syphonic action can be prevented by ventilation pipes (anti-syphon pipes) on the drain side of the sanitary fitting to equalize air pressure within the main stack. Alternatively special anti-syphon re-sealing traps which let air into the system without complete loss of trap seal can be used, or more usually today, the whole drainage system can be designed with careful regard to sizes, lengths, slopes, and positioning of pipework. Design rules are based upon empirical and scientific analysis of water flow and air pressure characteristics.

When the activities within a building are complex and it is not possible to group sanitary fittings around a centralised stack – as in hospitals, schools, and certain office layouts – a satisfactory drainage system can be accomplished by grouping fittings in 'islands', and by providing extensive horizontal pipework connection to strategically located vertical soil and/or waste stacks. This system is expensive and more complicated than the centralised system because the necessary prolification of horizontal pipes not only makes concealment by 'false' ceilings necessary, but also further technical problems again associated with syphonage must be overcome. Overlong horizontal lengths of pipework are liable to cause the discharged water from a sanitary fitting to remove the trap seal by *self-syphonage*. This occurs when a horizontal pipe flows full and prevents a build-up of positive pressure on the drain side of the trap until the last of the water has been drawn away. Fortunately baths and large flat-bottom sinks do not suffer self-syphonage as the last of the discharging water moves very slowly allowing the seal to re-settle, and the waste pipe of a wc is too large to flow at full bore. For other sanitary fittings with extensive horizontal lengths of pipework, larger sizes of pipe and depth of seals

must be incorporated, and also probably an elaborate ventilating system.

11.4 Rainwater collection

The planning of a drainage system for the removal of waste should be done in conjunction with the drainage system required for the collection of rainwater. This involves collection of rainwater falling on a building *and* perhaps around a building. The collected rainwater must then be conveyed in a similar manner as waste water to a local authority *surface water drain*. If this drain is not available, rainwater can be taken to a purpose-built *soakaway* situated away from areas likely to be detrimentally affected by excessive amounts of water, or to a conveniently located *water course* or a *storage vessel* for subsequent use as a water supply (figure 74).

Water falling on a building at roof level is collected and discharged into a rainwater pipe system by means of strategically located rainwater outlets. If a building is lower than about five storeys, easier means of access and maintenance generally make a system of collection involving guttering more acceptable. The water is conveyed to the vertical rainwater pipes which discharge either directly into the drain below ground or over trap seal gulleys at ground level. Sizes of pipes for the design of rainwater drainage systems in the UK are dependent upon the expected risk involved: 50 mm/hour of rainfall for flat roofs and other open paved areas; 75 mm/hour for sloping roofs; or 150 mm/hour when very occasional overflowing of rainwater outlets/gutters cannot be tolerated. (This amount likely to fall only during short periods of heavy storms.)

When considered together, there are three basic drainage systems approved by local authorities for the disposal of soil, waste *and* rainwater (figure 75).

(a) *Combined system* where both drains discharge into a common sewer. This is a simple and economic system because it involves less pipework and is easy to maintain. However, it has the disadvantage that vast amounts of liquids must pass through the sewage treatment works – particularly during periods following heavy rainfall.

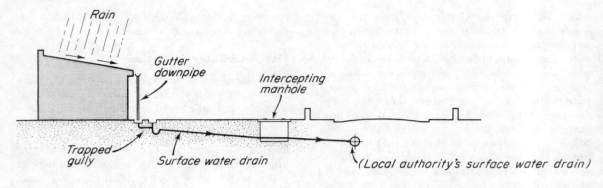

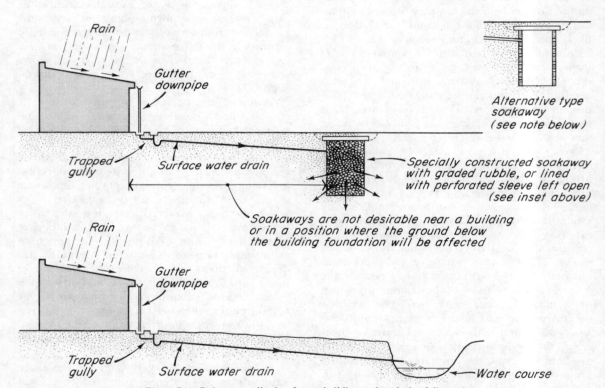

Figure 74   Rainwater collection from a building and methods of disposal

(b) *Separate system* where two drains are provided, and one of them receives the collected rainwater (or surface water) and conveys it direct to a suitable outfall without treatment, eg nearby water course or river. The second drain takes the soil and waste discharge and conveys it to the sewage treatment installation. This obviously involves more drainage pipes, but avoids the risk of overcharging the sewage treatment plant during periods following heavy rainfall.

(c) *Partially separate system* where a combined drain is used for soil, waste and rainfall. However, a second drain is also available to regulate the amount of rain/surface water discharging into the combined drain according to the capacity of the sewage treatment installation.

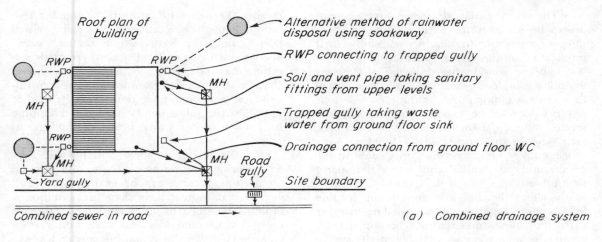

Roof plan of building

Alternative method of rainwater disposal using soakaway

RWP connecting to trapped gully

Soil and vent pipe taking sanitary fittings from upper levels

Trapped gully taking waste water from ground floor sink

Drainage connection from ground floor WC

Site boundary

Combined sewer in road

(a) Combined drainage system

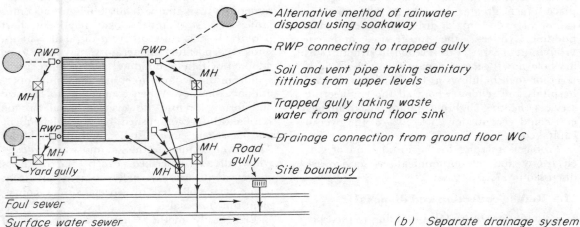

Alternative method of rainwater disposal using soakaway

RWP connecting to trapped gully

Soil and vent pipe taking sanitary fittings from upper levels

Trapped gully taking waste water from ground floor sink

Drainage connection from ground floor WC

Site boundary

Foul sewer

Surface water sewer

(b) Separate drainage system

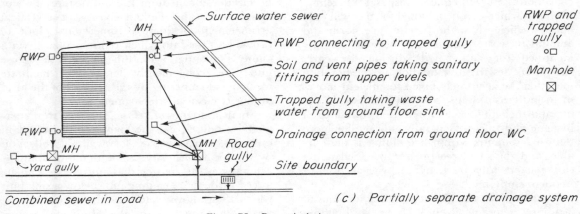

Surface water sewer

RWP connecting to trapped gully

Soil and vent pipes taking sanitary fittings from upper levels

Trapped gully taking waste water from ground floor sink

Drainage connection from ground floor WC

Site boundary

RWP and trapped gully
o□

Manhole
⊠

Combined sewer in road

(c) Partially separate drainage system

Figure 75   Domestic drainage system

89

Whatever internal drainage arrangements are to be incorporated within a building it is always important to consider carefully their implication on the precise system to be adopted. Not only can considerable economy be made by adopting simple systems, but also the visual impact on a building can be quite considerable – pipework can be exposed, or concealed in service walls, floors or ducts, and false ceilings, etc. When requirements for a building were much simpler, drainage specialists could arrive on site and, after initial competition for spaces with other trades, install their services. The new approach to the design of installation leading from a more sophisticated knowledge, together with the standardisation of components for greater economy, now requires designers to place much greater thought to the incorporation of even simple drainage systems in a building. Indeed, the integration and co-ordination of services in relation to structures as a whole now becomes a paramount design criteria for such buildings as blocks of flats, offices, hospitals and schools where all the engineering services (heating, lighting, plumbing, drainage, etc) could account for 25%–50% of the overall capital costs.

Problems relating to the installation of gas, electricity and telecommunications service are discussed in 17.10 *Services*.

11.5 Refuse collection and disposal

It is usually necessary for a building to incorporate adequate arrangements in its design for the collection and subsequent disposal of refuse. Methods which can be adopted for this will only operate efficiently if the precise type, form and amount of waste produce has been successfully identified (or anticipated) during the early stages of design investigation. Consideration may then be accurately given to efficient movement patterns of refuse about a proposed building and their influence on required standards of hygiene and safety. Generally, storage for more than a few hours within a building is undesirable, and the small receptacles in which refuse can conveniently be temporarily placed need frequent emptying.

Unobstructed and direct circulation routes to facilitate refuse collection from a building should also be thoroughly planned, thereby fur-thering the desire for protection from the pollution caused by unpleasant smells, visual horrors and noise. Refuse which is not destroyed at source is taken to local sites where crude selection may take place to permit incineration, consolidation and/or transportation to centralised tips. Some non-toxic refuse may be used to backfill areas subsequently needed for building sites, or used for other forms of land reclamation.

The design criteria for a building will vary when considering refuse collection and disposal methods applicable to, for example, medical, commercial, industrial, or domestic activities. A building designed to accommodate complex or multi-purpose activities (hospitals and certain factories) can generate many forms of waste – sometimes toxic, individually bulky, or accumulating in vast quantities over relatively short periods. In this case, different disposal systems, some incorporating incinerators, may have to be adopted within a building, and require great skill from the designer to ensure maximum operational efficiency. When convenient to the size and form of refuse, however, disposal can be satisfactorily accomplished by an independent *water-borne pipe system* installed within a building. This is very similar to the soil and waste *drainage system* already described, except the refuse is conveyed to an externally located pit from which it is collected by 'specialists' at convenient time intervals.

Less costly methods are available, one of which involves the use of a *dry chute*, and would be suitable for medium rise multi-storey flats or maisonettes. The method consists of a vertical arrangement of jointed impervious pipes to provide a tube into which refuse is placed via a chute. The outlet of the tube deposits the refuse into bins conveniently located to facilitate mechanical emptying by the vehicles of the local authority cleansing department.

Small amounts of refuse, such as are generated by houses, can be conveniently stored in dustbins or plastic bags for eventual collection by the local authority. Some domestic refuse – generally waste matter that putrefies – can be forced through a grinding unit below a sink unit which then allows it to pass through a normal waste pipe and into the drainage system beyond.

Further specific reading

Mitchell's Building Series

Environment and Services: Chapter 9   *Water supply*
                              Chapter 10 *Sanitary appliances*
                              Chapter 11 *Pipes*
                              Chapter 12 *Drainage installation*
                              Chapter 13 *Sewage disposal*
                              Chapter 14 *Refuse collection and storage*
Structure and Fabric Part 2: Chapter 7 *Chimneys, shafts, flues and ducts* (Ducts for services)
Components: Chapter 11 *Demountable partitions* 11.2 Integration of services
                 Chapter 12 *Suspended ceilings*
                 Chapter 13 *Raised floors*
                 Chapter 14 *Roofings* 14.2 Weather exclusion

Mitchell's Professional Library

Water, Sanitary and Waste Services for Buildings, Alan Wise

Building Research Establishment Digests

BRE Digest 81:   *Hospital sanitary services: some design and maintenance problems*
BRE Digest 83:   *Plumbing with stainless steel*
BRE Digest 98:   *Durability of metals in natural waters*
BRE Digest 151:   *Soakaways*
BRE Digest 205:   *Domestic water heating by solar energy*
BRE Digest 248:   *Sanitary pipework* Part 1: Design basis
BRE Digest 249:   *Sanitary pipework* Part 2: Design of pipework
BRE Digest 288:   *Dust explosions*
BRE Digest 292:   *Access to underground drainage systems*
BRE Digest 308:   *Unvented domestic hot water systems*

12 Security

Like many other performance requirements, the *security* aspects of a building involve the immediate physiological and psychological well being of the occupants. The main areas of concern relate to *unauthorized entry* into a building; *vandalism*; protection against *disasters*; and the reduction of *accidents*. The degree of risk associated with a particular building must be established during initial design stages in order that appropriate security measures can be incorporated which will permit the activities accommodated to be maintained without hindrance. In most cases, the primary and often most economical defence lies in the fabric of a building and, therefore, involves the use of compatible construction methods (materials and techniques of assembly).

12.1 Unauthorized entry

Unauthorized entry can be achieved either *visually* or *physically*. Both involve invasion of the 'private zones' of a building. But, whereas the former may be no more than inconvenient, the latter often involves violence towards people, and/or damage and theft of furniture, fabrics, machines, personal belongings and livestock. The defence systems involved for the two areas, therefore, must be considered separately, although they can be resolved together to provide a combined security.

Visual privacy can be achieved through external planning arrangements of a building which provide an acceptable degree of *remoteness* between observer and observed. When adequate distances for this are not available (see 17.2 *Site considerations*), or it is required to augment the remoteness for even greater privacy, the design of a building and its immediate surroundings play a more important role. The location of the more 'private zones' of a building can be away from the general fields of vision. Alternatively, high perimeter walls, trees and plants, landscaping, projecting wings and outbuildings can be employed as visual barriers.

Although less used today owing to energy conservation requirements, large windows which let in more daylight make room interiors more visible from outside. Cross lit rooms also reduce privacy levels by silhouetting figures. Whilst net curtains or slatted blinds may be considered a satisfactory solution by some people, it should still be possible to obtain the levels of daylights which the window was designed to provide, and for the occupants to enjoy uncurtained views out of a building without loss of visual privacy.

Figure 76 illustrates some of the principles involved when trying to provide reasonable visual privacy between dwellings based on the *Design Guide for Residential Areas* published by the County Council of Essex. They involve the inter-relationship between acceptable 'eye-to-eye' remoteness and the use of screening devices above eye level.

As an initial means of defence against unauthorized *physical* entry, the approach routes to a building should be carefully organized to prevent, or at least limit, the possibility of uncontrolled usage. This principle is always considered during the initial design stages of a building accommodating strongrooms, security stores, and prisons, etc. However, there has been a certain amount of complacency about the planning of a building having less dramatic risks. This has led to the development in the use of numerous elaborate security measures or costly security patrols, often added to a building as an afterthought in an attempt to justify incomplete initial design investigations. Nevertheless, although entrances can be positioned to provide good, well lit visibility and control, problems may still arise in a building requiring a high degree of security but where easy access by the public is necessary (banks, hotels, offices, police stations, museums, exhibitions, etc). Also, of course, the final defence against unauthorized entry into a building must rely on some form of locking mechanism.

A conflict often occurs, however, between the provision of secure doors/windows, and those which will permit the legal occupants of a build-

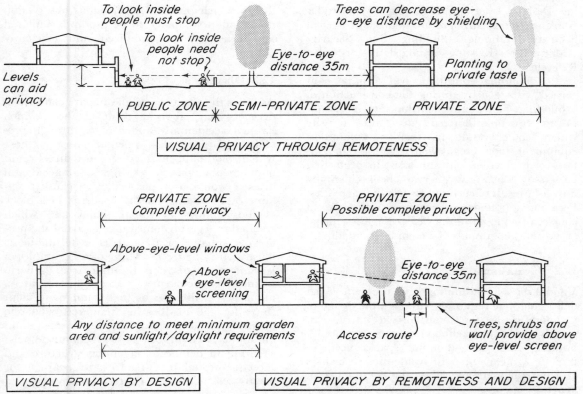

To look inside
people must stop

To look inside
people need
not stop?

Trees can decrease eye-
to-eye distance by shielding

Eye-to-eye
distance 35m

Levels
can aid
privacy

Planting to
private taste

PUBLIC ZONE   SEMI-PRIVATE ZONE   PRIVATE ZONE

VISUAL PRIVACY THROUGH REMOTENESS

PRIVATE ZONE
Complete privacy

PRIVATE ZONE
Possible complete privacy

Above-eye-level windows

Above-
eye-level
screening

Eye-to-eye
distance 35m

Any distance to meet minimum garden
area and sunlight/daylight requirements

Access route

Trees, shrubs and
wall provide above
eye-level screen

VISUAL PRIVACY BY DESIGN   VISUAL PRIVACY BY REMOTENESS AND DESIGN

Figure 76   Visual privacy provided by building location and design. (From Design guide for residential spaces, *Essex County Council)*

ing to escape without hindrance in the case of an outbreak of fire. A compromise is sometimes required when selecting locks (ironmongery), particularly for doors, although locks are now available which appear to satisfy both criteria. A similar form of compromise is often required in the design of fitments which must freely exhibit goods whilst protecting them from shoplifters.

Methods of construction should augment the initial design considerations which prevent *illegal* entry. For example, as certain lightweight constructions can be easily dismantled – boards removed from wall cladding systems or tiles lifted from roofs – care must be taken in their detail design to ensure their use if compatible with the security risk of a building. Windows, and doors, normally regarded as a means of giving access, or light and ventilation, should also be considered as a means of admitting a

criminal and must be carefully designed and protected accordingly.

Once the design of a building fabric has provided the maximum amount of security possible, further protection can be provided if necessary by electronic devices, alarm bells, and even by patrols or night watchmen. The Crime Prevention Officer (CPO) of the Police Force and/or Insurance Companies will advise at an early stage about the degree of risk involved and recommend suitable measures to be incorporated in the design and construction of a building.

The Avon and Somerset Constabulary have produced a booklet entitled *Security in Design* which gives an excellent checklist of important considerations under the following headings: Assessment of Crime Risk; Layout of Site; Building Design (Access from Exterior); Building Design (Access–Windows and Glass); Building Design (Interior); Ironmongery; Intruder Alarm Systems; Strongrooms and Safes; Special Security Planning; Building Site Security (Contractor); Existing Standards on Security; References.

The building contractor will also be concerned with security during the construction of a building. Fences and hoardings are necessary to prevent unauthorized entry into the site which could result in theft of materials, equipment, tools, etc, as well as vandalism. This form of protection will also assist in protecting passers-by who may otherwise be accidentally hurt as a result of certain construction activities. Permanent boundary fences and walls can, of course, provide an initial defence against illegal entry into the grounds of a completed building.

12.2 Vandalism

Vandalism is a continuing social problem which defies complete resolution although the design of a building, as for unauthorized entry, can greatly lessen the likelihood of its occurrence. A building and the adjoining spaces need to provide a means of positive identity to owners and normal users as well as the community in general.

The problem ranges from graffiti on accessible surface finishes to physical damage of a building fabric involving defacement, breakage, or complete destruction by demolition or fire. Graffiti can be avoided if not eliminated by the judicious selection of surface finishes; but physical damage is a serious problem which may result in a building employing 'fortress-like' construction methods. Everything must be robust and secure from attack where wilful damage is likely to occur and, of course, the appearance and other performance requirements of a building may suffer as a result. The 'ordinary' use of materials will have to be avoided – glazed areas should be of limited size and reinforced; doors of solid construction; walls of dense robust materials which are not easily ignitable, etc. Under certain conditions, suitably selected materials may require further protection by barriers or screens. Accidental damage can also be avoided by these devices.

To some extent the effects of vandalism can be lessened by a building owner through a design which provides conspicuous observation points, allows damage to be promptly repaired and litter to be minimized and efficiently cleared.

12.3 Disasters

As far as the effects on a building are concerned, disasters can take the form of those which happen accidentally and usually through previous ignorance of a phenomenon, or those which could happen as a direct or indirect result of planned actions. An example of an accidental disaster would be a gas explosion causing collapse of a building such as occurred in a block of flats known as Ronan Point and which resulted in legislative measures to avoid the progressive collapse of a construction technique as illustrated in figure 77. Many other examples of learning from disasters resulting from hitherto unknown occurrences are, unfortunately, numerous – the use of high alumina cement (hac) for the loadbearing reinforced concrete members in a building where warm humid atmospheric conditions exist (swimming pools) resulting in deterioration of the structure and subsequent collapse; inadequate bearing for prestressed reinforced concrete beams causing

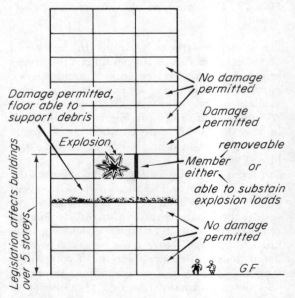

Figure 77 Legal requirements to avoid possibility of progressive collapse in tall buildings

their collapse; the lack of fire barriers in cavity constructions allowing the rapid spread of fire from one part of a building to another. Failures such as these and the Ronan Point disaster can cause serious loss of life and cost enormous sums of money to rectify.

Experience of disasters, together with the continuous research by investigative authorities (Building Research Establishment, Princes Risborough Laboratory, TRADA, etc) and testing organizations (British Board of Agrément) help to reduce the likelihood of their recurrence. However, there is always need for a designer to proceed cautiously by seeking out maximum information regarding the performance criteria with new products, materials, and untried methods of construction.

12.4 Terrorism

Acts of terrorism or war often involve the use of explosives to cause damage to buildings. For economic reasons, there can be few defensive measures taken to avoid complete or even partial destruction of a building constructed using normal techniques. However, when necessary, a building can be specially designed to withstand a certain degree of anticipated damage. Today, it seems that the ultimate form of this type of construction is one which must resist the light and heat, blast wave, tremors, and fall-out from a nuclear explosion. Construction methods can take the form of relatively simple 'do-it-yourself' techniques or those which involve housing large structures almost entirely below ground and which incorporate complicated life-support systems.

Specific forms of construction are also available to lessen the effects of natural disasters resulting from earthquakes, floods, hurricanes, etc.

12.5 Accidents

A building must be designed to ensure that the human activities it accommodates are carried out with the maximum amount of comfort, safety and efficient enjoyment. Also, of course, it is important that the construction of a building is carried out with reasonable comfort, high degree of safety, and hopefully, enjoyment (see 15.7 *Construction team*).

Mention has been made under *dimensional suitability* (appropriate size) regarding the need to manufacture components to sizes which are sympathetic with building operations and use. However, even appropriately sized components must be used wisely since it is not uncommon for people either to cause damage or to be damaged, quite accidentally, as a result of ordinary daily activities (figure 78). Careful consideration must be given to work surface heights in kitchens, offices or workshops, juxtaposition of conflicting activities within confined spaces, suitable clearances around specific activities, etc. The study of the relationship between people and their environment is known as *ergonomics* and applies physiological and psychological reasoning to anthropometric data. Figure 79 gives typical examples.

The design of spaces in and around a building should take into account the appropriately safe features and dimensions for stair flights, treads, risers and landings; slopes of ramps; heights and profiles for handrails and guardrails, etc. Circulation areas must be planned to give efficient movement patterns which avoid mixing of opposing activities. In this respect, special consideration must be given to the problems associated with the circulation of disabled people through a building, and into the various spaces it accommodates. Gas and water services should be separated from electrical services to avoid risks of explosion, fire and electrocution. If the opening parts of any windows above ground floor level are lower than 1020 mm from the inside floor, additional protection will be required against people falling out. Beyond third storey height, the minimum internal height of the window board should be increased to 1120 mm. Whenever large areas of clear glass are incorporated in the construction of walls, precautions must be taken to ensure that people are made aware of its presence. This can be done by physical and/or visual barriers which indicate proximity of a transparent obstruction.

Ideas of design for safety must be extended to appropriate surface finishes for a specific activity – non-slip floor tiles in swimming pools; non-combustible wall surfaces along fire escape routes; anti-static finishes in operating theatres.

The *use of colour* can play a very important role in helping to prevent accidents, although it

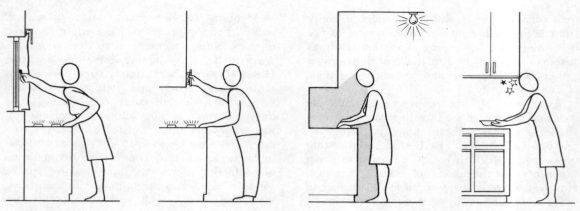

Dangerous designs of kitchen fitments which cause accidents

 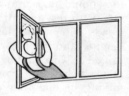 

Dangerous practices in cleaning the outside of windows from inside building

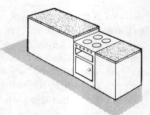

Unequal levels of working planes liable to cause accidents

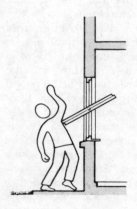

Dangerous window design

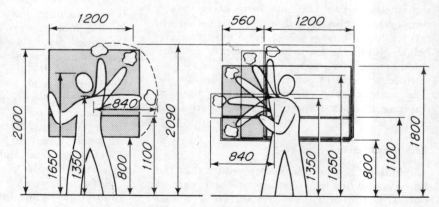

Cleaning inside face of window from inside building without steps and/or extension aids

Cleaning outside face of window from inside building. Maximum reach determines size and shape of fixed windows (shaded)

Figure 78   *Design factors influencing safety in a building (AJ)*

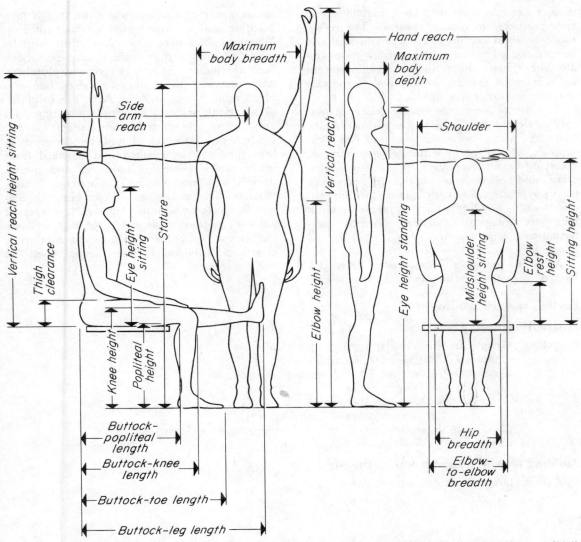

Figure 79   Human measurements used by designers. (From Human Dimension and Interior Spaces *by Panero and Zelnik, AP 1979)*

should never be used just to justify the safety of a badly designed feature. Apart from adding further dimensions to the shape and form of objects and planes, the addition of colour can act as a form of safety language. For example, because red is universally regarded as a warm and arousing colour, it can be used effectively to highlight equipment required to be used urgently in the event of danger, eg outbreak of

fire. According to the Young-Helmholtz Theory, human beings use a minimum amount of energy when reacting to red which results in a quick response time. Green requires slightly more energy and blue even more. When the safety precautions for certain areas of a building must be emphasized, the use of varying half tones of greys in the colour scheme should be avoided as these produce slow reactions to danger because they tend to camouflage real conditions due to lack of contrast. The reflectance value of certain colours should similarly be investigated to avoid dangers from glare, and also the effects which tungsten or fluorescent lights have on certain colours used to indicate safety precaution requirements.

The use of colour as a *safety coding device* is used to clearly identify electrical wiring circuits, hot/cold water services, and various gas supplies, as well as in industrial environments involving machines, steel plants, refineries, cranes and boilers. Some notable designers have used the resulting aesthetic qualities in the production of a building which represents modern high technological attitudes towards performance requirements (see chapter 2 *Appearance*).

Mention has also been made under 3.3 *Intended life span* to the need for 'specialized'

maintenance equipment when the design of a building makes conventional methods difficult. Whatever form necessary, every effort must be made to ensure that maintenance personnel, such as window cleaners, electricians, plumbers, can carry out their work in safety (see also Health and Safety at Work Act). For example, above the third storey height of a building (and also below where access for ladders is not convenient), windows should be designed so that they can be cleaned and reglazed from inside, unless there are balconies and special devices are incorporated for external maintenance. The maximum human reach is 550 mm for cleaning windows through or across an adjacent opening; side hung opening windows should have easy-clean hinges which produce a clear gap of 95 mm between frames when the window is open.

Further specific reading

Mitchell's Building Series

Components: Chapter 5   *Doors*   5.2(f) Strength and stability
            Chapter 7   *Glazing*   7.2(f) Security
            Chapter 9   *Ironmongery*   9.2 Performance requirements
                               9.4 Hinges
                               9.6 Locks and latches
                               9.8 Window stays and fasteners
           Chapter 10   *Balustrades and barriers*   10.2 Performance requirements
           Chapter 14   *Roofings*   14.2(c) Sub-structure

Building Research Establishment Digests

BRE Digest 289:   *Building management systems*

13 Cost

The design of a building must be judged not only by its appearance and the way it performs, but also by how much it costs. However, a true financial value is often difficult to assess because of the complexities of the building industry (figure 80). Once most of the other industries know what is to be produced, they can reasonably accurately formulate the precise processes involved in manufacture, and stipulate under what conditions the work will be carried out. The production of a building often does not rely on such entirely rational decisions. The form of a building can be influenced by many inter-related factors (location, planning constraints, soil conditions, availability of resources, etc), and the balance between appearance, comfort and convenience is inevitably finally resolved through subjective criteria. Methods of construction depend upon the knowledge and skill of the designer as well as the availability of materials and specific technical ability. The functional performance can be expressed in terms of cost of renewing or replacing fittings and fabric, methods required for adaptation to

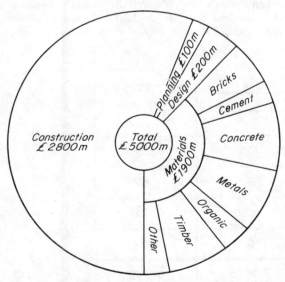

Figure 80   Apportionment of money spent within building industry

meet changing needs including those of heating, lighting and other services; and all these factors will be affected by the day-to-day use of a building.

13.1 Acquisition costs

Acquisition costs cover the financial aspects involved in the creation of a building – investment negotiations, professional fees, cost of land, building design and construction, etc. Further comments regarding most of these are made in Part B and only a few aspects are elaborated here.

The *cost of land* varies widely according to location, quality, size, and its value in terms of suitability for a particular building type, ie factory, office, hospital, housing, etc. Obviously, the consequences of these factors need to be carefully considered so that cost comparisons can be made with alternative sites. The value of the site after a building has been erected on it usually remains in proportion to the value of a particular development unless subsequent political or physical phenomena are likely to result in a change in its original characteristics, eg motor-way construction, tunnel formation, aircraft flight paths, compulsory purchase, change in planning zone designation, etc. The cost of land is high and growing faster than the other factors associated with the requirements for a building, and it is, therefore, essential that the land is used to its maximum efficiency (see also 17.2 *Site Considerations*). A more intensive use of land can be obtained by using a greater number of floors. Usually, however, the cost of suitable construction methods modify potential savings, and the necessity to incorporate lifts or mechanical ventilation systems will also add considerably to the subsequent running cost of a building.

There are, of course, a large number of financial considerations associated with the *design processes* of a building. Firstly, it should be remembered that government grants and subsidies are available which go towards the acquisition costs of certain types of building, eg currently those accommodating manufacturing activities. Available funds are usually given for initial construction rather than towards subsequent running costs and are usually fixed in relation to size of the proposed project. Unfortunately, this often does not encourage economic design because large, rather than efficient, forms of building provide better initial investment. Having partly financed such projects, unscrupulous developers or clients often leave their tenant users to run a building as efficiently as they may – any 'improvement' being at their own expense.

The cheapest form for a building is based on a compact cube, with no changes in wall, floor or roof planes. This form is difficult to achieve in truly optimum functional design terms, and may not be practicable or architecturally desirable (figure 81). Ideally, the spaces which a building provides should be the optimum usage (floor area and room height), and each activity accommodated planned to minimise the areas devoted to circulation as this inevitably contributes towards the time and cost necessary for communication. The siting of a building must give the best relationship with regard to access and with public zones – and also provide the best position which allows the most pleasing composition. Planning for maximum daylight penetration into a building and minimum exposure to the effects of adverse climatic conditions, can effectively reduce the initial and subsequent running costs for sources of artificial lighting and heating. The use of deep or interior spaces will also require mechanically produced environments. Noise, particularly in urban areas, is another important environmental factor related to cost. In city centres, tolerable noise levels can only be obtained by using sealed double glazing for windows which again necessitate the use of artificial ventilation, and perhaps, air conditioning.

The cost consequences of the *construction methods* used for each part of a building must be examined in a similar way to the cost consequences associated with the planning. Mention has already been made of the theoretically unlimited availability of material and technical resources for construction, and the limitations imposed by current economic constraints (1.3 *Performance requirements*) and other practical considerations (3.3 *Intended life span*).

During very early development stages of a building, the designer must be aware of the constraints that may be imposed by resources upon otherwise economic planning arrangements. Whenever possible, acquaintance should be made with available materials and technical ability, and a building produced which reflects these, rather than construction methods which are beyond economical means. Certain building arrangements are financially viable because they are based on the mobility of cranes and/or gantries. Also, the lifting capacity of certain

Single storey building with identical plan area			
Foundations	100	105	125
External walls	100	125	150
Floor	100	100	100
Flat roof	100	105	120
Overall	100	108·75	123·75

Figure 81   Comparison of elemental and overall cost per unit for buildings of different proportions but same plan area

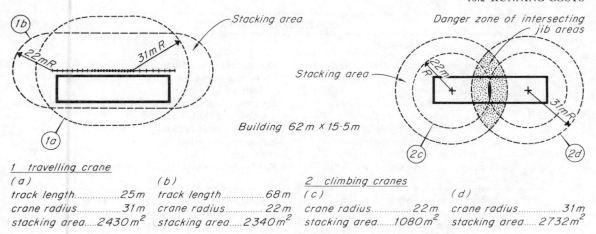

Figure 82   Radii of tower cranes and the affect on building shapes

1 travelling crane		2 climbing cranes	
(a)	(b)	(c)	(d)
track length............25m	track length............68m	crane radius...........22m	crane radius...........31m
crane radius............31m	crane radius............22m	stacking area......1080m²	stacking area....2732m²
stacking area.....2430m²	stacking area.....2340m²		

radii of tower cranes will limit the size of components which can be placed at the outer edges of a building under construction (figure 82).

One aspect which needs amplification concerns the relationship between *traditional* construction methods which involves the shaping of materials on site by skilled and semi-skilled labour, and *industrialized* construction involving the erection of large virtually finished factory-made components on site, usually by semi-skilled labour (see also chapter 4: *Dimensional Suitability*). Both forms of construction use the two basic resources, materials and labour; but the inter-related cost implication associated with each form affects their economical suitability.

In theory, the use of factory-made components for a building should result in considerable savings due to the possibility of large production runs. Such savings, however, may not always be available. The cost of basic materials accounts for about 50% of the total cost of materials used in traditional construction methods. After manufacture, the factory-assembled component must be moved for storage, and then eventually transported to site, hoisted and fixed in position. These processes often result in the incorporation of additional construction features (and materials) for handling during transportation and protection against damage during erection. The made component also needs storage or shelter facilities on site, as well as special plant (cranes, etc) for

its manipulation into position in a building. For example, a precast reinforced concrete wall component will need additional reinforcement to withstand the pressures of general handling and transportation; shuttering for *in situ* concrete connections to the structures of a building, or some form of bolted plate connection device; and a fairly sophisticated jointing system (gasket and mastics) for connection with adjoining components. If the component is delivered in a finished condition, particular care must be taken when it is being positioned, and afterwards while other trades are in close proximity. This adds considerably to the costs of handling and to the cost of site organisation and supervision.

The ultimate cost of a factory-made component, therefore, could be considerably higher than a similar form of traditionally constructed 'on site' component. On the other hand, traditional methods are subject to weather conditions unless special, and costly, precautions are taken; working conditions are far less congenial than in the factory, amenities poor and safety precautions sometimes primitive. A brief comparison is given in figure 83.

13.2 Running costs

Running costs include expenses incurred for general maintenance, cleaning and servicing of a building, and for renewing or repairing the fabric and fittings, as well as payments for heating, lighting, ventilating and services.

CONSTRUCTION METHOD		
	Traditional	*Industrialized*
Advantages	Well tried construction principles Greater possibility for design experimentation Users of building generally familiar with construction method Generally easily capable of adaptation Flexibility in tolerances during construction High degree of quality control not required	Standardized specifications/drawings Social conditions in factory better than on site Elements and/or components fabricated under controlled conditions Great degree of accuracy possible Site erection period likely to be shorter Raw materials storage on site reduced
Disadvantages	Specification/drawings vary for each project Work would take longer due to sequential nature of trades Work on site dependent upon material availability and storage New materials and components on site could be damaged Work on site often undertaken in poor conditions Social condition on site often below standard	System requires production in large quantities and therefore subject to market forces including government economic policies Designs largely dependent on manufacturing processes and aesthetic appeal often of secondary importance Factory manufacturing processes monotonous to workers Extra cost for transportation of large elements/components Usually 'specialist' site contractor required Accuracy of elements/components requires careful setting out, assembly and site control May not be cheaper than traditional construction method although less labour force employed Building usually costly to adapt

Figure 83   Comparison between traditional and industrialized construction method

Allowances for taxation rates and insurances should also be made to establish total running costs.

The implications which siting, design and construction (acquisition costs) have on running costs have already been commented upon, and further points are made in each chapter concerned with Performance Requirements. As mentioned in chapter 3 *Durability*, it is generally wisest to select more costly components, construction methods, and servicing systems which require low maintenance. The alternative use of low-cost components, etc, using high maintenance takes no account of the inconvenience, disruption and loss of earning power which arises with the need for repair or replacement. Techniques and materials which are only of

value in minimizing construction costs therefore fall short of what is really required if the designer is to be in a position to provide a building which offers the best value for money. It is also very important that the designer informs prospective users about matters concerning running costs of a building he has designed – see 16.8 *Maintenance manuals*.

13.3 Operational costs

Operational costs are generally (although not exclusively) associated with non-domestic buildings and include the cost of salaries for employees; provision of amenities which give congenial working conditions (bars, canteen, gymnasium, etc); machinery, power, and

Summary

Ground floor area: 625 m^2.
Total floor area: 1165 m^2.
Type of contract: JCT 1963 edition,
July 1976 revised (with quantities).
Tender date: 7 March 1983.

Work began: 31 May 1983.
Work finished: 3 October 1983.
Cost of foundation, superstructure,
installation and finishes
including drainage to collecting
manhole: £294 488.
Cost of external works and ancillary
buildings including drainage beyond
collecting manhole: £60 000.
Total: £354 488.

Cost analysis

All costs per m^2 based on estimated
final account.

Preliminaries and insurances
(Apportioned between all elements.)
This element is 7 per cent of the
remainder of the contract.

Work below lowest floor finish   £19·21
Pad and strip concrete foundations
integral with 125 mm thick ground
slab. Foundations 1 m deep bearing on
undisturbed lower chalk. Department
of Transport type 2 granular fill under
ground slab to make up levels from top
of chalk. Three coats liquid asphaltic
compound on top surface of ground
slab. (Suspended chipboard access floor
on stools off concrete slab giving
finished floor level 150 mm above
structural slab level included in
floor finishes.)

Structural elements

Frame   £29·29
Structural steelwork frame with a
typical column spacing of 6·75 × 6 m.
Primary 356 × 171 mm beams spanning
6 m between columns. Secondary
356 × 171 mm beams spanning 6·75 m
alternately onto primary beam or
column. Steelwork braced by diagonal
steel flats hidden in the inner lining in
cross-bracing bays. Spray applied
intumescent paint.

Upper floors   £9·83
Structural floors at first floor and roof
span 3 m between secondary beams,
540 m^2: £21·22/m^2.
Corrugated steel permanent
shuttering 51/1·2 mm gauge with
25 N/mm^2 dense concrete to overall
thickness of 100 mm. Design imposed
load of 5 kN/m^2.

Roof and rooflights   £23·05
Roof slab is same construction as first
floor—structure is designed for building
to have an additional storey in future.
Large central rooflight will become
stairwell. High performance felt
roofing in two layers on 40 mm
polyurethane insulation board on felt
vapour barrier on primed concrete slab.
Total roof area: 625 m^2; £42·78/m^2
average.
6 m × 2·6 m rooflight, continuous
barrel section, double skin, outer
polycarbonate, inner unwired diffusing
UPVC, glazing bars extruded
aluminium fixed to timber kerbs,
15·6 m^2: £106·00/m^2.

Staircases   £7·26
1 number internal, straight flight, steel
staircase with welded tubular steel
handrails, total rise: 3 m; width overall
tread: 1·4 m
2 number external, galvanised steel
spiral escape stairs, total rise: 3 m;
width overall tread: 1·1 m.

External walls   £30·15
Profiled metal cladding spanning
horizontally, coating PVF2, silver
colour; cold formed galvanised steel
vertical cladding posts; 50 mm
mineral fibre insulation; 12·7 mm
wallboard inner lining with painted
vapour barrier, 565 m^2:
£62·17/m^2.

Windows and external doors   £14·17
Anodised aluminium single glazed
vertical sliding, 169 m^2: £97·68/m^2.
Two pairs anodised aluminium glazed
doors; 4 number solid core timber fire
escape doors.
Total external door area: 18 m^2;
£152·13/m^2 average.

Partitions   £6·41
Metal stud partition system, 70 mm
studs, 12·7 mm plasterboard, dry lined
finish, 148 m^2. 100 mm blockwork
partitions to plant rooms, 38 m^2.
Total partition area: 186 m^2;
£40·15/m^2 average.

Internal doors and ironmongery   £0·94
Ply faced solid core flush doors in
softwood frames, 19 m^2: £57·64/m^2.
Number of single doors: 11.
Anodised aluminium lever handles,
mortice locks, door-closers and
panic bolts.

Finishes and fittings

Floor finishes   £30·15
Partial access raised floor with carpet
tiles over. Access floor comprises
30 mm tongue-and-groove chipboard
supported on 115 mm high stools at
600 mm centres, 180 number access
panels, 1165 m^2: £15·79/m^2.
Looped pile anti-static bitumen-backed
carpet tiles, 1123 m^2: £11·93/m^2.
Sheet vinyl in toilets and kitchen,
42 m^2: £16·95/m^2.
Total floor finish area: 1165 m^2.

Ceiling finishes   £7·68
Ceiling of exposed painted steelwork
and soffit with applied acoustic panels
of mineral fibre tiles, 666 m^2:
£4·50/m^2. Acoustic panels installed on
concealed grid, 499 m^2: £12·03/m^2.
Total ceiling finish area: 1165 m^2.

Decoration   £2·31
Emulsion paint to walls; oil paint to
exposed steel and ceilings; gloss paint
to skirtings and architraves.

Fittings   £2·31
Kitchen fittings and appliances,
domestic standard.

Services

Sanitary fittings   £0·92
6 number wall-hung washbasins,
vitreous china; 6 number washdown
WCs, fireclay; 4 number urinals,
vitreous china.

Waste, soil and overflow pipes   £2·76
UPVC solvent welded traps, branches,
stacks and overflows.

Cold water services   £0·46
Rising main only. All fittings feed
directly from mains. (Small feed and
expansion tank for heating system
included in heating costs.)
Number of cold draw-off points: 16.
Includes builders' work at £0·16/m^2.

Hot water services   £0·36
Packaged electrical water heater
feeding taps in kitchen and WCs (unit
includes 114 litres storage).
Number of hot draw-off points: 8.
Includes builders' work at £0·10/m^2.

Heating services   £15·61
Fuel: natural gas.
System: two pile radiator system with
thermostatic valves. Steel panel
radiators.
Boiler: high efficiency modular boiler.
Features: auto by-pass valve for
constant boiler flowrate. Reverse-
return pipework for easy balancing.
Space for expanding output to 200 kW.
Total heat load: 150 kW.
Includes builders' work at £0·05/m^2.

Ventilation   £0·18
Natural vent ducts to toilet lobbies.

Gas services   £0·09
Connection to existing meter in
adjacent building. One outlet.

Electrical services   £40·83
A 400A supply is provided by the
Eastern Electricity Board and
terminated in a cubicle switchboard.
Power distribution by flush floor socket
outlets in an access floor system.
Lighting consists of a fluorescent
uplighter system to a purpose-made
channel suspended below the ceiling
containing twin 1·5 m fluorescent
batten fittings.

Breakdown of electrical services	£/m^2
Meter and switchgear	10·13
Lighting installation	5·58
Power installation	12·19
Lighting fittings	12·39
Builders' work	0·54
Total	40·83

Total number of lighting outlets: 125.
Total number of power outlets: 177.
Total electrical load: 183 kW.

Drainage   £8·81
100 mm surface water drains to four
1·8 m diameter soakaways. 100 mm
foul connection to existing manhole.

Total per m^2 of floor area   £252·78
£294 488 (excluding external works)
1165 m^2 (inside external walls)

External works

Precast concrete paving around
building; car park for 120 cars finished
in tarmacadam; farmtrack; chain link
fencing; reinstatement of surrounding
land; £60 000.

Figure 84   Cost analysis (AJ)

materials used by the processes accommodated in a building; and the costs of adapting a building to meet changing user requirements. These costs, therefore, must be supplied by, or worked out in close conjunction with, the client and his advisers.

Sometimes the layout and design of a building dominates the way certain operations are carried out, the number and type of staff employed, and hence the overall operational costs. The type of activities mostly affected in this way involve transport systems or product manufacture/handling, but even the efficiency of these will be finally determined by the calibre (and salaries) of supervision and management staff.

When possible, the prospective building owner must be made aware of the effects which other cost factors of a design have on operational costs. For example, the choice of temperature control in offices can have an affect on the efficiency of employees. During sudden cold spells a slow room heating response may produce errors or temporary cessation of clerical performance giving rise to subsequent financial losses. To ensure continuous internal heating levels in a building it is possible to increase the acquisition cost of construction (say by double glazing or extra thermal insulation) and/or increase running costs for fuel (say by using relatively expensive forms of quickly responding temperature controls).

13.4 Cost evaluation

The true cost evaluation of a building should involve a close examination of viable alternative acquisition, running and operational costs. The depth of investigation possible and, therefore, the accuracy in achieving the 'true value' of a building, will depend on the information available and skill of interpretation during the inception, preliminary and detail design stages. Although cost analyses (figure 84) are available through technical publications for numerous schemes, which may be similar to one in hand, they are normally not able to supply information regarding operational costs or the effects of solely financially orientated design decisions on building users. When pertinent, this information allows the historical cost analysis data to be used for the future creation of a building specifically orientated towards definable long-term financial goals.

Various techniques, therefore, exist for the cost analysis of a proposed building. Simply expressed, these include a *cost-in-use* method which analyses the relationship between acquisition and running cost in order to achieve optimum performance requirement criteria; a *cost-effectiveness* method which analyses the relationship between acquisition, running and operational costs so as to permit adaption of certain performance requirements to allow for *specific* user attitudes; and a *cost-benefit* method (the most difficult) which analyses economical building performance criteria in terms of *variable* user-benefits.

The whole area of cost evaluation is known as *terotechnology*. BS 3811:1974 *Glossary of Maintenance Terms in Terotechnology* states: 'Its (terotechnology) practice is concerned with the specification and design for reliability and maintenance of plant, machinery, equipment, buildings and structures; with their installation, commissioning, maintenance, modification and replacement; and with feed-back of information on design, performance and costs.' In other

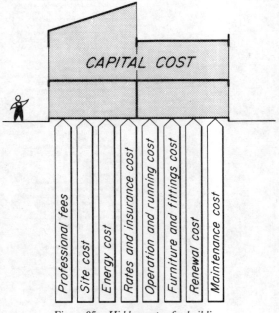

Figure 85   Hidden costs of a building

words, the central idea involves the pursuance of economic *life cycle costs* for a building through combined design, technical, financial and managerial skills. Very few building owners and users calculate the true costs of their building (figure 85). Clients (and designers) need to be made aware of life cycle costs so that they apply their minds to the best available data in a structured and systematic way in order to obtain a clearer understanding of the future cost implications for a proposed building. Today the high costs of materials, labour and energy imply that the running costs of buildings are of greater importance relative to acquisition costs. Despite possible short-term reductions in prices, the long-term trend is likely to continue upwards, and operational costs must be used more fully to buffer the effects of over-costly capital investments. A true understanding of this will facilitate the preparation of estimates to give the realistic cost consequences of building designs against which the more subjective judgements of appearance, comfort and convenience can be set. For this purpose comparative tables can be prepared which set out the costs and values arising from each design alternative: preferred appearance, comfort and convenience factors can be compared against the monetary figures which sum up the cost consequences of the design.

Further specific reading

Mitchell's Building Series

Environment and Services: Chapter 7 *Thermal insallations*
Materials: Introduction (Economics), and within each chapter on materials
Structure and Fabric Part 1 Chapter 2 *The production of buildings*
(Economic aspects of building construction)
Structure and Fabric Part 2 Chapter 1 *Contract planning and site organization*
(Planning consideration)
Components: Chapter 2 *Economic criteria*
Chapter 3 *Industrialized system building* 3.2 Economic and social factors

Building Research Establishment Digests

BRE Digest 247: *Waste of building materials*
BRE Digest 259: *Materials control to avoid waste*

PART B

An analysis of a building in terms of the processes required, the building team which implement them, and the methods used in communicating information

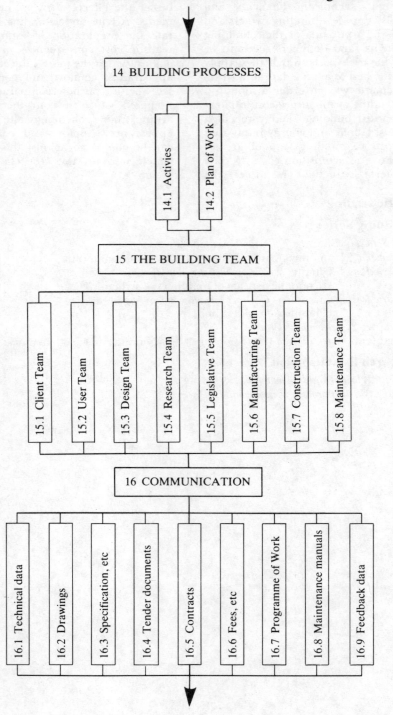

14.1 Activities

The processes currently involved in the erection of a building are very complex. This is primarily because it concerns many different categories of personnel, and form of communication, resources, and operational skills. Activities include:

- the initial decision to build
- securing of financial resources
- selection of appropriate location
- appointment and briefing of suitable member to be involved with design and construction operations
- definition of precise functional requirements
- the design process and decision on how to build
- implementation of erection
- operations necessary to maintain building in the state of continuous performance for which it was intended
- operations necessary to adapt building to new functions.

All these activities may be affected by an elaborate system of approvals, controls, checks and cross checks which not only involve nearly all the members of the entire building team, but also a variety of outside bodies in varying administrative, technical, financial, and fiscal capacities. Furthermore, whereas the creation of a building was formerly a leisurely occupation ultimately dependent upon craft-based skills, the whole process is now greatly influenced by the desire to achieve profit on financial investments as soon as possible and the exploitation of machine technology. Therefore, although most buildings of today are far more complex and sometimes much larger, the time scale from the realization of the need for a building to its use after erection is generally far shorter than at any previous period in the development of building.

In view of these complexities, a great deal of attention has been devoted to methods of achieving clarity of communication between the participants. The Tavistock Institute reports, *Communications in the Building Industry* (1965) and *Interdependence and Uncertainty* (1966) are typical examples which stress the need for long-term investigation. Increased effectiveness means better organization and control of all the complex operations involved in building which affects clients, designers and builders.

It is often stated that the only common factor underlying the whole process is the economic one and this should, therefore, provide the basis of organizational procedures. Economics are significant in assessing the nature of the individual contributions of the members of the building team and often explain the reasons for the elaborate system of mutual controls, checks and doubts usually blamed for delays and 'inefficiency' within the building process.

14.2 Plan of work

Figure 86 indicates a *Plan of Work* which was originally published in the first edition of the *Royal Institute of British Architects' Handbook* in 1964. Its intention was to provide a model procedure for methodical working of the design team employed on projects which have sufficient common factors to make it widely relevant. The *Plan of Work* has subsequently become widely known among other professions concerned with the design and construction of buildings as is if capable of being used in a variety of ways. It can assist the planning of projects and be adapted to form the basis for control of organizational procedures. In developing the model certain assumptions were made:

(a) that it relates to a building costing £300,000 (1964 price) which uses a full design team;
(b) an architect is the principal designer and leader of the design team;
(c) the architect is appointed at an early stage of the building process;
(d) each stage follows sequentially;
(e) the cycle of work in *each* stage involves
  - stating objective and assimilation of relevant facts
  - assessment of resources required and setting up of appropriate organizations

Outline plan of work

Plan of work diagram 1

Stage	Purpose of work and Decisions to be reached	Tasks to be done	People directly involved	Usual Terminology
A. Inception	To prepare general outline of requirements and plan future action.	Set up client organisation for briefing. Consider requirements, appoint architect.	All client interests, architect.	Briefing
B. Feasibility	To provide the client with an appraisal and recommendation in order that he may determine the form in which the project is to proceed, ensuring that it is feasible, functionally, technically and financially.	Carry out studies of user requirements, site conditions, planning, design, and cost, etc., as necessary to reach decisions.	Clients' representatives, architects, engineers, and QS according to nature of project.	
C. Outline Proposals	To determine general approach to layout, design and construction in order to obtain authoritative approval of the client on the outline proposals and accompanying report.	Develop the brief further. Carry out studies on user requirements, technical problems, planning, design and costs, as necessary to reach decisions.	All client interests, architects, engineers, QS and specialists as required.	Sketch Plans
D. Scheme Design	To complete the brief and decide on particular proposals, including planning arrangement appearance, constructional method, outline specification, and cost, and to obtain all approvals.	Final development of the brief, full design of the project by architect, preliminary design by engineers, preparation of cost plan and full explanatory report. Submission of proposals for all approvals.	All client interests, architects, engineers, QS and specialists and all statutory and other approving authorities.	

Brief should not be modified after this point.

Stage	Purpose of work and Decisions to be reached	Tasks to be done	People directly involved	Usual Terminology
E. Detail Design	To obtain final decision on every matter related to design, specification, construction and cost	Full design of every part and component of the building by collaboration of all concerned. Complete cost checking of designs.	Architects, QS, engineers and specialists, contractor (if appointed).	Working Drawings

Any further change in location, size, shape, or cost after this time will result in abortive work.

Stage	Purpose of work and Decisions to be reached	Tasks to be done	People directly involved	Usual Terminology
F. Production Information	To prepare production information and make final detailed decisions to carry out work.	Preparation of final production information i.e. drawings, schedules and specifications.	Architects, engineers and specialists, contractor (if appointed).	
G. Bills of Quantities	To prepare and complete all information and arrangements for obtaining tender.	Preparation of Bills of Quantities and tender documents.	Architects, QS, contractor (if appointed).	
H. Tender Action	Action as recommended in NJCC *Code of Procedure for Selective Tendering* 1972.*	Action as recommended in NJCC *Code of Procedure for Selective Tendering* 1972.*	Architects, QS, engineers, contractor, client.	
J. Project Planning	To enable the contractor to programme the work in accordance with contract conditions; brief site inspectorate; and make arrangements to commence work on site.	Action in accordance with *The Management of Building Contracts** and Diagram 9.	Contractor, sub-contractors.	Site Operations
K. Operations on Site	To follow plans through to practical completion of the building.	Action in accordance with *The Management of Building Contracts** and Diagram 10.	Architects, engineers, contractors, sub-contractors, QS, client.	
L. Completion	To hand over the building to the client for occupation, remedy any defects, settle the final account, and complete all work in accordance with the contract.	Action in accordance with *The Management of Building Contracts** and Diagram 11.	Architects, engineers, contractor, QS, client.	
M. Feed-Back	To analyse the management, construction and performance of the project	Analysis of job records. Inspections of completed building. Studies of building in use.	Architect, engineers, QS, contractor, client.	

*The publications Code of Procedure for Selective Tendering (NJCC 1972) and Management of Building Contracts (NJCC 1970) are published by RIBA Publications Ltd for the NJCC.

Figure 86   RIBA Plan of Work

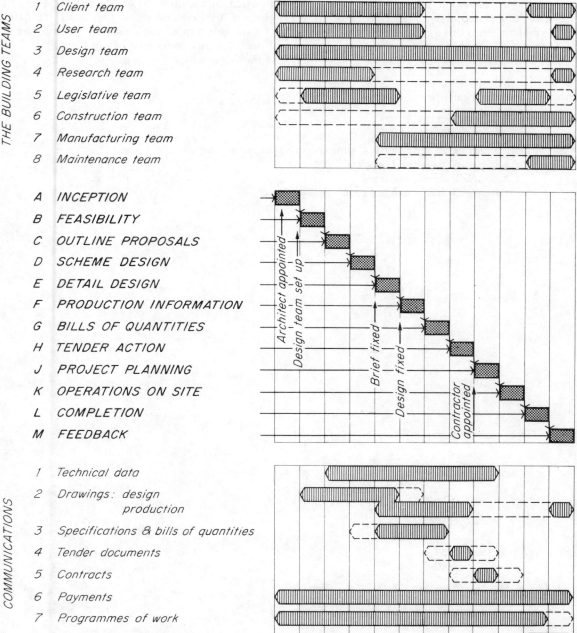

THE BUILDING TEAMS

1 Client team
2 User team
3 Design team
4 Research team
5 Legislative team
6 Construction team
7 Manufacturing team
8 Maintenance team

A INCEPTION
B FEASIBILITY
C OUTLINE PROPOSALS
D SCHEME DESIGN
E DETAIL DESIGN
F PRODUCTION INFORMATION
G BILLS OF QUANTITIES
H TENDER ACTION
J PROJECT PLANNING
K OPERATIONS ON SITE
L COMPLETION
M FEEDBACK

COMMUNICATIONS

1 Technical data
2 Drawings: design
    production
3 Specifications & bills of quantities
4 Tender documents
5 Contracts
6 Payments
7 Programmes of work
8 Maintenance manual

Figure 87 Relationships between RIBA Plan of Work and the involvement of the Building Team and the Methods of Communication

– planning the work and setting timetables
– carrying out work
– making proposals
– making decisions
– setting out objectives for next stage.

This *Plan of Work* will be used as a basis for the text in chapter 15 *The Building Team*, and chapter 16 *Communications*. Figure 87 indicates the relationship of each subject to the *Plan of Work*. Further reference should be made to *Architect's Job Book*: Fifth edition.

Further specific reading

Mitchell's Building Series

Structure and Fabric Part 1: Chapter 2 *The production of buildings*
Structure and Fabric Part 2: Chapter 1 *Contract planning and site organization*
Components: Chapter 1 *Component design*
Chapter 2 *Economic criteria*
Chapter 3 *Industrialized building systems*

15 The Building Team

CI/SfB (A1m)

Building is a group activity and its success depends on a good understanding and co-operation between a large number of people. The participants involved can be conveniently arranged into groups or teams according to their particular interest and/or involvement as follows:

Client Team
User Team
Design Team
Research Team
Legislative Team
Manufacturing Team
Construction Team
Maintenance Team

Figure 87 associates each team to the RIBA Plan of Work for a medium size building project as described in chapter 14. In this way the activities of each can be approximately related to the various stages of the design and building process. However, this represents only one method of sequential timetabling and also indicates the principal designer of a building to be the leader of the building process. Various permutations can be adopted involving different 'leaders', or no 'leaders' at all, as well as employment of teams at different stages according to the precise circumstances. Figure 88 indicates the inter-relationship of the various teams in the sequence to be considered in this chapter.

15.1 Client Team

The client, or prospective building owner, has the responsibility for defining the building to suit needs, establishing and providing the necessary finances, agreeing design and construction phases, timetabling, and, of course, fulfilling the management and running of the completed project. This implies that not only must the client's *brief* (or list of requirements) for a building be represented accurately and clearly, and that he should maintain strategic knowledge through the building process, but also he must be able to make prompt decisions when requested. However, many of these are usually resolved by the client with the aid of his professional advisers.

The type of client varies but can include a domestic couple, a director of a small company, or a carefully constituted project team of a large industrial complex consisting of departmental heads controlling finances, building skills and other specialisms. Large organizations very often incorporate a committee of laymen backed by consultants, such as exist in government administrative departments. Nevertheless, whatever the combination or form of the client organization, their responsibility is always to provide clear and concise instructions to the building professionals.

The British Property Federation is an association devoted to the interests of property owners and others with a major concern for property. Their main objective is to promote a better understanding between property owners (ranging from large development companies to owner-occupiers of single dwellings) and the public, government and local authorities. The Federation provides general advice on all matters relating to the law, management and administration of property, and on taxation, housing and rating problems.

The necessity to build

A potential client must establish whether to build or not to build, and it is first necessary that he should carry out a comprehensive appraisal of his, or his company's needs. Having decided that a new building is necessary to provide additional or alternative space, it is important that consideration is then given to when the space will be needed. For people who are not connected with the building industry, the time scale for building often seems surprisingly (and unnecessarily) long. A minimum of six months will be required between making an appraisal and the start of even a quite modest size building. Larger buildings can take several years in resolving the problems associated with land acquisition, establishment of

111

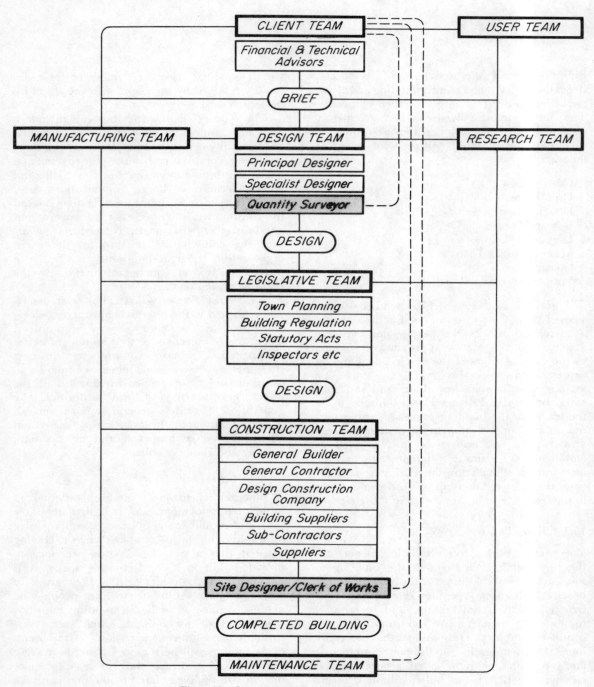

Figure 88   Sequential relationship of Building Team

rights, development permits, planning permission, building approval, contractor selection and subsequent erection. Initial delays may also be encountered while seeking out and appointing a suitable team of professionals (lawyers, financiers, designers, etc) to advise on these activities.

There are various options open to a person or organization requiring more space other than commissioning a new building. For example, there may be the possibility of the purchase or lease of existing property which can be altered, adapted, extended, or renovated in a suitable manner. Alternatively, there is the possibility of the purchase or leasing of a precisely suitable building. Often developers will build office blocks and shopping areas speculatively, and many town councils will build factory units for sale or rental. These buildings usually, however, require fitting out to greater or lesser amounts by the eventual owner or lessee. Clients wishing to investigate the potential use of an existing building, or one which is under construction having been commissioned by others, should seek the advice of the **Estate Agents** dealing with the particular interest. These are professionally qualified people involved with the buying, selling and managing of property generally. They may belong to the Royal Institute of Chartered Surveyors (RICS), to the Auctioneers and Estate Agents Institute (AEAI), or to the Incorporated Society of Auctioneers and Land Property Agents (ISALPA).

Financial resources

Most building is undertaken from money made available in the form of a loan – therefore, interest rates (loan charges) are important. In this respect, the government has direct influence and can use the building industry as a regulator for the economy of the country. Taxation and/or adjustment to the bank rate on loans can cause boom or slump periods. Funds for a public building are made available from local or national taxation and there is concern to spend as little as possible with due regard to the value for money. Financial control is implemented by the use of the 'cost yardstick' in housing; cost per place in schools, or cost per bed in hospitals, etc. In the private sector, there is a little more flexibility in that the client will spend in propor-

tion to stated aims – which may be as diverse as a temporary-use building, or a prestige building. Private funds arise from the profits of industrial undertakings or the savings from individuals which again are closely linked with investment and borrowing potential, mortgage rates, and grants to charities, etc. Figure 89 indicates the approximate apportionment of money spent on building in 1983.

Some building types which the Government in power considers should be encouraged for the benefit of a particular region or the country in general may be eligible for financial support from national taxation. Formally industrial premises and hotels have been supported in this manner. Indeed, often the reason for a building is implemented as a result of this financial aid or *grant*, but designers must always confirm the opportunities which may exist for the client. For smaller scale work involving the domestic sector, the Housing Acts specify grants which are available for the provision of adequate sanitary appliances where they do not exist, and for essential repairs or building extensions which bring privately-owned homes to a satisfactory standard of habitation. Currently, there are even government grants towards the costs of providing thermal insulation to roof spaces of domestic buildings.

Once money becomes available for a building, the client will require speedy action for its design, construction and subsequent use so that the lost interest, which would have been gained through alternative financial investments, may be speedily recouped. Indeed, today, speed of use is even more essential to ensure that the value of the original sum is not extensively reduced by the effects of monetary inflation in the cost of labour and materials. The total cost of a building must include the professional fees of the Design Team which the client appoints. In conventional terms the contractor's 'fees' and those associated with initial local authority services (planning, byelaws, water for works, etc) are included in the costs for the building (see chapter 13 *Cost*).

15.2 User Team

As the majority of building is for the direct employment of people or involves people for its continued function, ie maintenance, a building

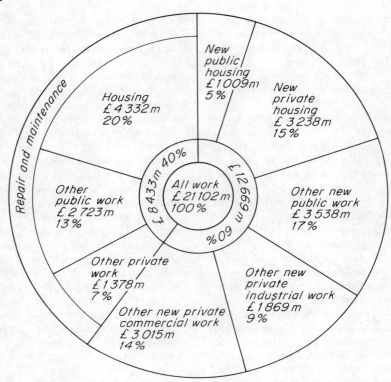

Figure 89   Approximate construction output in Great Britain for 1983 by all agencies at constant 1980 prices seasonally adjusted. Percentages relate to total of all work undertaken (£21 102 m). Source: D o E

User Team forms a vital link between design concepts and built reality. Formulating user requirements may be simplified when client and occupants are the same. However, this is not always the case, and it is the initial responsibility of a non-user Client Team to establish user requirements when formulating a brief with their professional advisers.

Typical User Teams which can supply information include, Tenants' Associations, Medical Associations, Consumer Associations, Tourist Board, Unions, etc, and their membership may include professional advisers as well as lay people. Often, the Design Team must draw independently upon the expertise, experience and research of these bodies when advising on a client's own interpretation of user requirements for a proposed building. In this way the designer can modify certain views and provide for the psychological and physiological well-being of future occupants.

The performance requirements for a building are sometimes derived from feed-back information supplied by the users of similar existing designs. This applies particularly to experience gained from designs which resulted in loss of security or accidents (see chapter 12 *Security*). Where a novel project is being contemplated, detailed research must be carried out to ascertain the effects which certain designs or construction methods are likely to have on potential occupants – if necessary a life-size prototype of the proposed enclosure must be made for users to experience and test.

15.3 Design team

The design of a building can no longer be the total responsibility of one person. There are a great many people concerned with supplying the design expertise which will make a building possible and these people are known collectively as the Design team.

Within this team there are two types of building designers – *Principal Designers* with the responsibility for the overall design of the project, and *Specialist Designers* who provide expertise concerning certain aspects of a building and whose requirements are often co-ordinated by the principal designer. Cost control and financial advice to client, principal and specialist designers is generally provided by a **Quantity Surveyor**. The fees for the professional services of principal designers, specialist designers, and the quantity surveyor are paid by the client (see 16.6 *Fee accounts and Certificates of Payment*).

Principal designers

Professional principal designers generally include architects, interior designers, and building surveyors. Each is capable of fullfilling the function of a building designer, although, because of their training, they may provide different emphasis in the approach to the problems associated with the production of a building.

Architects design and prepare the production information for most building projects, and on small general purpose buildings their expertise permits them to be the sole designer. They will also inspect the construction work on site. The title 'architect' is protected under the Architects' Registration Council of the United Kingdom (ARCUK), and only persons appropriately qualified and registered can use it.

Most architects are also members of the Royal Institute of British Architects (RIBA) and are governed by its Charter, Byelaws, Regulations, and Code of Professional Conduct under, and in addition to, the general law. The purpose of the RIBA as described in its Charter is 'the advancement of Architecture and the promotion of the acquirement of the knowledge of the Arts and Sciences connected therewith'. The object of the Code is to promote the standards of professional conduct in the interest of the public and consists of three Principles:

1 '*A member shall faithfully carry out the duties which he undertakes. He shall also have proper regard for the interests both of those who commission and of those who may be expected to use or enjoy the product of his work.*'
2 '*A member shall avoid actions and situations inconsistent with his professional obligations or likely to raise doubts about his integrity.*'
3 '*A member shall rely only on ability and achievement as the basis for his advancement.*'

Each of these Principles contains a number of Rules giving instructions about specific aspects of practice; for example, under Principle 1, an architect must act impartially when interpreting the clauses of a building contract between client and contractor; under Principle 2, an architect must not take bribes or allow his name to promote any service or product; under Principle 3, an architect must not offer financial inducements to obtain work.

Interior designers can also prepare design and production information for a building, and provide supervision of work, but, as their title implies, they may be specifically concerned with the interior of a building and need additional advisers in order to deal with all the design and construction processes involved in a total building. Interior designers can be members of the Chartered Society of Designers. This organisation has Codes of Professional Conduct which reflect ideals equivalent to those of the RIBA.

Building surveyors are sometimes responsible for the design and supervision of certain building work although they more usually carry out surveys of structural soundness, condition of delapidation or repair, alterations/extensions to existing buildings and market value of existing buildings. However, it must be borne in mind that the emphasis of their training lies in the technical aspects of building rather than the aesthetic aspects. They may be members of the Royal Institute of Chartered Surveyors (RICS) and will, therefore, be governed by its code of conduct – (see also comments under 15.1 *Client Team, Estate Agents*).

It is also possible for both architects and surveyors to become members of the Incorporated Association of Architects and Surveyors (IAAS) whose purpose is to facilitate collaboration between the two professions.

Specialist designers

These include civil and structural engineers; services engineers; and those concerned with specific aspects of architecture, including landscape, interiors, office planning, etc. **Civil and structural engineers** are employed to assist principal designers on building projects which contain appreciable quantities of structural work, such as reinforced concrete, complex steel or timber work, or foundations which are either complex or abnormal.

Civil engineering is defined as the construction of roads, bridges, tunnelling and motorways, etc. There is often a certain amount of building work in civil engineering construction and, likewise, many building projects contain civil engineering construction. *Structural engineering* deals especially with the calculation of the structural parts in a building. When projects consist mainly of structural work, the civil or structural engineer will be principal designer. They can be members of the Institute of Civil Engineers (ICE) or the Institute of Structural Engineers (IStructE): the Association of Consulting Engineers (ACE) issue codes of conduct defining professional responsibility.

Services engineers work with other designers and are concerned with environmental control – lighting, heating, air conditioning, and sound modulation; electrical installations, plumbing and waste-disposal systems; and mechanical services, such as lift installations and electrical conductors. The Chartered Institute of Building Services (CIBS) was formed from an amalgamation of the Institute of Heating and Ventilating Engineers (IHVE) and the Illuminating Engineering Society (IES) with the object of promoting the science and practice of service engineering as well as advancement of education and research in the field. Electrical engineers are governed by the regulations of the Institution of Electrical Engineers (IEE), and mechanical engineers by the Institution of Mechanical Engineers (IMechE). Senior members of the CIBS, IMechE, and IEE can also become members of the Association of Consulting Engineers (ACE). Figure 90 indicates some of the service engineers involved in the design of buildings.

Acoustic	– modulation and audio
Air conditioning	– heating and refrigeration
Communications	– lifts, hoists, escalators, paternosters and conveyors
Catering	– food preparation and service
Drainage	– above and below ground, and water-borne refuse
Electrical	– heating, air conditioning, refrigeration and lighting
Gas supply	– heating including industrial applications
Heating	– gas, electric, oil, solid fuel, solar and ambient
Fire protection	– escape and fire fighting equipment
Lighting	– natural and artificial
Plumbing	– hot and cold water services
Refrigeration	– preservation and installations
Refuse disposal	– storage, chutes, water-borne and disposal
Sanitation	– pest control, appliances and drainage
Security systems	– protective, anti-theft and pilfering devices
Telecommunications	– telephone, security, cable and tv
Thermal insulation	– fabric design and installation
Ventilation	– natural and artificial
Water supply	– availability and storage

Figure 90 Some Service Engineers involved as specialist designers for the environmental control aspects of buildings

Other specialist designers such as landscape architects, interior designers, graphic designers, space planners, acoustical and production engineers are employed according to the type and complexity of the building project. Each discipline is represented by a professional Institute or Association.

Quantity surveyor

An essential part of any design process involves cost control. The costing services for smaller, less complex building projects are generally provided by the principal designer working in conjunction with the client and specialist designers. However, for larger or more complex projects it is usual that a *quantity surveyor* is employed to give cost advice and sometimes a cost control service (see chapter 3 *Cost*). Chartered quantity surveyors are governed by the Codes of Professional Conduct issued by the Royal Institute of Chartered Surveyors (RICS), and like other members of the Design Team, their fees are paid by the client. The RICS is the professional body for certain other members of the building profession (Building Surveyor, Estate Agents, etc) and some quantity surveyors prefer membership of the Institute of Quantity Surveyors (IQS) which deals solely with their profession.

Until recently, no country, other than the United Kingdom, regularly employed a quantity surveyor for building projects. His primary role is to prepare a *Bill of Quantities* from the drawings and specifications supplied by the design team. This itemises the type, form and amount of materials to be used in the construction project. The Bill (see 16.3 *Specification and Bill of Quantities*) will also define the legal requirements for the project, including the form of contract to be adopted between the client and the contractor. The whole document, together with relevant drawings and specifications supplied by the Design Team will, therefore, provide a good basis for obtaining prices for the project from a number of interested contractors (competitive tendering).

Prior to the work on the Bills of Quantities, the quantity surveyor may be expected to give cost advice on any alternative solutions that may be considered during the various stages of the design of a building. Also, during the actual construction period for a project, he must measure and value the work carried out at regular (monthly) intervals and submit details to the overall financial administrator (usually the principal designer) for payments from client to contractor. The quantity surveyor also advises on the use of sums of money listed in the Bill of Quantities for contingency or provisional items, the cost of making variations in areas originally described in the Bills or indicated on the drawings, and settlement of the final account for the finished project.

Student assistants and technicians

Principal and specialist designers, as well as quantity surveyors, often employ students of their individual profession to assist them in their work. The students are mostly post first-degree standard and, as such, their work will require close supervision by suitably qualified staff within an organization. If a part-time or 'sand-wich' mode of professional education is not available, the use of this type of assistant may be limited to university or polytechnic vacation times, or the periods designated as 'time out' for *practical training* such as is required by the RIBA and CSD Examining Boards.

Greater continuity of assistance may, however, be provided by *technicians* who are specially trained in a number of areas concerned with the design features of a building and the subsequent inspection of construction or installation work. According to qualifications and experience, technicians may be responsible for carrying out surveys, technical feasibility studies, as well as the preparation of designs, and working drawings, and models. They can also become involved in cost analysis, obtaining legal approvals, contract administration and assisting with site inspections. The RIBA played a major role in setting up the Society of Architectural and Associated Technicians (SAAT) which represents the specially trained architectural, structural engineering and quantity surveying technicians. Nevertheless, as far as architecture is concerned, it is important that technician skills are not abused by allowing them to influence the appearance of a building in areas which the principal designer neglects either through choice or ignorance.

Methods of operation

Principal designers, specialist designers, and quantity surveyors can work in private practice either as part of a single discipline firm, or as part of a larger multi-discipline firm. They can also be employed by public bodies such as central or local government, by major and industrial firms, and by building contractors. As mentioned earlier, the fees of the designers and quantity surveyors are paid by the client at prescribed intervals related to the stage of development in a project. Where they are employed as part of a large organization, however, these fees will be absorbed as part of the overall charges and paid to the consultants as a regular salary. Depending on the precise nature of a project, and as rough guide, the *combined cost* of these professional fees, including cost planning advice, will vary from between 12% and 20% of the final construction costs (see 16.6 *Fee accounts and Certificates of Payment*).

	Principal designer	Specialist designer	Quantity surveyor
A Inception	Agree extent of work to be done, financial arrangements and appointment with client		Agree extent of work to be done, financial arrangements and appointment with client
B Feasibility	Elicit all information by questionnaire, etc from client Carry out studies on site-user information, eg boundaries, rights of way, rights of light, easements, services, etc Make particular enquiries to local authority regarding planning approvals, preservation orders, site lines, availability of services, etc Prepare feasibility report and present to and discuss with client	Provide initial guide about possible factors influencing design and cost of proposals Assist in preparation of feasibility report and if required help in presentation to client Agree fee for services	Produce cost area guides, ie cost per m^2 or m^3 Advice on alternative cost factors Discussions on financial aspects of specialist designer's discipline
C Outline proposals	Study circulation and space problems for site and proposed building Try planning/massing solution and investigate alternative costs Produce diagrammatic analysis and discuss problems Try out various general solutions indicating critical dimensions and mean-space allocation Contribute to, prepare and present report to client	If not already done, agree condition of appointment with client Provide more detailed information on design and cost problems Assist in preparation of outline proposal	Estimate cost of project based on outline plans and brief specification Provide cost evaluation of elements for designs as they develop Provide comparative cost information relative to possible alternative materials and construction techniques Prepare elemental cost at sketch design stage and estimate of building cash flow
D Scheme design	Complete outstanding user studies Consult other members of design team Prepare full scheme design, taking individual and group advice Receive and discuss proposals of specialist designers and QS Prepare presentation drawings and obtain client and other members of design team approval Prepare more detail drawings for local/central government approvals	Formulation of accurate information regarding space and dimensional requirements Detailed negotiations with principal designer, other specialist designers and QS Preparation of preliminary drawings, specification and schedules indicating proposals: assist with presentation to client Negotiations for local/central government approvals	Evaluate design within cost plan limits Prepare preliminary measures of quantities Advice on alternative cost factors Discussions on financial aspects with specialist designers Advise on running cost for proposed building

	Principle designer	Specialist designer	Quantity surveyor
E Detail design	Complete user studies Complete final design Close co-operation with specialist designers and QS	Finalize drawings, specification and schedules Close co-operation with principal designer, other specialist designers and QS	Detail consultation with principal and specialist designer
F Production information	Prepare drawings, specifications and schedules Agree drawings, specifications and schedules of specialist designers Close consultation with QS, agree sub-contractors and suppliers	Assist with incorporating requirements into principal designer's documents Advise on suitable sub-contractor, etc	Assist with cost information for principal and specialist designer's drawings, specification and schedules Checks on cost factors, agree sub-contractors and suppliers
G Bills of Quantities	Complete all information for QS including contractual arrangements Assist in completion of Bills of Quantities	Assist with preparation of Bills of Quantities Advise on form of contract, methods of payment, etc	Preparation of Specification and Bills of Quantities Advise on form of contract, methods of payments, etc
H Tender action	Agree list of suitable contractors with client and other interested parties In conjunction with QS issue tender documents and answer queries arising Receive tenders and consider with QS If necessary interview short listed contractors Report to client with recommendation of appointment of contractor, adjustment or re-tendering Draw up and arrange for client and contractor to sign contract documents	When appropriate, advise on selection of contractor and/or sub-contractors Assist with recommendation for appointment of contractor	Assistance with list of suitable contractors and issue of tender documents Check arithmetic and content of tenders from contractor Prepare detail report on tenders for principal designer and give recommendations regarding acceptance, adjustment or re-tendering Assist in drawing up contract documents
J Project planning	Prepare list of critical dates for contractor's programming and pass on sets of contract and construction information documents to contractor for use on site Arrange start of work and inform site inspectorate	Assist in preparation of critical dates for contractor's programming and supply all necessary information for contractor through principal designer Brief contractor when necessary	Supply cost information relative to site organisation Agree programme of certification with contractor Advise on contract interpretation

Figure 91 Sequence of events for small building project following RIBA Plan of Work format

119

	Principle designer	Specialist designer	Quantity surveyor
K Operations on site	Hold regular site meetings using agenda and minutes Give general supervision and negotiate with sub-contractors Issue instructions and certificates for payment in consultation with QS Keep client informed of progress and financial reports Obtain approval for increased costs and time for completion	Attend regular site meetings Provide general supervision and negotiate with sub-contractors Issue instruction and certificates for payment through QS and principal designer Prepare progress reports Obtain approvals for increases costs and delays in contract	Attend regular site meetings Measure and value work in progress for interim payments Provide cost advice on and measure and value instructions and variations Agree costs involved in delays in building programme
L Completion	Agree completion of work with QS and contractor Agree final costs and prepare statement Arrange for any defects to be rectified Arrange for final payments	Agree completion of work with contractor, principal designer and QS Agree final costs for their work and prepare statement Arrange for any defects to be rectified Notify principal designer and agree final payments	Agree completion of work with contractor, principal and specialist designer Prepare final account in agreement with contractor Agree final payment following rectification of any defects Prepare elemental cost of finalized building based on final account
M Feedback	Prepare building owner's manual Arrange for storage of contract documents	Assist in preparation of building owner's manual Arrange for storage of contract documents	Assist in preparation of building owner's manual Advise on running costs of building

Figure 91 Sequence of events for small building project following RIBA *Plan of Work* format—*continued*

Figure 91 briefly summarizes the sequence of events for a building project, where the principal designer is the team leader. For small uncomplicated contracts – say less than 500 m^2 – principal designers may perform all the functions of the Design Team. They will establish the brief, appraise the building requirements, obtain approvals, provide sketch schemes, develop these in detail, present them for evaluation to the client, provide drawings and provisional assessments, compile the production information and supervise the job on site. A short-list of builders will price from detailed specifications and key drawings, and the successful builder will provide a break-down of the price and the schedule of rates for the main activities. For medium sized contracts – say 500 to 2000 m^2 – the principal designer will probably work in conjunction with the quantity surveyor. The principal designer is normally selected first and is responsible for the design and management of the project. The quantity surveyor can be appointed on the recommendation of the designer or chosen separately. The designer and quantity surveyor together undertake a general appraisal of the building requirements. For large projects – say over 2000 m^2 the Design Team will comprise the principal designer(s), quantity surveyor(s) and specialist designers – the skills of the specialist designers being related to the particular complexity of the projects. Where circumstances are such that the principal designer is not the leader of the team in so far as management function is concerned, then various divisions of responsibility should be made to all members of the team at the out-set of a project (see pages 131–134).

Professional liability

Principal designers are the conventional leaders of the Design Team because, following the client's instruction, they normally produce the design of a building. In this case, as stated earlier, it is also their responsibility to supervise and co-ordinate the production of all drawings, schedules, and specifications, thereby ensuring that the project is carried out in accordance with their aims for the design. Principal designers should make certain that adequate information is available to other members of the Design Team so that decisions relative to the

continuity of work can be made at the right time. They must also try to foresee, as far as is practicable, the problems likely to arise and take action on unplanned eventualities. Being leader of the Team, therefore, principal designers must take the ultimate responsibility for the project. Accordingly, any deficiencies which may occur as a result of any design and construction decision, or mis-management of associated contracts often make the designers liable to legal claims for negligence. This may appear to be perfectly reasonable, but must be considered with some caution depending upon current legal interpretations of what constitutes the *reasonable* 'skill and care' which should be exercised by designers in the performances of their duties. Whilst certain claims for negligence are just, it must be remembered that design and construction processes involve innumerable value judgements when balancing the demands imposed by various performance requirements for a modern building. Each demand can only be satisfied at some expense to others. Also, the liability of principal designers often extends into the work of others, in that there is a responsibility for '*inspection*' and, therefore, acceptance.

The Civil Liabilities (Contribution) Act makes provision for a writ to name all and any party who might conceivably have any degree of liability for negligence in connection with a claim. Furthermore, the designers are currently liable *for life* for the cost implications (including inflation costs) arising from the errors in decisions made. In the event of a belated claim, the identifiable personal estate of the principal designer may even be penalised after his death. It is, therefore, necessary to take out *professional indemnity insurance* against possible claims, and when unlimited liability periods are required the premiums are very high. Furthermore, it is recommended that designers retiring from private practice continue this insurance cover, although, because they will no longer be practising, it will be given at a gradually reducing premium every year after cessation of practice.

15.4 Research Team

Any synopsis of the Building Team, however brief, must include reference to those members who make understanding and development of

121

current construction methods (materials and technical ability) possible, ie the *researchers*. The process of building has moved from a craft-based art towards a science-based technology over a period of about the last 50 years. This is particularly true in the environmental science areas of energy usage, thermal comfort, sound control, and lighting, where quantifiable criteria have found application in the design of the external envelope of a building.

The function of building research through government agency was begun in Britain in 1921 when the Building Research Station now known as the *Building Research Establishment*, was founded. The aim of the research is to discover facts by means of scientific study and, in matters concerning building, covers a very wide area of knowledge requiring controlled programming of critical investigation of chosen subjects. The BRE regularly publish information, current papers and monthly digests (*Building Research Establishment Digest*) about their work.

The BRE has numerous fields of study and provides an *Advisory Service* to the Building Team providing information which includes the following:

(a) *Building materials*
Research into old materials and new materials to produce an advanced scientific knowledge of their structure and behavioural characteristics so as to produce economic advantage in use.

(b) *Structural engineering*
Concern with the problems of design, service ability and safety related to cost.

(c) *Geotechnics*
Provision of a rational set of principles and analysis to help structural engineers to design a wide range of structures in and on the ground with a calculated risk and with economy.

(d) *Mechanical engineering*
Investigations into the mechanisation of the building process and the development of new production equipment and methods and study of new building plant.

(e) *Environmental design*
Study of the physical environment in and around building including research on human needs, the physics of building and design and development of engineering services bearing in mind the relevant cost implications.

(f) *Building production*
Concern with the use of resources of capital, materials and labour for new building, the maintenance of old buildings, and life-cycle costs.

(g) *Urban growth*
Study of problems associated with urban growth, either in new and expanded towns, or in sectors of existing towns.

Also, separate bodies associated within the BRE make special study of:

(h) *Fire*
Studies concerning the behaviour of people subjected to the effects of a building fire and smoke; evaluation of fire statistics; undertaking fire tests; identifying and quantifying fire and explosion hazards and methods of reducing them; evaluating the performance of structures in fires in relation to future design guidance and the compilation of legislation.

(i) *Timber technology*
Investigation into the properties and performance of wood, wood-based products and similar materials used in lightweight components; problems relating to the processing of home-grown timbers and jointing components; the understanding of the causes of deterioration of wood and wood products with the aim of reducing loss by the evaluation of preventative, remedial and preservative treatments; timber grading by quantifiable stress criteria and the effective design of timber structures.

There are many other advisory bodies allied to the construction industry which are prepared to supply informed opinions on specific problems. Like the BRE Advisory Service, they charge a fee and include the Construction Industry Research and Information Association (CIRIA) which often helps to finance research into aspects of construction by allocating funds to universities, other research bodies and industrial organizations. The Building Cost Information Service (BCIS) provides a cost analysis service and issued reviews of building costs; the

Royal Institute of British Architects Services (RIBAS) provides architects with practice information and includes a legal advisory service; and the Society for the Protection of Ancient Buildings (SPAB) advise (without charge) on possible listing of old buildings as well as give recommendations concerning technical aspects of conservation.

Research and development organizations also exist which are sponsored by industries to promote the scientific and aesthetic use of particular materials. These organizations include the Brick Development Association (BDA); Clay and Tile Federation (CTF); Timber Research and Development Association (TRADA); British Woodwork Federation (BWF); Constructional Steel Research and Development Organization (CONSTRADO); and the Cement and Concrete Association (CCA). Other organizations promote the use of data concerning the use and specification of suitable and tried construction methods as well as many other aspects, including documentation methods, testing procedures, and the legal requirements associated with the building process. These include the National House Builders Council; National Building Specification; Concrete Society; Fire Precaution Associations; Chartered Institute of Building Services; Property Services Agency; Building Advisory Service. These, and the many other organizations too numerous to mention here, can supply information which assists in making critical decisions during design and construction processes.

More detailed reference must be made to the importance of the British Standard Institute and the British Board of Agrément. The basic function of the BSI lies in the preparation and promulgation of *British Standards* covering nearly all aspects of productive industry. These Standards represent the recommendations made by specialist committees of interested parties in a particular subject and, for the building industry, cover dimensions, quality, performance, safety, testing, and analysis, etc, of materials, components, and methods of assembly or construction. Standards relating to materials and components are known as 'specifications' and for methods of assembly as 'codes of practice'. Most designers, quantity surveyors,

and builders are conversant with the existence of Standards to meet specific requirements, although they are not always very clear as to just what particular recommendations are given. More and more products carry the now familiar 'kitemark' – the registered BSI symbol – which implies that the product concerned has been tested and found to meet the appropriate requirements. Furthermore, British Standards recommendations regarding appropriate methods of construction frequently form the basis of the mandatory minimum requirements contained in the Building Regulations.

The principal objective of the British Board of Agrément, which employs the resources of the BRE and other organizations, is to help to bring into general use in the building industry new materials, products, components and processes. Accordingly, the Board offers an assessment service based on examination, testing and for other forms of investigation. The aim is to provide the best technical opinion possible within the knowledge available, and those innovations satisfying critical analysis relative to their intended performance are issued with an *Agrément Certificate*. Products or processes already covered by a British Standard do not normally fall within the scope of the Agrément Board – items to be subsequently covered by a British Standard will be issued with a certificate renewable after three years.

Agrément Certificates, therefore, not only supply a critical analysis of new materials, products, components and processes for members of the Design Team, but also furthers the production of new British Standards, and often provide a proof that certain innovations suit the requirements of the Building Regulations.

15.5 Legislative Team

After the client's brief for a building has been established, it will be necessary for the design team to start negotiation with the local authority in which it is to be situated in order to clarify certain legal requirements. The location and design of a building are controlled by the Planning Authority, and the construction by the Building Regulation Authority. These are supported by many other legal requirements concerning fire precautions, clean air, highways, factories, offices, shops, railways, etc.

Although in theory a designer must be conversant with all the legal requirements affecting a building, this can be an impossible task. There are probably more than a thousand Acts of Parliament which make reference to the building process and at any time during the course of construction (and even after occupation) a building is liable to be inspected by such diverse officials as factory inspectors, health inspectors, water inspectors, petroleum officers, planning and building control officers, firemen and policemen. While not necessarily guaranteeing the avoidance of demands being made during a late stage of construction which cause additional expense and loss of building time, it is the responsibility of a designer to be aware of the relevant legislation influencing the design of a particular building. At least this awareness means that contact can be made with the appropriate controlling person or organisation for specific advice during the various design stages of a building.

The **Town Planning Acts** are administered through the Local Authority by professional planners who are members of the Town Planning Institute (TPI) and, perhaps, the Royal Institute of British Architects. In dealing with the creative, social and administrative aspects of planning, the Planning Officers frequently employ specialist advisers, ie sociologists, ecologists, statisticians, economists, planning technicians, geographers, architects, graphic designers, and landscape architects.

Initial advice will be given on the suitable location of a building in a selected area (*zones*); amount of building permitted on a particular site (plot ratios and densities); the influence of previous planning decisions affecting the site; tree preservation; car parking; building lines; general public circulation; road widths and proposed widenings; and changes of use from one activity to another. Advice will also be given when work is to be carried out on existing buildings of special architectural or historical interest. Such buildings will have become '*Listed*' by the local planning authority in an effort to safeguard heritage, and any proposed demolition, extensions, alterations or modifications require special consent. A Building Preservation Notice can also be served on any individual when a building of special interest is proposed for demolition. Indeed, whole areas of countryside and urban development can be designated or zoned as a *conservation area*, and either no development permitted or only that permitted which is necessary for overall maintenance of features required to be preserved. Some planning authorities also issue guide lines about the appearance of a proposed building in an attempt to control the general character of a neighbourhood. (See DOE circular 8/87 Historic Buildings in Conservation Areas: Policy & Procedures.)

Having incorporated the requirements of the Planning Officer (or negotiated compromises) a formal application can be made which consists of a set of drawings indicating proposals, and documentation giving details of ownership, land usage, densities, etc. If the application is successful, permission will be granted for the proposals to commence construction *as far as the provision of the Town Planning Act are concerned*. Planning applications can be made in two consecutive stages, a fee being paid to the local authority for each according to the size of the proposed projects. *Outline planning consent* gives permission to the principle that a building, as yet not designed in detail, may be erected in a certain locality, eg a shop or office in an area or zone of an urban area designated for commercial usage; *full planning consent* gives permission for a specific building to be erected when precise details are finalised and proposals are known.

Before making a planning application for an industrial building it may be necessary to obtain an *Industrial Development Certificate* from the Department of Trade and Industry. This procedure attempts to control the effects of indiscriminate development and also focus industrial activities on areas of high unemployment. For similar reason, in highly congested urban areas such as London, request for planning permission to erect an office building may have to be accompanied by an *Office Development Permit* which is issued by the Department of the Environment.

The Town Planning Act stipulates that a decision of approval or rejection of a project should be given within two months of receipt of application, but most authorities currently request an extension of this period rather than formally reject a proposal owing to the lack of sufficient time for detailed consideration. Appeals against rejections, or any conditions imposed on a consent, can be made to the Secretary of State.

Building Control legislation in England and Wales (Scotland an Northern Ireland have separate systems) is covered by the *Building Act 1984* and the *Building Regulations 1985*. The *Act* is a consolidating statue drawing together requirements previously found in such documents as the Public Health Acts 1936 to 1961, the Health and Safety at Work Act 1974, the Fire Precautions Act 1971 and the Housing and Building Control Act. The Building Act calls for series of *Approved Documents* to give practical guidance on some of the ways of meeting building control requirements and these combine to form the basis of the **Building Regulations 1985**. A *Manual*, giving general guidance in the form of explanatory notes, accompanies the Approved Documents. The main purpose of the Regulations is to ensure the health and safety of people in or about a building, although they are also concerned with energy conservation and access to buildings for the disabled.

Building work controlled by the Regulations includes new buildings, extensions and material alterations to existing buildings, as well as the provision, extension or material alteration of controlled services or fittings and the work required for a change of use. Certain small buildings and extensions (normally under $30\,\text{m}^2$) and buildings used for special purposes are exempt from the Regulations.

To ensure compliance with the provision of the Regulations, the Client Team (acting on the advise of the Design Team) can have the work approved and supervised under construction by the local authority *or* by a privately employed Approved Inspector acting with the local authority. The two methods are independent and effect the way in which initial building control approval is sought.

Under full local authority control, two main options are available. One is to deposit working drawings for approval, to be given either conditionally or in stages within five weeks or, by agreement, within two months. Although having the drawings passed gives some protection in the event of subsequent problems, there is no need to wait this long before starting work. The alternative option involves giving a *Building Notice* to the local authority who may then ask for drawings to help them when inspecting the work. This option is not permitted for the construction of shops or offices. For both options, 48 hours notice must be given before work commences and, if the local authority considers that the Regulations are contravened at any stage, they must serve a notice requiring rectification unless they have approved the drawings. There are procedures for challenging the views of the local authority and there is no need for certification upon completion of the work.

When the local authority is fully involved in this way, Building Control Officers, Inspectors or District Surveyors will inspect the work during the various stages of construction to ensure compliance with the Regulations. These inspectors are usually members of the Royal Institute of Chartered Surveyors (RICS), or have qualifications of other professional bodies concered with the construction of buildings. Some building designers may wish to discuss proposals for use of materials or certain constructional strategies with these officials prior to the submission of drawings. A fee is payable to the local authority for this whole approval process as prescribed in *The Building (Prescribed Fees) Regulations*. Instead of seeking local conservation, which involves checking calculations, etc, certificates of compliance can be supplied instead. These are prepared by 'Approved Persons' who are professionals in those areas of design (members of CIBS or ICSE).

The alternative method of obtaining approval under the Building Regulations 1985, using the privately employed *Approved Inspector*, requires the submission of an *initial notice* to the local authority. This notice must be accompanied by certain drawings and evidence of insurance. The local authority must accept or reject the initial notice within ten working days, and once accepted, their powers to enforce the Regulations are suspended. It becomes the duty of the Inspector to notify the Client Team if the work contravenes the Regulations. If the defective work is not remedied within three months, the initial notice must be cancelled. On satisfactory completion of work the Inspector issues a certificate to the local authority and the Client Team.

The fee payable to an Approved Inspector is a matter of negotiation with the Client Team. Currently most Approved Inspectors derive from the National House Builders Council (NHBC). However, here is increasing interest from many professional bodies, including the Incorporated Association of Architects and Surveyors (IAAS) Faculty of Architects and Surveyors (FAS), and the Institute of Building Control Officers (IBCO), as well as the RIBA, CIB, ICE, ISE, and RICS.

In addition to the Building Act and Regulations, a designer may need to consult the inspectors who administer the many other ACTs directly affecting the building process. These include the Office, Shops, and Railway Premises Act; Factories Act; Clear Air Act; Chronically Sick and Disabled Act; Civil Amenities Act;

Noise Abatement Act; Licensing Act; Housing Acts; and the Health and Welfare Act, etc. There are also peripheral 'laws' in respect of cost, dimensional co-ordination or certain building types, and costs allowances (yardsticks). Account must also be taken of the recommendation and requirements of the service authorities, ie gas, water, electricity, telecommunications, drainage, fire brigade, and police. The rights of the building or land owners adjoining the site of a proposed building are also protected by legislation and require attention from a designer, ie Rights of Lights; Rights of Way; etc.

A building site can be a place of danger if care is not taken or there is a lack of experience at the management level. There is a vast amount of legislation related to this problem, but currently the principal Acts are the Factories Act 1961; the Office, Shops, and Railway Premises Act 1963; and the Health and Safety at Work etc, Act 1974. The Factories Act incorporates many other major Regulations, including the Construction (General Provisions) Regulations 1961, but all will eventually be superseded as new regulations come into force under the Health and Safety at Work etc, Act. These Acts require a builder or contractor to ensure that the building site maintains safe and healthy conditions for employees, and that the general public should be adequately protected from dangers resulting from site operation. The regulations should be on display for employees to read, and adequate instruction must be given to ensure safety consciousness. Factory Inspectors have power to enter, inspect and examine a building site at all reasonable hours and to make examinations and enquiries to ensure compliance with the regulations. He must be informed when serious accidents occur, or when disasters happen such as when cranes or lifting appliances collapse, or if there is any fire or explosion.

The builder is required to employ a competent person to inspect and supervise certain construction activities. Most of the larger building organizations have a Safety Officer who may be either employed or under contract from a Group Safety Scheme. This officer will visit site offices, workshops, and site works, liaise with site personnel and where necessary report to the organization's safety committee and to the local authority Factory Inspector.

The Design Team also has responsibilities under the Health and Safety at Work etc, Act 1974, because their projects must not create hazards for building operatives during construction. Although this may influence a change between initial concepts and the final design, the safety requirements are entirely reasonable. A building organization has the right in refusing to allow his employees to become involved in construction methods which are liable to be dangerous.

15.6 Manufacturing Team

The Manufacturing Team supply the materials, components and equipment which are used during the construction processes of a building, and, therefore, incorporate many organizations and interests. Traditionally, construction processes relied on the supply of readily available materials easily converted into manoeuverable forms or sizes which could be adapted further by craftsmen on site to suit a particular design. With the need to economise in labour and reduce costs, building procedures became more rationalized. Materials were formed into readily usable components during manufacturing processes, and were assembled with little adaptations after delivery to site. This *rationalized traditional* construction procedure reduced the number of separate operations and saved time on site.

However, the continual advancement of technology, and increases in complexity and size of buildings today results in even more complex construction processes. Manufacturers must extend their services from the supply of single components, to the supply of much larger parts of a building (elements) and indeed whole buildings. Site operations are reduced to a minimum using mechanical plant, and methods of building become largely concerned with the organisation of the systematic supply and assembly of pre-fabricated items, ie *system building* (see also 13.1 *Acquisition costs*).

Some manufacturers produce items which will not normally fit with the components of other manufacturers, and the resulting method of building is commonly referred to as *closed system building*. When component design is co-ordinated between the manufacturers of different products so that they can be used together without alterations, or become interchangeable, the building method is called *open system building*.

The purpose of this brief account of an aspect of the evolution of building processes is to indicate the vital role which the manufacturing team have on the design and development of a building (figure 88). In many respects, they should be members of the Design Team. Generally, however, manufacturers are often only concerned with the entire suitability of their particular product as it leaves the factory, and it is up to the Design Team to assess their performance relative to other criteria. (Mention has already been made to the influence of the Research Team in assisting the work of the Design Team in this area.) Whether producing materials, equipment, components or building systems, Manufacturing Teams often incorporate their own research and development organization to test products. Manufacturing Teams will also employ public relation organizations to produce information about their products for circulation to members of Client, Design, Research, Legislative and Construction Teams.

15.7 Construction Team

A study of building workers carried out by the Building Research Establishment lists over 50 separate occupations associated with construction. Therefore, the erection of a building depends on an industry where total reliance is placed on the diverse attitudes, abilities, and adaptability of its workers. Conventionally, these workers were grouped under 'trade' headings according to their skills (figure 92), and 20 years ago most were employed by a *main contractor* managing and directing all works on a site using a general foreman to co-ordinate the work of each trade or *sub-contractor* foreman.

Today, most specialist trades are employed as *nominated sub-contractors* by the client or principal designer on behalf of the client; a relatively few key tradesmen being employed directly by the main contractor as *ordinary sub-contractors* (known also as non-nominated sub-contractors or domestic sub-contractors).

Nominated tradesmen or sub-contractors may be required to design and provide specialist

Asphalter	– roof, floor and wall (basement) finishes
Bricklayer	– laying brickwork
Carpenter	– structural and carcassing timber work
Concretor	– placing concrete
Drainlayer	– providing below ground drainage
Electrician	– electrical installation
Excavator	– levelling site and digging drain/foundation trenches
Floor tiler	– internal floor finishes
Gas-fitter	– gas installation
Glazier	– fixing glass
Joiner	– timber work to finished components
Metal worker	– sheet metal applications (roofing)
Painter and decorator	–finishing components
Paver	– external paths/road finishes
Plasterer	– plastering walls/ceilings, screeding and rendering
Plumber	–plumbing installation, flashing and gas pipes (interior)
Scaffolder	–erecting scaffolding and working platforms
Steel erector	– erecting steel columns and beams
Steel fixer	–cutting, shaping and positioning steel re-inforcement
Tiler and slater	– roof finishes

Figure 92   Basic list of trades which would be employed for erection of a simple building

elements within a building from a statement of performance requirements, but the main contractor is still entirely responsible for the satisfactory completion of the work involved. It is also quite common for the client or principal designer on behalf of the client to employ *nominated suppliers* for certain specialist materials, components or equipment which are to be used or fixed into position by the main contractor.

Although this system of site organization remains normal with most small and many medium size building firms, there is an increasing tendency for the larger main contractors to become *building managers*, responsible for the co-ordination of the erection of a building using *only* nominated sub-contractors or suppliers. Perhaps the main reason for this lies in the fact that the continuous employment of their own tradesmen cannot be guaranteed during periods of economic recession. Enforced redundancies are sometimes contractually difficult and likely to prove expensive.

Available skills

The training of building trade craftsmen has traditionally been very much a matter for employers and unions. The City and Guilds of London Institute is the examining body for building craft workers and produce syllabuses which form the basis of construction courses taught at technical colleges. In order to compete with the other employment areas in terms of time-related earning power, the basic apprenticeship period for building tradesmen was reduced and has resulted in a gradual but continuous 'de-skilling' of traditional crafts. Furthermore, the greater emphasis now being placed on academic subjects in secondary schools, together with the unattractiveness of adverse climatic conditions on exposed UK construction sites, has resulted in far less recruits than previously. Government attempts through the Construction Industry Training Board (CITB) and the Training Services Agency (TSA) have endeavoured to boost training interests in skilled manpower. So has the trend towards providing an artificial climate around a building site by transparent protective sheeting, or moving most construction processes into a factory producing preformed units for final and speedy erection on site. Nevertheless, the reduction of practical tests in favour of the theoretical assessments of craft skills and knowledge have produced building managers rather than tradesmen.

The importance of the move away from the traditional skilled craftsmen on site lies in the need for building designers to concentrate earnestly upon the selection (or implementation) of suitable construction methods which is known to be realistic, relative to the depth of *practical expertise* likely to be available at the time when a project is to be erected. There is, therefore, an even greater need than before for close consultation between Design (including quantity surveyor) and Construction Teams – especially if hitherto untried materials and/or techniques are being contemplated. If this is not possible,

the Design Team must include sufficient expertise (perhaps through Specialist Designers) to be able to supply a potential main contractor with more detailed specifications and drawings of the chosen construction method than perhaps was necessary for corresponding innovations in the past.

Notwithstanding the above comments concerning the role of trade skills, it is important to realise that mechanization based on the development of petrol, diesel, and electric engines, pneumatic and hydraulic engineering, has influenced building methods considerably. In recent years mechanization has become universal on site – excavators have replaced hand digging of foundation trenches; mechanical hoists have largely replaced the necessity to carry materials up ladders by hand; pumping techniques make concrete more manoeuverable without the need for wheelbarrows; cranes transport large building components; hand-power tools have replaced hammer and chisel; mobile heaters have been introduced for drying out, etc. Indeed, certain building techniques rely on the use of equipment specially designed to carry them out. The increased use of plant or machines means that 'semi-skilled' operatives are often in the majority on a building site. However, completely 'un-skilled' activities still remain – mainly to service skilled and semi-skilled workers – and are fulfilled by 'labourers'.

The Building Employees' Confederation, (until recently the National Federation of Building Trade Employers) is the most important union of representing workers within the Building Industry and has affiliated organizations representing specific trades. The Confederation of British Industry (CBI) represents the interests of most building firms in the UK. The professional qualifications for builders involves membership of the Chartered Institute of Building (CIOB), which concerns itself with the education and management training with the industry.

Types of building organizations

The term 'main contractor' used earlier will now be investigated in greater detail since, for the efficient and economic construction of a building, it is imperative that the right type of organization is selected. Criteria generally relate to size and complexity of the project in hand, although the speed with which it can be erected and the special resources which may be provided by a particular builder play an important role in selection.

The sometimes loose separation between simple and complex project, relates more to the intended purpose (design) of the building rather than actual size: single or predominantly single-storey projects are not necessarily simple buildings, and multi-storey or large span projects not necessarily complex. The general configuration of a building, the amount and degree of complication of the services it accommodates, the characteristics of the site, and the complexity or otherwise of the construction method are more accurate divisions between simple and complex. The selection of a suitable builder may additionally be influenced by members of either the Client Team or the Design Team who may have preference for a builder with whom they are familiar. Generally, however, the available range of suitable builders can be narrowed by their ability to build a particular project to suit the intended financial outlay allocated to it.

Although complicated by the reason stated above, main contractors can be divided into three basic groups: *general builders*; *general contractors*; and *design and construction companies*. Currently, each could be co-ordinated by building

managers as explained earlier, but this role is mostly associated with the larger organizations of the latter two groups. All could, therefore, employ tradesmen as part of their regular staff, and use sub-contractors to deal with the specialist construction areas necessary for a particular project.

General builders (figure 93)

These will take on a wide range of work, but most concentrate on particular types and sizes of projects which are seldom in excess of £250,000. The *smallest* firms will be the 'jobbing builders' concerned with minor repairs of existing building where the principal workers are craftsmen (bricklayers or carpenters) who may employ sub-contractors for other trades such as plastering, electrical and plumbing installations, etc. The larger firms will be based on a 'family' organization, employing about 70 operatives (skilled and un-skilled), and six office staff. They normally operate in a particular area and therefore avoid excessive commuting distances. When their own building staff, plant and/or material resources are insufficient to carry out a particular trade or specialism, sub-contractors will be employed.

Some organizations with continuous building programmes, undertake part or even all of the construction processes themselves through the

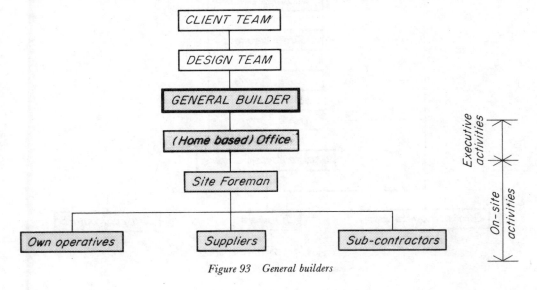

Figure 93   General builders

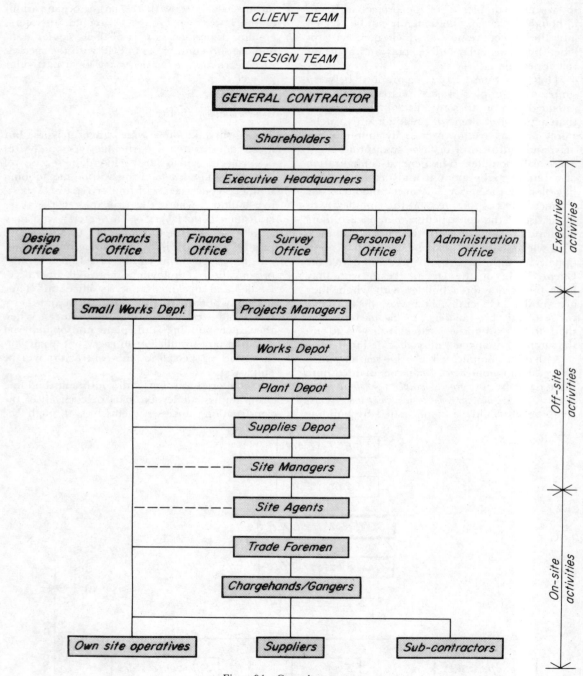

Figure 94   General contractors

agency of capital works departments which directly employ all the personnel concerned. These capital works departments are usually associated with maintenance and the building of small general purpose works, or sometimes, housing under the control of local government.

General contractors (figure 94)

Contracting firms will range in size and include multi-national organizations having international, national and regional headquarters. These firms will probably carry out both building and civil engineering projects of a very large size, ie multi-million pounds, but will also carry out much smaller projects requiring a special expertise they may have developed. For example, many general contractors specialize in particular types of work based on local traditions or availability of workers – shopfitting, specialist joinery work; special expertise in bricklaying, or concrete work resulting from particular skills in shuttering and form work. Some include research and development divisions, and have evolved individualistic construction methods based on the exploitation of certain materials such as pre-cast concrete units, steel frame components or load-bearing timber panels.

The general contractor's organization must necessarily be divided between office and site activities. Offices will often be concerned with estimating, tendering, site planning, construction process and planning, quantity surveying, cost control, and the bulk purchasing of materials and hire of plant. In large firms there will be separate transport, personnel, and employees' welfare sections.

The work on site will be under the control of a *site or project manager*, who may co-ordinate many other projects on different sites. The resident contact on a particular site will be the *general foreman or agent* who will be responsible for the operatives (contractor's own employees and hired sub-contractors) which work in teams under the control of the *trade foreman* or the *ganger* in charge of the skilled, semi- and un-skilled workmen.

A site office will employ staff to keep careful records of the work in progress, and assist in effective control and costing. This staff will also be concerned with time sheets, delivery and store records, weather records, and details of work progress.

Drawing office staff may also be employed, either at the main headquarters, or on a particular site. Their function is to make larger and more detailed working drawings when it is felt special information is required, such as the temporary work for shoring up excavations or methods of constructing shuttering for *in situ* concrete work required by a particular building design. It may also be necessary for this drawing office staff to confer with a designer regarding alternative methods of construction when greater efficiency can be achieved as a result of a general contractor's special resources.

A great deal of contracting work is carried out for and by government agencies and runs into several million pounds of work each year. This is true of both local government and of nationalised industries such as transport, gas, and electricity. These together with large commercial undertakings, have their own building works departments known as *direct labour organizations*.

Design and construction companies (figure 95)

These companies provide a service which is more elaborate than the previous two, as they will undertake the responsibility for *both* design and construction of a building project. This type of contractor is usually a specialist in specific form of building such as housing, factories or offices using a particular form of construction.

Design and construction companies, therefore, combine the services of the Design Team with those of general contracting, and are employed direct by a client for a particular project. In effect, the company provides a 'package deal' in which a contractor is responsible for all the major decisions on design and technical matters, prepares plans and specifications, obtains approvals, and carries out the contruction. They either employ salaried designers to prepare the design for projects, or pay a fee to independent principal designers and specialist designers. Arrangements are also made with building specialist and sub-contracting staff. There is, or course, an area of free enter-

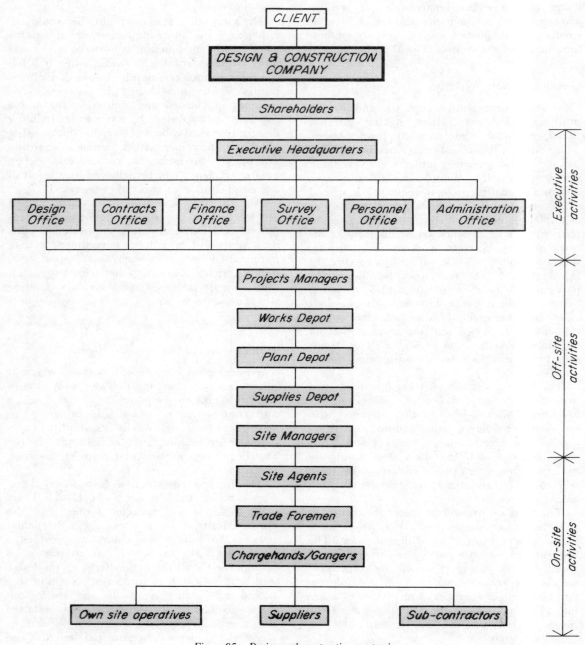

Figure 95   Design and construction companies

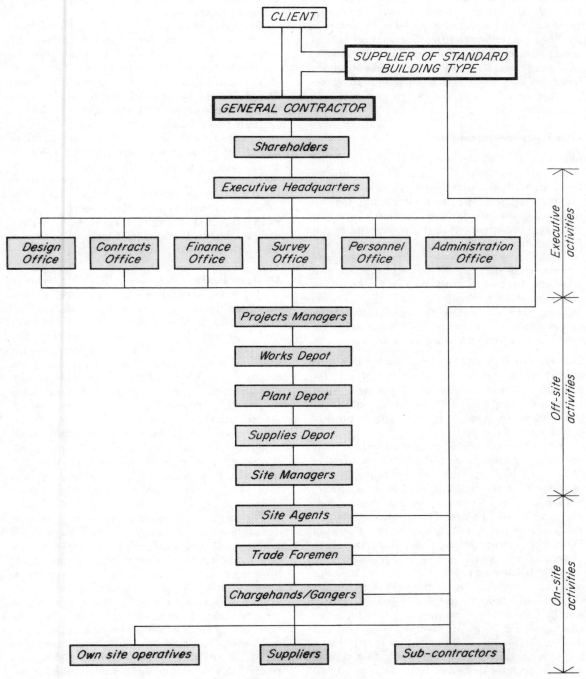

Figure 96   Package deal contractors

prise building in which contracting firms will seek out and acquire land, negotiate planning permission and develop a project to their own specification before setting out to find a customer for the project (figure 96). Conversely, with many overseas projects, fee paid consultant designers may work as a team with contractors to make an all-in offer for design and construction supervision.

Other personnel on site

During the construction period of a building site, other personnel who may not be directly involved with the practical work of a building organization will also be present, and the co-ordinating role of the site foreman, agent, or manager must also include responsibility for making sure their needs are recognised. This includes consideration for their safety and welfare while they are on site. Appropriate and adequate insurance must be taken out which compensates for accidents and the possible financial loss caused by delays resulting from any accidental damage they may cause while on site. These personnel include those from:

Client Team Visits may be made, but it is important that they only observe and not, for example, issue instructions. This will only confuse channels of communication and eventually cause the Design Team to lose control of the running of the project.

Nevertheless, when a client requires a regular check on the quality of the work, a *Clerk of Works* can be employed with responsibility of ensuring that the building organization strictly adhere to the documentation and instructions supplied by the Design Team (and agreed by the client). On small projects, the Clerk of Works would visit the site at regular intervals to inspect critical work, but on larger projects he would probably be resident on site and have his own office there. Although paid for by the client, the employment of a Clerk of Works usually follows the recommendation of the principal designer, especially when it is considered that a particular building project requires a fairly high degree of supervision. For similar reasons specialist designers may also recommend that a Clerk of Works specializing in their discipline should be employed when a building organization is lacking the necessary co-ordinating expertise.

Records will be kept by a Clerk of Works of all events, such as delay periods caused by inclement weather, strikes, or availability of materials, components and labour. This information is invaluable in assisting the Client and Design Teams to establish the need for extensions of construction time for a project which is requested by a builder or contractor.

Clerks of Works will have been trained in all aspects of construction and contract management, and are often tradesmen who have decided to turn their acquired experience of building into an overall supervisory skill. The professional body representing Clerks of Works, examining, and issuing their code of professional conduct is the Institute of Clerks of Works of Great Britain Incorporated (ICW).

Design Team (including QS) Visits will be made to deal with queries, supply information, issue additional instructions, establish financial criteria and monitor progress. For this purpose the principal designer usually organizes site meetings at regular and frequent intervals, the first one or two taking place before any work commences. All members of the Design and Construction Teams may not necessarily have to attend every meeting, but it is essential, of course, that individual members are present when issues are to be raised concerning their expertise. The agendas for site meetings must, therefore, be carefully planned in advance. Other and probably less formal meetings will take place as necessary to ensure the smooth running of a project.

For large or complicated projects it may be desirable to have a member of the Design Team permanently resident on site during construction activities. Depending on the nature of the project involved, he or she may be a representative of the principal designer, quantity surveyor, or any of the specialist designers. In certain cases a member from each may be necessary. The function of this resident designer or quantity surveyor is to answer and authorize action on any day to day queries which a contractor may have, and supply detailed information about areas of the project when the drawings, schedules, and specifications issued by his or her office require further clarification or amendment. In fulfilling these duties, the resident designer or quantity surveyor liaises with his or her office, who will provide the overall communication links between other members of the Building Team.

Research Team Visits may be made to monitor any work they have recommended, or to give advice concerning special problems which have arisen during the course of construction. Information gathered during these visits may form the basis of future useful research.

Legislative Team Most members of the Legislative Team are likely to visit the site, direct invitation or, often, in the form of spot checks on parts of the work relevant to their delegated power. The Building Control Officer or District Surveyor will regularly visit the site, usually by invitation of the builder, to inspect aspects of construction required by the provision of the Building Regulations (figure 97). When 'Approved Inspectors' have been commissioned to ensure compliance with the Regulation, they will also be visiting the site Highway Engineers, Public Health Inspectors and Town Planning

Officers will check compliance with approvals and investigate complaints made by the public concerning such matters as the builder's access into a site, mud on roads, noise obstruction, or other nuisances or infringements.

Although not strictly as members of the Legislative Team, the public utility services organizations (gas, water, electrics, and telecommunications, etc) will periodically check for possible damage to existing installations and/or organized work to new installations. Police may wish to inspect the site during the day or night to discuss security measures, or resolve problems concerning road obstruction caused by unloading lorries, etc; the Fire Officer will check for fire hazards and to establish that recommendations have been incorporated as the work proceeds; the Safety Officer and the Factory Inspector will examine the way that safety, health and welfare facilities are maintained on site.

Construction Team The personnel on a building site – either permanent or visiting – who are not involved in practical work may include contract administration staff, health offi-

---

LONDON BOROUGH OF ENFIELD

FOR OFFICIAL USE ONLY

For the attention of the Chief Building Surveyor

Date ....

BUILDING CONTROL STATUTORY NOTICE
(Building Regulation 14)

FEE PAID

£...................... TO PAY

Plan No. BC .... Nature of Works...................................

Address ....
I hereby give notice that:—
(a)   the above works will commence on .... at .... a.m./p.m.
(b)   the under-mentioned work or building
      will be ready for inspection on ....

Name and Address of Builder ....

Tel. No. ....   Signature of Builder ....
1. Excavations.
2. Foundations.
3. Damp Proof Course
4. Hardcore/Concrete Oversite.
5. Drains (before haunching or covering)

6. Drains (after haunching or surrounding and backfilling)
7. Occupation of Building.
8. Completion.

Note:— (a)   delete items not applicable.
       (b)   48 hours notice required under Building Regulations for item (a)
       (c)   24 hours notice required under Building Regulations for items
                    (b) 1,2,3,4 and 5.
       (d)   not less than 7 days notice required for item 7.
       (e)   not more than 7 days notice required for items 6 and 8

BC/15

AREA

Figure 97   Typical notification card used to inform local authority of progress in building work on site and the need for official inspection for compliance with 'Building Regulations 1985'

135

ing technicians, secretaries, canteen staff, and union officials. On larger projects, sub-contractors may have a similar range of staff on site, and will have regular visits from their management staff. Suppliers will deliver materials and components to designated storage areas.

Apart from a building site being inspected by members of the above Teams, often special visitors make application or are permitted by a building organization to observe the construction work during progress. These visitors include interested professional institutions, user teams, councillors, foreign visitors, and students of the building design professions, etc. With this type of visitor the site agent or manager must particularly ensure that proper insurance cover exists should they damage themselves or cause damage resulting in additional work having to be done which consequently causes delay in completion of a project.

15.8 Maintenance team

The chosen design and construction method of a building must take into account the effects which time will have on their performance (see chapter 3 *Durability* and chapter 13 *Cost*). Because of the complicated requirements, their inter-relationships, and the multiplicity of stylistic conventions which influence the selection of design and construction methods used today, it is sometimes necessary for the Design Team to consult certain specialists who, during the investigatory phases of design, can offer advice which goes toward the assurance of satisfactory performance standards for the intended life-span of a building. These specialists collectively form a consultative *Maintenance Team* who, although they may not necessarily become involved with the physical processes, can use their acquired experience and research to advise a designer on a suitable solution to a particular problem, especially where the more normal procedures of maintenance are impracticable. For example, very tall tower blocks often present external cleaning problems which could be solved by incorporation of one of several special items of equipment – cantilevered gantry devices and specially profiled curtain wall mullions to allow safety clips to be inserted for external manual cleaning; or strategically positioned sparge

pipes which allow water to be sprayed on façades, thereby eliminating the need for external manual cleaning altogether. Inevitably the appearance of the building will be affected by such devices and it is, therefore, very important that the Design Team and consultative Maintenance Team work in close harmony (figure 98). The precise methods adopted for subsequent maintenance and cleaning will also be influenced by the attitude of the Client Team towards the running costs of a building. The Maintenance Team will, therefore, give valuable advice which affects the ultimate cost evaluation of a particular project. Three-dimensionally profiled glass façades often present particular financial problems associated with both maintenance and cleaning. Additional costs can be incurred in the design, manufacture, and installation of special climbing gantries to facilitate accessibility, and additional costs are also sometimes involved in

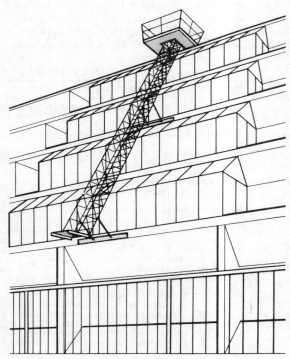

Figure 98   Window cleaning apparatus necessary for complicated facades which became a permanent feature of a building affecting appearance

special training programmes for maintenance and cleaning personnel. Nevertheless, for certain prestige building types the aesthetic desirability of three-dimensional glass façades may more than offset the additional costs involved.

Normal maintenance procedures can usually be formulated using the expertise of the Design Team without the need for special consultation, ie, methods of access to services, dimensions of ductwork and crawlways within ducts to permit space for repair or alterations, the routines necessary for continued or adjustable service supply (heating, lighting, ventilating, drainage, etc), and the care needed to maintain finishes, furniture and fitments. On completion of a project, the Client Team must be presented with a Maintenance Manual compiled by the Design Team which incorporates the advice of the consultative Maintenance Team. This manual describes how a building can be expected to perform, what measures have been taken to ensure it does, and what action must be taken in the future (see 16.8 *Maintenance manuals*).

Further specific reading

Mitchell's Building Series

Structure and Fabric Part 1 Chapter 2 *The production of buildings*
Structure and Fabric Part 2 Chapter 1 *Contract planning and site organization*
Chapter 2 *Contractors' mechanical plant*
Components: Chapter 3 *Industrialized system building*

Mitchell's Professional Library

Architectural Practice and Procedure, Philip H P Bennett
The Supervision of Construction, John W Watts
The Supervision of Installation, John W Watts

Building Research Establishment Digests

BRE Digest 289: *Building management systems*

The operational procedures and other management activities associated with the design, construction and subsequent performance of a building rely a great deal upon quite complex information being transferred between the various participants of the Building Team. Ideas and technical data must be dispersed to a wide range of people at both professional and non-professional, skilled and un-skilled levels. For this reason, methods used for communication should not only clarify issues, but also attempt to bring harmony to the work processes involved in the creation of a building, and foster the co-operation which ensures maximum contributions from all those with tasks to perform.

Communication methods have developed in type and sophistication to meet the needs of the disparate parties increasingly becoming with the creation of a building. Traditional methods involving memos, letters, reports, minutes, schedules, diagrams and drawings, are gradually becoming universally standardized in format to simplify understanding and speed production. Telecommunication systems, recording tapes and closed circuit television makes instant communication easy.

The introduction of micro-chip technology – the modern computer – has changed attitudes towards efficiency and communication within the building industry today. Design aims, methods of communicating design information and technical data can be instantly dispersed between designer, specialist designer, client, and contractor by coding through the universal digits of the computer. This process raises possibilities of absolute understanding and, in the face of ever-increasing building cost, the concept of keeping options open until the last minute. For example, it is now possible to design a school and produce all the drawings, schedules, specifications and other contract documents in less than three days when using *Computer Aided Design* (CAD) technology. This process obviously involves a modification in the way the various Teams operate when creating a building and deal with design and construction prob-

lems. The Designer will no longer only ponder over a drawing board in search of a solution to a problem through reliance on his intuitive skills and ability to interpret technical facts through memory or prolonged research (figure 99). Instead he will have multifarious solutions presented to him by a computer, all with equal technical and economical competency: his problem becomes one of selection. However, this selection process must include a realization of aesthetic ideals, and although the computer can incorporate ergonomic and other data about physical comfort it cannot (as yet) design a building to give full psychological delight to a human being. The techniques applicable to computer technology involve the production and manipulation of information, information storage, and information transmission. The extent to which these activities can be carried out depend upon the capacity of the computer – its size, cost, and programme capability and complexity. Information technology computers

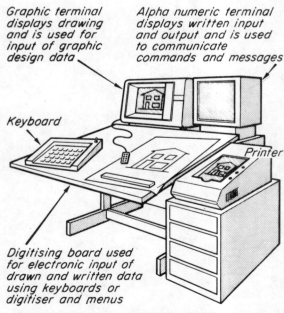

Graphic terminal displays drawing and is used for input of graphic design data

Alpha numeric terminal displays written input and output and is used to communicate commands and messages

Keyboard

Printer

Digitising board used for electronic input of drawn and written data using keyboards or digitiser and menus

Figure 99   Computer Aided Design (CAD) work station

range from pocket size dicataphones and calcu-
lators, to larger, but fully portable equipment;
then desk top equipment and large processors,
through to 'intelligent' machines connected with
the terminals of other remotely located machines.
Needless to say the competence of the operator is
of paramount importance.

Figure 87 illustrates the RIBA Plan of Work
for a building of moderate cost and which
involves the activities of a full team of designers
with a principal designer as leader, as described
in chapter 15 *The Building Team.*

The following areas of communications have
been identified, against the progress of the work:

– Technical data
– Drawings
– Specification of Works and Bills of Quantities
– Contracts
– Tender documents
– Fee accounts and Certificates of Payments
– Programmes of work
– Maintenance manual

16.1 Technical data

The time has long past when members of the
Building Team were able to rely on only the
technical data contained in a few text books for
the design and construction of a building. The
development of new uses for materials and of
new techniques of construction, together with
continual research into all aspects of building,
has undoubtedly led to an 'information explosion' in the building industry. Technical data is
now readily available from research organizations in the form of Research Bulletins and
Papers, and from manufacturers of materials,
equipment, components, and building systems
in the form of trade literature and/or actual
samples of products – mostly distributed by the
travelling representatives of a company.

But the vast amount of new and important
information supplied through these organizations makes accumulation and categorization
of the knowledge difficult, and thorough assimilation virtually impossible. It was necessary,
therefore, for the industry to establish suitable
methods of presenting technical data to the
Building Team, and recommend a method of
cataloguing documents, etc, so that they could

be readily found when required. The International Council for Building Research (CIB)
published a list of headings for the guidance of
authors of technical literature which classified a
method of giving information based on sequential performance criteria. From 1961 onwards a
method of classification for technical information started to be adopted which developed into
the currrently used *CI/SfB system.*

This system was first created in Sweden by
what was then known as *Samarbetskommittën for
Byggnadsfrågor* (The Co-ordinating Committee
for Building) and related more to the specific
requirements of the building industry than the
UDC system (Universal Decimal Classification)
employed in UK libraries. However, in order to
make the SfB system relate precisely to the UK
building industry, the RIBA instigated modifications resulting in the adoption of the CI/SfB
system (CI = Construction Index). Figure 100
gives the four tables of this modified system
(tables 1 to 3 constitute the original SfB
system), and each table represents a major
group of subjects or concepts which is now
available in the building industry. Broadly
speaking, the tables represent the design process
for a building as it proceeds through levels of
increasing detail – the construction work proceeds through these same levels but in reverse
order. Therefore, CI/SfB is a common language
for communicating information to all members
of the Building Team. The Tables are shown in
more detail in Appendix A, and figure 101 illustrates how the system may be used.

Whether the CI/SfB system is used in total or
only partially, depends on the size of organization requiring to classify information, or the
degree of cataloguing necessary for easy
retrieval by an individual. For example,
members of the Client Team may only require
information relating to Table 0 (Physical
Environment); or a manufacturer to Tables 1,
2, 3 (Elements and Construction Form and
Materials); whereas Design and Construction
Teams may need to make use of all Tables.

A selection of critical technical data can be
obtained from a *Building Centre* located in one of
many areas in the UK, and which form the
recognized meeting places for manufacturers,
users and designers. The Building Centre,
London, published a short, but useful, guide to

CI SfB

Table 0
Physical environment

Planning areas

Planning areas

0 Planning areas
01
02 International, national scale planning areas
03 Regional, sub-regional scale planning areas
04
05 Rural, urban planning areas
06 Land use planning areas
07
08 Other planning areas
09 Common areas relevant to planning

Facilities

1 **Utilities, civil engineering facilities**
11 Rail transport
12 Road transport
13 Water transport
14 Air transport, other transport
15 Communications
16 Power supply, mineral supply
17 Water supply, waste disposal
18 Other

2 **Industrial facilities**
21–25
26 Agricultural
27 Manufacturing
28 Other

3 **Administrative, commercial, protective service facilities**
31 Official administration, law courts
32 Offices
33 Commercial
34 Trading, shops
35–36
37 Protective services
38 Other

4 **Health, welfare facilities**
41 Hospitals
42 Other medical
43
44 Welfare, homes
45
46 Animal welfare
47
48 Other

5 **Recreational facilities**
51 Refreshment
52 Entertainment
53 Social recreation, clubs
54 Aquatic sports
55
56 Sports
57
58 Other

6 **Religious facilities**
61 Religious centres
62 Cathedrals
63 Churches, chapels
64 Mission halls, meeting houses
65 Temples, mosques, synagogues
66 Convents
67 Funerary, shrines
68 Other

7 **Educational, scientific, information facilities**
71 Schools
72 Universities, colleges
73 Scientific
74
75 Exhibition, display
76 Information, libraries
77
78 Other

8 **Residential facilities**
81 Housing
82 One-off housing units, houses
83
84 Special housing
85 Communal residential
86 Historical residential
87 Temporary, mobile residential
88 Other

9 **Common facilities, other facilities**
91 Circulation
92 Rest, work
93 Culinary
94 Sanitary, hygiene
95 Cleaning, maintenance
96 Storage
97 Processing, plant, control
98 Other ; buildings other than by function
99 Parts of facilities ; other aspects of the physical environment ; architecture, landscape

Table 1
Elements

(−−) Sites, projects, building systems

Substructure

(1−) **Ground, substructure**
(10)
(11) Ground
(12)
(13) Floor beds
(14)−(15)
(16) Retaining walls, foundations
(17) Pile foundations
(18) Other substructure elements
(19) Parts of elements (11) to (18) Cost summary

Structure

(2−) **Primary elements, carcass**
(20)
(21) Walls, external walls
(22) Internal walls, partitions
(23) Floors, galleries
(24) Stairs, ramps
(25)−(26)
(27) Roofs
(28) Building frames, other primary elements
(29) Parts of elements (21) to (28) Cost summary

(3−) **Secondary elements, completion** if described separately from (2−)
(30)
(31) Secondary elements to external walls ; external doors, windows
(32) *Secondary elements to internal walls ; internal doors
(33) Secondary elements to floors
(34) Secondary elements to stairs
(35) Suspended ceilings
(36)
(37) Secondary elements to roofs ; rooflights, etc
(38) Other secondary elements
(39) Parts of elements (31) to (38) Cost summary

(4−) **Finishes** if described separately
(40)
(41) Wall finishes, external
(42) Wall finishes, internal
(43) Floor finishes
(44) Stair finishes
(45) Ceiling finishes
(46)
(47) Roof finishes
(48) Other finishes to structure
(49) Parts of elements (41) to (48) Cost of summary

Use for doors generally if required

Services

(5−) **Services, mainly piped and ducted**
(50)−(51)
(52) Waste disposal, drainage
(53) Liquids supply
(54) Gases supply
(55) Space cooling
(56) Space heating
(57) Air conditioning, ventilation
(58) Other piped, ducted service
(59) Parts of elements (51) to (5 Cost summary

(6−) **Services, mainly electrical**
(60)
(61) Electrical supply
(62) Power
(63) Lighting
(64) Communications
(65)
(66) Transport
(67)
(68) Security, control, other services
(69) Parts of elements (61) to ((Cost summary

Fittings

(7−) **Fittings**
(70)
(71) Circulation fittings
(72) Rest, work fittings
(73) Culinary fittings
(74) Sanitary, hygiene fittings
(75) Cleaning, maintenance fitt
(76) Storage, screening fittings
(77) Special activity fittings
(78) Other fittings
(79) Parts of elements (71) to (Cost summary

(8−) ***Loose furniture, equipmen**
(80)
(81) Circulation loose equipme
(82) Rest, work loose equipmen
(83) Culinary loose equipment
(84) Sanitary, hygiene loose equipment
(85) Cleaning, maintenance loc equipment
(86) Storage, screening loose equipment
(87) Special activity loose equipment
(88) Other loose equipment
(89) Parts of elements (81) to (Cost summary

External, other elemen

(9−) **External, other elements**
(90) External works
(98) Other elements
(99) Parts of elements Cost summary

Use only (7−) if preferrec

Table 2
Constructions

Constructions, forms

Cast in situ work

Block work ; blocks

Large block, panel work ; large blocks, panels

Section work ; sections

Pipe work ; pipes

Wire work, mesh work ; wires, meshes

Quilt work ; quilts

Flexible sheet work (proofing) ; flexible sheets (proofing)

Malleable sheet work ; malleable sheets

Rigid sheet overlap work ; rigid sheets for overlapping

Thick coating work

Rigid sheet work ; rigid sheets

Rigid tile work ; rigid tiles

Flexible sheet work ; flexible sheets

Film coating and impregnation work

Planting work ; plants, seeds

Work with components ; components

Formless work ; Products

Joints

Used for special purposes eg specification

Table 3
Materials

a* Materials
b*
c*
d*

Formed materials e/o

e Natural stone
f Precast with binder
g Clay (dried, fired)
h Metal
i Wood
j Vegetable and animal materials
k
m Inorganic fibres
n Rubbers, plastics etc
o Glass

Formless materials p/s

p Aggregates, loose fills
q Lime and cement binders, mortars, concretes
r Clay, gypsum, magnesia and plastic binders, mortars
s Bituminous materials

Functional materials t/w

t Fixing and jointing materials
u Protective and process/property modifying materials
v Paints
w Ancillary materials

x
y Composite materials
z Substances

*Used for special purposes
eg resource scheduling by computer

Table 4
Activities, requirements

Activities, aids

(A)   Administration and management activities, aids
(Af)   Administration, organization
(Ag)   Communications
(Ah)   Preparation of documentation
(Ai)   Public relations, publicity
(Aj)   Controls, procedures
(Ak)   Organizations
(Am)   Personnel, roles
(An)   Education
(Ao)   Research, development
(Ap)   Standardization, rationalization
(Aq)   Testing, evaluating

(A1)   Organizing offices, projects
(A2)   Financing, accounting
(A3)   Designing, physical planning
(A4)   Cost planning, cost control, tenders, contracts
(A5)   Production planning, progress control
(A6)   Buying, delivery
(A7)   Inspection, quality control
(A8)   Handing over, feedback appraisal
(A9)   Other activities ; arbitration, insurance

(B)   Construction plant, tools
(B1)   Protection plant
(B2)   Temporary (non protective) works
(B3)   Transport plant
(B4)   Manufacture, screening, storage plant
(B5)   Treatment plant
(B6)   Placing, pavement, compaction plant
(B7)   Hand tools
(B8)   Ancillary plant
(B9)   Other construction plant, tools

(C)   *

(D)   Construction operations
(D1)   Protecting
(D2)   Clearing, preparing
(D3)   Transporting, lifting
(D4)   Forming : cutting, shaping, fitting
(D5)   Treatment : drilling, boring
(D6)   Placing : laying, applying
(D7)   Making good, repairing
(D8)   Cleaning up
(D9)   Other construction operations

*Used for special purposes

Requirements, properties

(E/G)   Description [2]
(E)   Composition [2.01/2.03]
(F)   Shape, size [2.04/2.06]
(G)   Appearance [2.07]

(H)   Context, environment [3]

(J/T)   Performance factors [4/5]
(J)   Mechanics [4.01]
(K)   Fire, explosion [4.02]
(L)   Matter [4.03/4.06]
(M)   Heat, cold [4.07]
(N)   Light, dark [4.08]
(P)   Sound, quiet [4.09]
(Q)   Electricity, magnetism, radiation [4.10]
(R)   Energy [4.11]
        Side effects, compatibility [4.12/4.13]
        Durability [4.14]
(S)
(T)   Application [5]

(U)   Users, resources

(V)   Working factors [6]
(W)   Operation, maintenance factors [7]

(X)   Change, movement, stability factors

(Y)   Economic, commercial factors [8/10]

(Z)   Peripheral subjects ; forms of presentation ; time ; space [11]

Note : Codes in square brackets above are from *CIB MasterLists for structuring documents relating to buildings, building elements, components, materials and services* (CIB Report No 18, 1972).

CI/SfB is a flexible system eg free codes throughout the tables can be used for special purposes on information which is not for general publication ; for Tables 0 and 1, headings and codes in colour will often provide sufficient breakdown

Figure 100   CI/SfB Construction classification (RIBAS)

CI/SfB New edition (1976): how to use it

1 CI/SfB is a common language for communicating information in the building industry. Like any other language, it has to be learnt. After a time (quicker than with most languages) the key words and codes which make it up can be used without conscious effort.

2 It consists of four 'tables'. Each of these represents a major group of subjects (concepts) about which information is passed in the building industry:

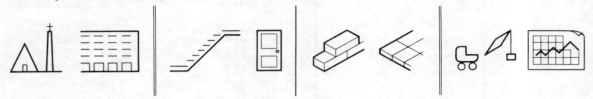

3 Each table consists of key words and codes to go with them:

Numbers		(Numbers in brackets)		Letters		(Letters in brackets)	
RELIGIOUS	6	SERVICES	(5−)	BLCKWK/BLOCKS	F	OPERATIONS	(D)
Cathedrals	62	Drainage	(52)	Concrete	Ff	Storing	(D1)
Churches	63	Gas	(54)	Clay	Fg	Transporting	(D3)

4 The tables are shown in more detail in the Construction indexing manual, and on the RIBA wall chart.

5 Classification (numbering for filing) is done in three stages:
  1 Decide what the document is about, in terms of CI/SfB (difficulty arises when it is about several subjects).
  2 Select most appropriate key word(s) or code(s) (or both).
  3 Write these down, sometimes in a classification 'box':

Any code from Table 0   Any code from Table 1   Any code from Tables 2, 3   Any code from Table 4

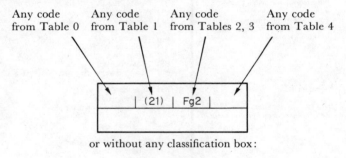

or without any classification box:

(21) Fg2

or without any classification box, using words only:

external walls – bricks, blocks – clay

6 When classifying, make a careful decision as to whether or not any code is required from each of these tables – 0, 1, 2, 3, 4, in that order. Any other 'citation order' may be adopted if there is a good reason for it. The order, once decided upon, should be used consistently.

7 File in a 'filing order' which may also be – 0, 1, 2, 3, 4 – or any other order used consistently.

Project information

8 In the case of project information the principal designer, specialist designer or quantity surveyor is the originator of the document, whether drawn or written and is therefore able to use the classification as a set of definitions to decide what should go on each drawing or in each section of the specification etc. The classification is used *before* the document is produced and *not, as in library classification, after it has been produced.*

Figure 101   Method of using the CI/SfB classification system (RIBAS)

manufacturers of trade literature for the building industry which stated: '*Trade literature may be defined as information which enables the user to select, specify and utilize a product in service. This information also helps him to compare similar products and services. He may thus elect the ones that best suit his requirements. The fully informative piece of trade literature is a great aid in the preparation of drawings, specifications, and bills of quantities.*' Building Centres incorporate libraries of building and engineering product literature (including samples) classified under the CI/SfB; they provide an information service which includes a telex link to European sources of reference. Specifiers of non UK products must however ascertain that testing data complies with BSI Standards or have an appropriate Agrément Certificate.

Some Building Centres may provide contact offices for professional and research bodies such as the RIBA, ICE, BRE, and TRADA. Whereas Building Centres primarily provide permanent information resources, manufacturers and trade organizations also frequently sponsor temporary national building exhibitions. Well designed UK products may be exhibited at a *Design Centre* run by the *Council of Industrial Design* whose chief purpose is the promotion of British design.

Instead of collecting and classifying technical information, or in order to supplement an established technical library, some organizations prefer to employ the services of companies specializing in the dissemination of information. As far as the building industry is concerned the major distributors of technical information are *Barbour Index Limited*, and *RIBA Services Limited* (including *Services to Specifiers Ltd*). These organizations provide a basic library of literature (Barbour Index include a microfilm presentation) which is periodically updated with new information. RIBA Services Ltd currently provide technical information in the forms of Product Data and Practice Data. The former is linked to a Product Selector which provides simple computer programmes as well as listing firms able to supply products and services; the latter gives information on legislation, practice, building failures, design, British Standards and information technology. Less elaborate collective information is available from such sources

as the Architects Standard Catalogue, Building Commodity File, Specification, etc.

16.2 Drawings

Drawings of many types provide the main method of communication between all the members of the Building Team. As the information required at any one stage of the implementation of a building will vary, and also be at different levels of complexity according to user requirements, the different types of drawing are divided for convenience into two broad categories.

Design drawings communicate the form of a building in terms of shape, colour, and texture; *Production drawings* communicate the technical, physical and economic aspects of a building which are associated with its construction, subsequent use, and maintenance. The information conveyed on both types of drawing are generally supplemented by reports, schedules, samples, models, discussions, etc. In reality there should be no firm division between design and production drawings, just as in reality there should be no division between design ideas and construction. However, certain drawings need to convey more about appearance to less technically minded parties, while other drawings are needed to convey technical information to parties whose priorities are other than overall appearance and function of a project. Though all the Building Team would undoubtedly benefit from studying a comprehensive range of drawings for a particular project, such a need is of most value to the principal designer who is concerned with the all embracing quality of the project. It is the responsibility of the overall co-ordinator of the project, (whether principal designer, contractor, or some other party) to ensure that the most useful drawings are distributed to the appropriate members of the Building Team at the precise time they are required.

Design drawings

There are two types of design drawings; those concerned with the preliminary investigation processes for a design, and those concerned with the presentation of a design solution. Both are produced during the 'design' stages of a project.

COMMUNICATION

Design drawings for *investigation purposes* communicate information between designers, quantity surveyors, etc, (Design Team), and if involved during early stages of negotiations, the builder or contractor (Construction Team). In addition, this information about the project will also be passed to the Client, and eventually form the basic information necessary to produce a preliminary or *sketch design*.

The earliest form of investigation drawing is that which provides information about the site (and existing building if a conversion is involved), and the immediate environment including adjoining structures, roads, services, etc, likely to influence a project. This is known as a *site survey*, and will enable the Design Team to commence with the design. Government produced Ordnance Survey maps are useful initial references (figure 102).

As the design process continues, the Design Team may require specific information about the cost of different solutions for a potential scheme, and supply *sketch drawings* to a quantity surveyor for comment. A contractor may also be able to help in establishing economic and efficient construction methods by suggesting the suitable type and availability of labour and mechanical plant for a particular design.

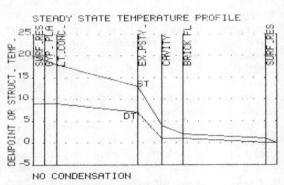

```
U VALUE & CONDENSATION ANALYSIS
**********************************

WRITTEN BY:
TIM VENN DIP.ARCH. MSC. RIBA

POLYTECHNIC OF NORTH LONDON
DEPT. OF ENVIRONMENTAL DESIGN
APRIL 1983

WALL CONSTRUCTION FOR ANALYSIS
MATERIAL        RSTY  THIK  RES.CE
SURF.RES.IN.     0     0     .13
GYP. PLASTER     2     15    .03
LT.CONC.BLOCK    5.3   100   .53
EX.PSTY.HI/D     27.7  30    .83
CAVITY           0     25    .18
BRICK FLETTON    1.78  102   .18
SURF.RES.OUT.    0     0     .05

TOTAL RESISTANCE       =     1.93
U VALUE=  .52 WATTS/M2 DEGC
STEADY STATE CONDENSATION ANAL.

CONDITIONS:
OUT.TEMP= 0 C OUT.VP= 6 MB
IN.TEMP= 20 C INSIDE VP= 11.3 MB
```

Figure 103   Computer analysis of specific forms of construction (Brick/block cavity wall)

Specialist designers usually prepare sketch drawings of their proposals for similar reasons and communicate vital information which enables the principal designer to incorporate them in the design. Drawings for these purposes need not be elaborate when information is being communicated between parties having a common aesthetic and/or technical language.

Computer 'print-outs' are increasingly used for the production of technical data (figure 103) to develop concepts outlined by sketch design decisions. Using this method, speedy and accurate information is made available for assessing lighting levels created by different shape and size windows, heat losses for different constructions,

Figure 102   Section of Ordnance Survey map

and economic spanning methods for alternative structural solutions.

Design drawings for *presentation purposes* are prepared by the designer to illustrate to the client the appearance of a project, the general disposition of the accommodation to be provided, and the effects the overall scheme has on the environment, including details of landscaping and the immediate physical surroundings. Drawings should use techniques which are readily understood – preferably three dimensional representations such as perspectives, (figure 104), axonometrics, or isometrics – and two-dimensional representations (orthographic) should be clear and provide easily identifiable features, eg people, furniture, trees, etc. The drawings range from simple pencil sketches to highly finished fully coloured and mounted drawings. Very little technical information is usually given, although this depends on the expertise and requirements of a particular client: those with a technical expertise influencing the design will probably need information to show how their interest has been incorporated.

Whenever possible, information about the design should be supplemented by scale models and/or sample boards of materials to be employed. Some principal designers prefer to use the skills of '*visualisers*' specialising in artistic techniques so that a design is most effectively presented. Also, line perspectives can be produced by computers (figure 105). After being fed the appropriate information, computers are also capable of producing a video presentation of a 'walkabout' through a proposed building. This preview can also assist a designer during sketch design stages.

With a few minor additions, design drawings can form a suitable submission to the local authority for Town Planning consent, although generally for this they need not be as elaborately presented.

Production drawings

These are often referred to as 'working drawings', and are produced by designers in order to communicate technical information throughout the Building Team. The nineteenth century architect, Sir Edwin Lutyens once commented '. . . *a working drawing is merely a letter to a builder telling him precisely what is required of him – and not a picture to charm an idiotic client.*'. The building

Typical perspective drawing (Artist: Malcolm Carver)

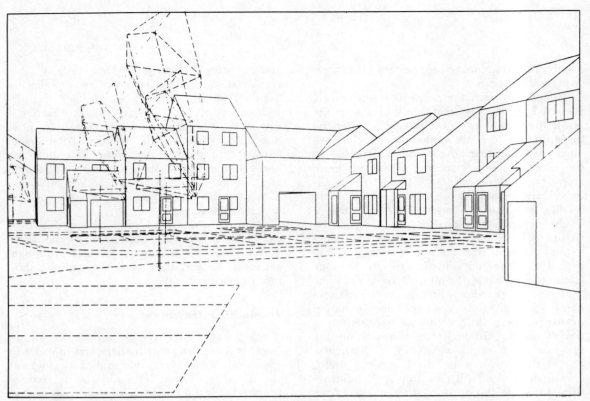

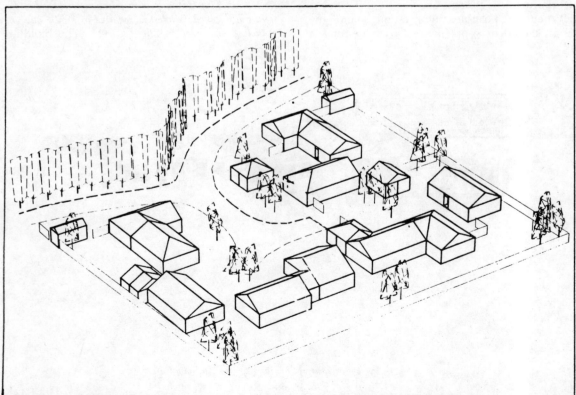

Figure 105(a) and (b)   Three-dimensional representation of different aspects of a building complex produced with the aid of a computer (Diamond Redfern & Partners and GMWC)

process has become much more complex since, and figure 106 lists the uses to which production drawings may currently be placed. Generally, the principal designer of a project prepares information and incorporates the data supplied by others (including specialist designers) to producing a master set of production drawings. For larger projects, certain specialist designers (structural and mechanical services engineers, etc) prepare independent production drawings, but this must be done in close collaboration with the principal designer in order to ensure unified .intentions and maintenance of professional responsibility.

Means of obtaining official consents and statutory approvals
Analysis of cost factors
Establishes use of materials
Informs extent of sub-contractors' work
Source of information for other contract documents
Provides details for tendering
Indicates contractual commitments
Basis for ordering materials and components
Establishes type and amount of labour required
Demonstrates construction detailing
Forms part of documentation during site meetings
Indicates degree of supervision
Provides check for variations from the contract
Assists with the measurement of progress
Guides on amount of interim financial payments to contractor
Basis for agreeing completed works and final payment
Provides check on defects in site construction method
Records work completed
Indicates factors to be included in maintenance manual

Figure 106 Functions provided by production drawings during the realisation of a building project

Together with the written descriptions of a project (eg Bills of Quantities, Schedules and Specification of Works) production drawings are a vital part of the legal documentation upon which contractual arrangements are based. It is important that any alterations which are necessary after the formal issuing of production drawings are communicated to all members of the Building Team as they could result in changes in legal arrangements. Amendments influencing

cost and/or uses of materials and labour need particular attention.

A typical basic list of items to be incorporated on production drawings for a small project is indicated in Appendix B. From this it will be realised that they must provide an accurate record of the principal designer's intentions at all stages of the construction process, and the information must be interpreted by many people with different sets of priorities, eg building control officers, quantity surveyors, salesmen, foremen, tradesmen, labourers, etc. It is important, therefore, that production drawings are clearly representative and easily understood; comprehensive, and sufficiently detailed for their purpose; and are produced in a format which enables them to be easily collated so that specific drawings can be found by particular users when required.

For this reason, a great deal of work has been carried out, initially by the British Standards Institute, in order to establish a system of co-ordinating production drawings and the information they communicate so as to avoid errors, inadequate data, and omissions. The need of the different parties involved were analysed, and recommendations made in BS 1192: 1984/7 *Building Drawing Practice*, that information can best be communicated by arranging production drawings into sets which reflect the construction process. Members of the Building Team first need to know the shape, size and *location* building to be constructed and its constituent parts; then about the methods to be adopted for the *assembly* of the parts (type of material and labour required); and finally about details of *components* to be used. Figure 107 indicates in greater detail the type and purpose for each category of drawing together with the scales which are recommended for each. Schedules provide more detailed specialized information and are a collection of mostly written information about the repetitive parts of a building, such as doors, windows, finishes, etc, (figure 108).

This method of structuring information greatly assists individuals of the Building Team in identifying the group of production drawings which are able to give the particular information they require. However, consideration must also be given to how this information is presented to assist easy reference and understanding. Drawing sheets should be used which conform to a uniform series of sizes to facilitate handling or storage and the international 'A series' of paper sizes (figure 109) should be adopted for all drawings and written material, including trade literature, etc. Secondly consideration must be given to one of the standard methods of graphically representing materials, components, and dimensions so that a common grammar is established between the members of the Building Team, thus reducing ambiguity and speeding understanding. Figure 110 indicates the British Standard recommendations for dimensions on drawings. The recommendations of BS 1192 are in four parts, covering general principles; architectural and engineering drawings; symbols and conventions; and landscape drawings. Some of the conventions currently used are illustrated in figure 111.

In addition to the above factors it is important that the layout of drawing sheets should be done in a systematic manner: particular attention should be given to the title panel of the drawings, as ambiguity can cause a considerable waste of time when seeking out particular information. Figure 110 illustrates a title panel as recommended by the current BS 1192.

The messages conveyed by the production drawings are really a translation of the three-dimensional ideas of a designer, and therefore, perhaps the best form of graphical presentation should also be three-dimensional, ie isometric or axonometric. This form is particularly useful when new construction methods or difficult junctions between several components need amplification (figure 39). However, three-dimensional representations are very time consuming to prepare and, for the majority of building work, are not normally necessary. There are two basic methods of communicating information on production drawings. The *conventional method* consists of drawings containing many notes about the construction methods to be employed which are further developed by the clauses of Specifications and/or Bills of Quantities. There is also the *systematic method* which consists of outline drawings only, but makes frequent direct cross-reference (using CI/SfB system) to supplementary documents such as standard details, schedules, specification clauses, and Bills of Quantities. These forms are illus-

trated in figures 112 and 113: the selection of the most suitable one for a particular project depends upon the organisation of the Building Team, and on the complexity of the building work involved. The conventional form of

text continued on page 157

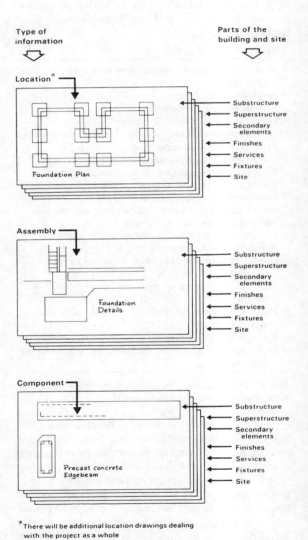

*There will be additional location drawings dealing with the project as a whole

Figure 107   Production or working drawings ... continued

... *continued*

Type of drawing and purpose			Appropriate Scale
Location	Site and external works:	General site plan	1 : 2500 1 : 1250
		To identify, locate and dimension the site and external works	1 : 100 1 : 200 1 : 500
	Building:	To identify, locate and dimension parts and spaces within the building and to show overall shapes by plan, elevation or section	1 : 50
		To locate grids, datums and key reference planes	1 : 100
		To convey dimensions for setting out	1 : 200
		To give other information of a general nature for which a small scale is appropriate (eg door swings)	
	Element:	To give location and setting-out information about one element, or a group of related elements	1 : 100 1 : 200
	Cross references:	To show cross-references to schedules, assembly and component drawings	1 : 100 1 : 50
Schedule	Element:	To collect repetitive information about elements or products which occur in variety To collect cross-references to assembly and component drawings	
Assembly	Element:	To show assembly of parts of one element including the shape and size of those parts	1 : 5
		To show an element at its junction with another element	1 : 10
		To show cross-references to other assembly and component drawings	1 : 20
Component	Element or sub-elements:	To show shape, dimensions and assembly (and possibly composition) of a component to be made away from the building	1 : 1 1 : 5 1 : 10
		To show component parts of an *in situ* assembly which cannot be defined adequately on the assembly drawing	1 : 20 1 : 50

Figure 107   Production or working drawings. Crown Copyright. Reproduced from Building Research Establishment Digest 172 by permission of HM Stationery Office

DOOR GROUP PREFIX LETTERS			INTERNAL DOORS													EXTERNAL DOORS coded XD												
DOOR NUMBERS	REFERENCE NUMBERS	AZA	1	2	3	4	5	6	7	8 DOUBLE	9	10	11	12	13	1 DOUBLE	2	3	4	5 DOUBLE	6	7 DOUBLE	8	9 DOUBLE	10	11	12	13
common suiting master keying.																												
UPRIGHT LOCKS	WITH ONE KEY	100																										
	LOCKS TO PASS	101	/	/	/		/					/	/	/														
	WITH TWO KEYS	102																										
	REBATED COMPONENTS	103																										
	ROLLER BOLT, ONE KEY	104																										
	ROLLER BOLT, LOCKS TO PASS	105																										
	ROLLER BOLT, TWO KEYS	106																										
	ROLLER BOLT REBATED COMP'S	107																										
UPRIGHT DEADLOCK	WITH ONE KEY	112																										
	LOCKS TO PASS	113								1.					1.		/	/		/		/		/		/	/	/
	WITH TWO KEYS	114																										
	REBATED COMPONENTS	115													/		/	/		/		/		/		/	/	/
LEVER HANDLES	PAIR ON ROSE	130																										
	PAIR ON BACKPLATE, KEYHOLE	131	/	/	/		/					/	/	/														
	PAIR ON BACKPLATE, NO KEYHOLE	132																										
	ESCUTCHEONS	133								2					2		2	2		2		2		2		2	2	2
PULL HANDLES	150 mm CENTRES FIXING	134																										
	225 mm CENTRES	135				/	/	/	/					/		/	/	/	/	/	/	/	/	/	/	/	/	/
	300 mm CENTRES	136																										
FINGER PLATES	300 x 75	140			/		/	/	/	/				/		/	/	/	/	/	/	/	/	/	/	/	/	/
KICKING PLATES	625 mm WIDE	141												1.														
	725 mm	142				/	/	/	2	2		/																
	775 mm	143	/	/	2						2		2							/	/					/	/	
	825 mm	144														/	/	/	/		/	/	/	/				/
	875 mm	145																										
FLUSH AND BARREL BOLTS	PAIR FLUSH BOLTS	150								/						/				/		/		/				
	SOCKET FOR WOOD	151																										
	SOCKET FOR CONCRETE	152																										
	PAIR BARREL BOLTS	153																										
OVERHEAD CLOSERS	FOR EXTERNAL DOORS OPEN OUT	160																										
	FOR INTERNAL DOORS	161				/		/	/				/	/														
	DOOR SELECTOR	162																										
	OVERHEAD LIMITING STAY	163														/	/	/	/	/	/	/	/	/	/	/	/	/
DOOR STOPS	FOR TIMBER	170																										
	FOR CONCRETE	171				/					/	/	/	/														
	POST MOUNTED DOOR HOLDER	172																										
	CABIN HOOK	173																										
LETTER BOX	LETTER BOX.	174																									/	
HAT AND COAT HOOKS	ALUMINIUM	180																										
	NYLON COATED SECRET FIX.	181																										
	PAIR NYLON COATED SECRET FIX	182																										
	NYLON COATED SCREW FIX.	183																										

Figure 108   Ironmongery schedule

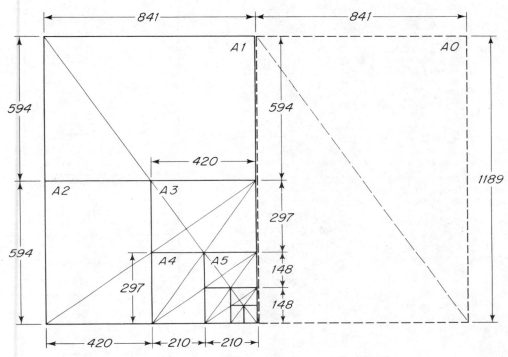

A-sizes retain identical proportions (1:√2),
each sheet size being half the size next above

The rectangle x × y has a
surface area of 1m²
for sheet size AO

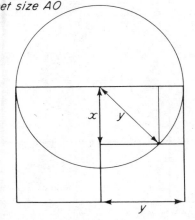

A-size	mm
A0	841 x 1189
A1	594 x 841
A2	420 x 594
A3	297 x 420
A4	210 x 297
A5	148 x 210
A6	105 x 148
A7	74 x 105
A8	52 x 74
A9	37 x 52
A10	26 x 37

Measurements represent trimmed sizes

Figure 109   The internationally agreed A-series format of paper sizes to be adopted for all written documents, drawings and trade literature

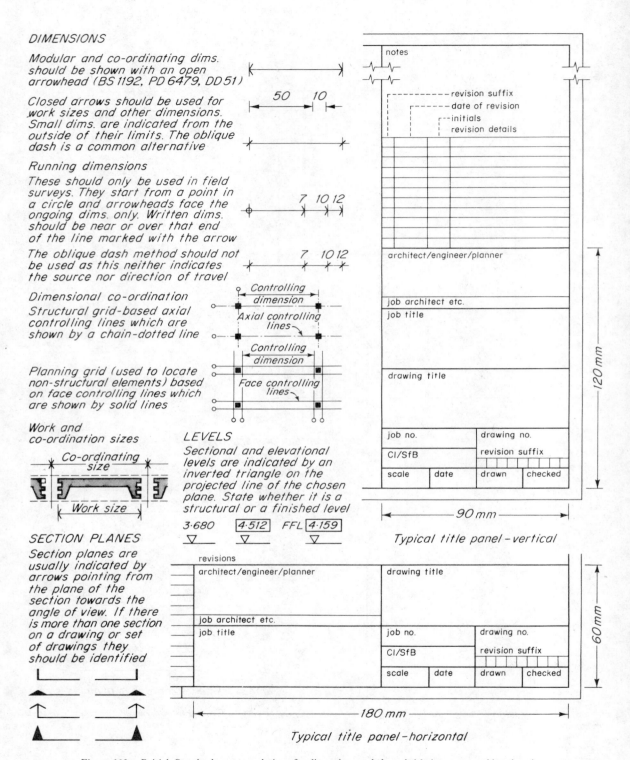

DIMENSIONS

Modular and co-ordinating dims. should be shown with an open arrowhead (BS 1192, PD 6479, DD 51)

Closed arrows should be used for work sizes and other dimensions. Small dims. are indicated from the outside of their limits. The oblique dash is a common alternative

50   10

Running dimensions

These should only be used in field surveys. They start from a point in a circle and arrowheads face the ongoing dims. only. Written dims. should be near or over that end of the line marked with the arrow

7 10 12

The oblique dash method should not be used as this neither indicates the source nor direction of travel

7 10 12

Dimensional co-ordination

Structural grid-based axial controlling lines which are shown by a chain-dotted line

Controlling dimension
Axial controlling lines

Planning grid (used to locate non-structural elements) based on face controlling lines which are shown by solid lines

Controlling dimension
Face controlling lines

Work and co-ordination sizes

Co-ordinating size
Work size

LEVELS

Sectional and elevational levels are indicated by an inverted triangle on the projected line of the chosen plane. State whether it is a structural or a finished level

3·680   4·512   FFL 4·159

SECTION PLANES

Section planes are usually indicated by arrows pointing from the plane of the section towards the angle of view. If there is more than one section on a drawing or set of drawings they should be identified

notes
revision suffix
date of revision
initials
revision details

architect/engineer/planner

job architect etc.
job title

drawing title

job no.   drawing no.
CI/SfB   revision suffix
scale   date   drawn   checked

90 mm

120 mm

Typical title panel – vertical

revisions
architect/engineer/planner   drawing title

job architect etc.
job title   job no.   drawing no.
                                 CI/SfB   revision suffix
                                 scale   date   drawn   checked

180 mm

60 mm

Typical title panel – horizontal

Figure 110   British Standard recommendations for dimension symbols and title boxes on working drawings

152

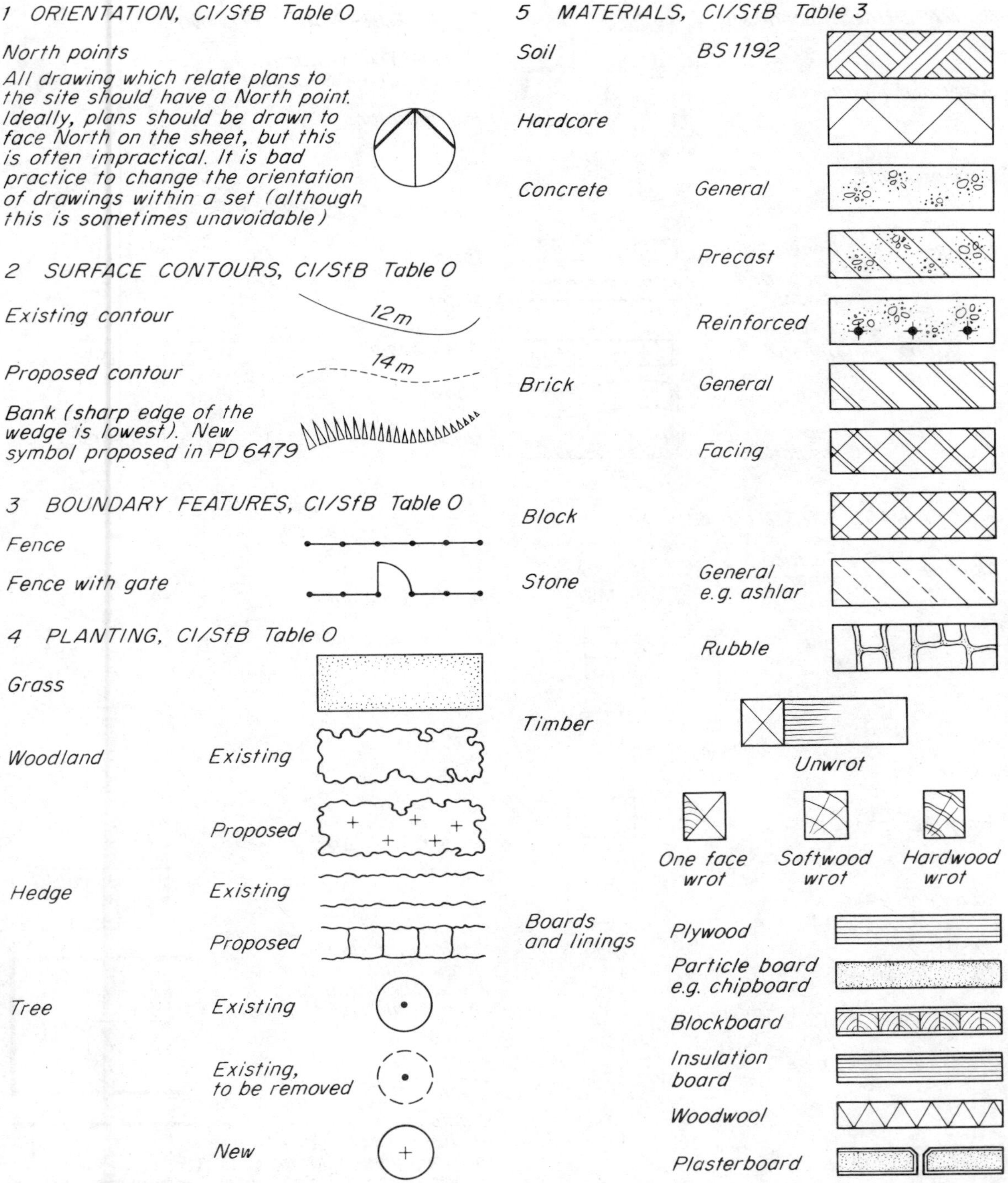

1 ORIENTATION, CI/SfB Table 0

North points
All drawing which relate plans to the site should have a North point. Ideally, plans should be drawn to face North on the sheet, but this is often impractical. It is bad practice to change the orientation of drawings within a set (although this is sometimes unavoidable)

2 SURFACE CONTOURS, CI/SfB Table 0

Existing contour

Proposed contour

Bank (sharp edge of the wedge is lowest). New symbol proposed in PD 6479

3 BOUNDARY FEATURES, CI/SfB Table 0

Fence

Fence with gate

4 PLANTING, CI/SfB Table 0

Grass

Woodland   Existing

                  Proposed

Hedge   Existing

                  Proposed

Tree   Existing

                  Existing, to be removed

                  New

5 MATERIALS, CI/SfB Table 3

Soil   BS 1192

Hardcore

Concrete   General

                  Precast

                  Reinforced

Brick   General

                  Facing

Block

Stone   General e.g. ashlar

                  Rubble

Timber

Unwrot

One face   Softwood   Hardwood
wrot   wrot   wrot

Boards   Plywood
and linings
                  Particle board e.g. chipboard

                  Blockboard

                  Insulation board

                  Woodwool

                  Plasterboard

Figure 111   British Standard recommendations for graphic symbols used on working drawings   Continued ...

5 MATERIALS (continued)

Steel and metalwork

Clay products

Plaster — General

Decorative render

Insulation — General

Quilt

Loose insulation

6 DRAINAGE, CI/SfB (U46)

Manhole — Surface water

MH

Foul water

MH

Gully

G

Rainwater shoe

RWS

Rainwater head

Rainwater pipe

RWP

Discharge (soil) pipe

DP

Vent pipe

VP

7 FIXTURES AND FITTINGS

Kitchen fixtures and equipment, CI/SfB (73)

Refrigerator

R

Sink

S

Sanitation, CI/SfB (74)

Bath

Bidet

BT

Wash basin

WB

Shower unit

S

Urinal

WC

Laundry fixtures and equipment, CI/SfB (75)

Washing machine

WM

Furniture, CI/SfB (82)

Bed

Table

Chair

8 STAIRS AND RAMPS, CI/SfB (24)

Stairs and steps:
arrows always point up.
Tread numbered from lowest
step to highest floor level

Ramp:
arrow indicates direction
of rise; give gradient and
levels at both floor levels

1:10

... *Figure 111   continued* ...

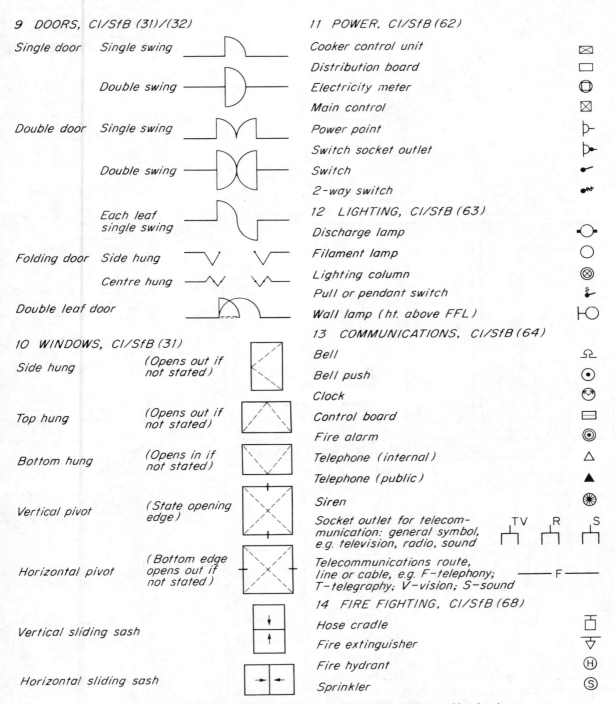

9 DOORS, CI/SfB (31)/(32)

Single door   Single swing

Double swing

Double door   Single swing

Double swing

Each leaf
single swing

Folding door   Side hung

Centre hung

Double leaf door

10 WINDOWS, CI/SfB (31)

Side hung   (Opens out if
not stated)

Top hung   (Opens out if
not stated)

Bottom hung   (Opens in if
not stated)

Vertical pivot   (State opening
edge)

Horizontal pivot   (Bottom edge
opens out if
not stated)

Vertical sliding sash

Horizontal sliding sash

11 POWER, CI/SfB (62)

Cooker control unit

Distribution board

Electricity meter

Main control

Power point

Switch socket outlet

Switch

2-way switch

12 LIGHTING, CI/SfB (63)

Discharge lamp

Filament lamp

Lighting column

Pull or pendant switch

Wall lamp (ht. above FFL)

13 COMMUNICATIONS, CI/SfB (64)

Bell

Bell push

Clock

Control board

Fire alarm

Telephone (internal)

Telephone (public)

Siren

Socket outlet for telecommunication: general symbol,
e.g. television, radio, sound   TV   R   S

Telecommunications route,
line or cable, e.g. F-telephony;   ——— F ———
T-telegraphy; V-vision; S-sound

14 FIRE FIGHTING, CI/SfB (68)

Hose cradle

Fire extinguisher

Fire hydrant

Sprinkler

Figure 111   *British Standard recommendations for graphic symbols used on working drawings*

155

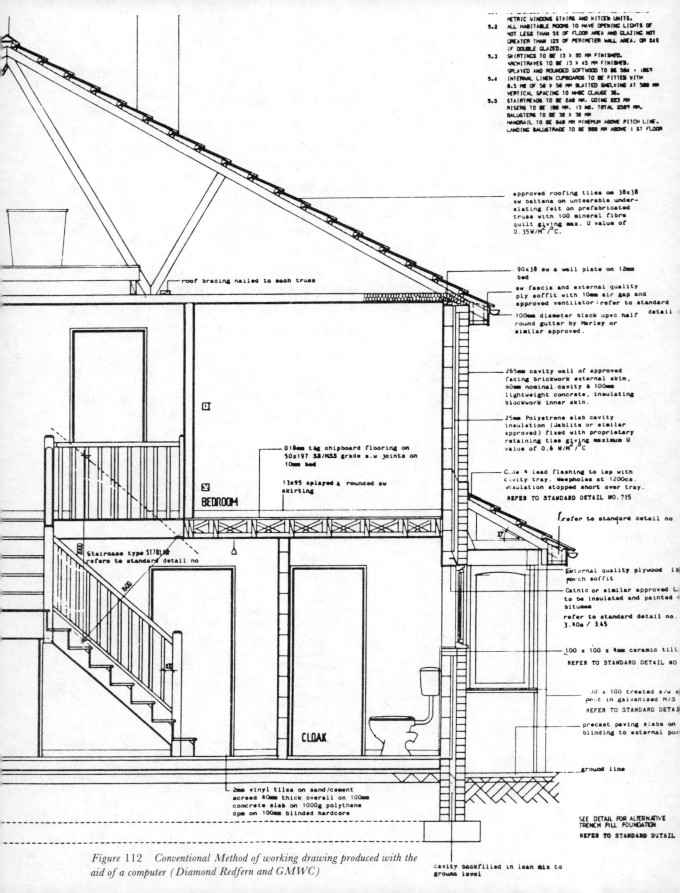

*Figure 112   Conventional Method of working drawing produced with the
aid of a computer (Diamond Redfern and GMWC)*

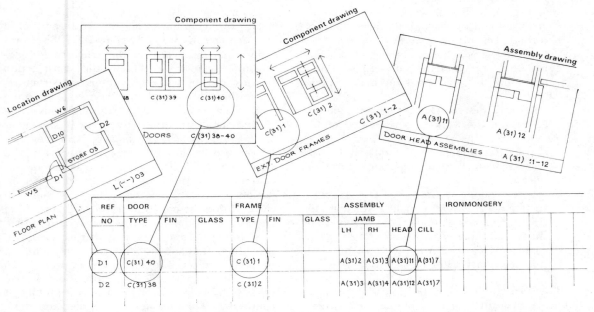

Figure 113 Systematic Method of production or working drawing. Crown Copyright. Reproduced from Building Research Establishment Digest 172 by permission of HM Stationery Office

production or working drawing shown in figure 112 has been produced by a computer: the hand drawn equivalent, which still forms the major technique of presentation today, is similar to most of the illustrations in this book. Further information concerning this subject can be obtained from **BRE Current Paper 18/73** *Working drawings in use.*

The use of a 'systematic method' for production drawings has developed with the use of computers, which are able to supply drawings for particular design and construction problems, eg repetitive details relating to columns and beams on structural grids.

16.3 Specification of Works and Bill of Quantities

These, together with the production drawings, combine to form the legal documents describing the totality of the construction process. A *specification* is a written document prepared by members of the Design Team and provides fundamental information which, for various reasons, cannot be incorporated on the production drawings for a proposed project. It provides an inherent part of the design process because it describes the quality of workmanship which is considered necessary during construction. Information for a specification, therefore, evolves during the preparation of the production drawings, and is subsequently issued in two main parts.

(a) *Preliminaries* which describe the legal contract through which the construction work is controlled including insurances; the facilities to be provided by members of the Building Team on site; the general conditions of the site and/or existing building to be converted including details of access routes, roads and local restrictions; and other details concerning the general running of a project, including information

157

about the quality of work required, nominated sub-contractor and suppliers, hours of working, generation of noise, dust, etc.

(b) *Trade Clauses* which describe the actual construction methods (materials and techniques) to be adopted and may be done by;

   – precisely describing the materials to be used and the work to be executed under each trade (see page 127);
   – or, by describing the materials and work required for each separate part or element;
   – or, by giving a *Performance Specification* (Statement of requirement) which accurately details the quality of work expected in terms of performance criteria for each part involved, without describing a method of achieving it.

The purpose of a Specification is to provide information to the potential builder or contractor of a proposed project which, together with the drawings, will enable a reasonably accurate price to be submitted or tendered for the work involved. The detailed information it provides could not be adequately indicated on the production drawings because these would become far too complicated, and extremely difficult to read. In order to avoid unnecessary confusion, it is important that notes on a drawing should not duplicate clauses of specification. Drawing notes should only provide a general comment necessary for an overall understanding, and which lead to the more detailed description in the specification. Only for very simple projects involving few trades, will drawing notation suffice for describing the work – the legal arrangements being left to the clauses of the formal building contract. The Co-ordinating Committee for Project Information (CCPI) sponsored jointly by the RIBA, BEC, RICS and the ACE, is concerned with guidance on the content and arrangement of drawings in conjunction with a common arrangement of work sections for specifications and measurement.

Writing specifications is often thought to be tedious and time consuming. Over the years attempts have been made to standardize phraseology, rationalize descriptions and provide a structure to the clauses in order to speed production whilst still providing an adequate means of communication. The latest edition of the '*Standard Method of Measurement*' (SSM) provides a system advocating the structuring of specification clauses to follow the construction process, starting with excavations and structure, proceeding through all the services to finishing trades and external works. Although this has a certain logic, an alternative system known as the *National Building Specification* (NBS) has particular advantages for some designers in also providing a CI/SfB coding system for specification clauses. It is ideally suited to those organisations already using this system of classification for collating technical information and for the production of drawings (figure 114).

A *Bill of Quantities* is also a written document providing fundamental information about a proposed project, but varies from a Specification in that it arranges the information into a form more suitable for direct pricing by a builder or contractor. They are used for larger and more complicated contracts, and are prepared within the organization of a Design Team by a quantity surveyor using the information supplied by production drawings and specification notes. A currently acceptable method of presentation for Bills of Quantities consists of three main parts:

(a) *Preliminaries* – as described for Specification,
(b) *Preambles to Trades* is a general specification and description of materials and workmanship,
(c) *The Quantities* is a description of the individual items to be priced and also the numbers, amounts or *quantities* of each required for the project.

The Construction Team's quantity surveyors or estimators generally prepare their own list of quantities, when they are not supplied, from information indicated on the production drawings and in the Specification. But as the Bills of Quantities prepared by the Design Team describes the work to be done, the condition relating to the work *and* measures the materials and components required (and, therefore, type and amount of labour), it is easier for a contractor's quantity surveyors or estimators to prepare a more accurate price for a project. (Furthermore, an individual contractor's estimated price is based upon standardized infor-

■ against the clause number denotes 'included in Small Jobs Version'.

N13 CLAY/CONCRETE SINGLE LAP TILING **N13**

To be read in conjunction with Preliminaries and General Conditions.

TYPE(S) OF TILING

If there is more than one type of tiling, because of variation in base, pitch, or tile size, the item can be repeated as type N13/2, etc.

Sub-items which do not apply should be deleted.

For guidance on completing the item(s) see notes on page 1.

TYPE(S) OF TILING

N13/ ■ _ _ _ _ _ _ _ _ _ _ TILING:

Base: _
Pitch: _ _ _ degrees

Tiles: _
Minimum head lap: _ _ _ mm
Fixing of general areas: _ _ _ _ _ _ _ _
Fixing of local areas (see clause 1351): _ _ _ _ _ _ _ _ _ _

Fittings/Accessories: _ _ _ _ _ _ _ _ _ _ _

Battens: Size: _ _ _ x _ _ _ mm

Counterbattens: Size: _ _ _ x _ _ _ mm, centres coinciding with rafters.

Underlay: _ _ _ _ _ _ _ _ _ _ _ _ _ _ _ _ _ _ _
Minimum horizontal lap: _ _ _ mm

TILES

N13: PRODUCTS

A210, A410
Single-lap tiles should not be laid at less than 35° pitch without evidence of satisfactory use at the lower pitch in conditions similar to those proposed.

A210
There is no British Standard for this type of tile.

A210 ■ CLAY SINGLE LAP TILES and fittings:
Type: _ _ _ _ _
Size: _ _ _ _ _
Finish: _ _ _ _ _
Colour: _ _ _ _ _
Manufacturer and reference: _ _ _ _ _

A410 ■ CONCRETE SINGLE LAP TILES and fittings: to BS 473, 550, Group B.
Type: _ _ _ _ _
Size: _ _ _ _ _
Finish: _ _ _ _ _
Colour: _ _ _ _ _
Manufacturer and reference: _ _ _ _ _

ACCESSORIES

H210 PREFORMED VALLEY GUTTERS:
Manufacturer and reference: _ _ _ _ _

H410
This clause is based on an amendment to BS 5534: Part 1 due to be published early in 1982.

Preservative treatment for battens is optional, see Binder 1, Appendix 3. Appropriate treatments include:

- Boron diffusion
- OS double vacuum
- OS Immersion, 3 minutes.

Preservatives containing copper must not be specified when aluminium nails are to be used.

H410 ■ BATTENS/COUNTERBATTENS: sawn softwood free from decay and insect attack except pinhole borers.
Wane: admissible on one arris only.
Wane, knots and knot holes: in total not exceeding one third of the width of the surface on which they occur.
Fissures: generally not exceeding 600 mm long, and not exceeding 60 mm long at ends of battens.
Moisture content: not more than 22% at time of fixing.
Preservative treatment: _ _ _ _ _

L310
Suitable for draping over rafters.

L310 ■ BITUMEN FELT UNDERLAY: reinforced felt to BS 747, Type 1F.

L410
Plastics underlays must be permeable to minimise the risk of condensation, and of adequate strength. Many suitable types are covered by Agrement Certificates. Check suitability for specified batten gauge.

L410 ■ PLASTICS UNDERLAY:
Gauge: _ _ _ μm
Manufacturer and reference: _ _ _ _ _

Full Version N13 Page 2

Figure 114 National Building Specification: standardised clauses based on either CI/SfB or SMM format which can be used to produce a full specification. Product clauses are followed by workmanship clauses. The guidance notes in the right hand column are particularly useful (Example from RIBAS)

mation, and this facilitates the Design Team when comparing it with the prices of any other contractor who may wish to do the work – see next section.)

16.4 Tender documents

The production drawings, specification and/or Bills of Quantities together form the initial *contract documents* which are submitted to the Construction Team for pricing. The principal designer may agree with the client to submit these documents to a builder or contractor who is known to be capable of executing the particular work involved. This will produce a *negotiated tender*, where a price for the work is prepared on the basis of discussion around the contents of the contract documents. This method of tendering has the advantage that a considerable amount of time is saved by not involving other builders or contractors, and the selected organization may well be able to suggest modifications to certain construction methods resulting in saving of capital.

As an alternative, the principal designer may advise the client to obtain *competitive tenders* from a suitable range of Construction Teams. As described earlier, Bills of Quantities are particularly useful for this purpose as they permit individual Construction Teams to submit tenders on a uniform basis. Equally, the Design Team is able to check and compare the figures of each competitor efficiently before making recommendations to the Client Team. Before this recommendation is given, however, the Design Team must also ascertain the suitability of resources and quality of workmanship available from each competitor. The period of time required for the construction work is also an important consideration. It is not uncommon for the organization submitting the lowest tender price to be rejected because of doubt in these areas.

Work which is being undertaken by competitive tender is often advertised, and with local authority work this is mandatory. On receiving the contract documents, contractors should visit the site of the proposed project in order to place their economic assessments on the actual condition likely to affect construction work. Each competitor will prepare their estimate for the

work and submit a tender, which gives prices for the individual items in the Bill of Quantities.

Matters to be considered when pricing work include the current cost of materials, machinery and transport, fluctuations in wage rates, and the overhead expenses involved in maintaining an administrative organization. To this must be added an allowance for waste of the materials, depreciation of equipment, labour on non-productive work, and, of course, a reasonable margin for profit. The profit of a Construction Team is not only realised through good organization of site work, but involves the efficient costing ability of their quantity surveyor or estimator at the time of tendering. Consideration must be given to how much work is to be done within the organization and how much is to be sub-let. In the latter case, prices must be obtained for materials and labour of sub-contractors or suppliers. The management personnel within the organization must discuss the estimate to establish a figure for *contingency items* resulting from likely increases in materials and labour before the contract ends. When the cost of certain items indicated in the contract documents cannot be predicted, such as the degree of excavation, provisional allowances are made on a '*day work basis*'. This incorporates the cost of material/equipment required and a *unit rate* for the time taken for the work, which includes the hourly sum paid to the workmen, as well as insurance, holiday, pay grants, travel and overtime expenses, and adverse weather payments, etc. The absence of information caused by poorly prepared or the non-existence of Specification or Bills of Quantities could mean that much of the work indicated on drawings will be priced by using this method, but with additional sums added to cover uncertainties. The lack of information concerning some constructional aspects of a project could result in massive extra costs once the work has proceeded and, therefore, can lead to serious repercussions.

If the price for the construction work is considered acceptable and the project proceeds, it is quite usual for the quantity surveyor of the builder or contractor to continue with the financial management for the construction work. This entails agreement of interim work, 'measurement' or accurate assessment of completed work and materials on site, and negotiation of payments with the client's quantity surveyor at intervals agreed in the building contract. Details are submitted to the principal designer for confirmation and payment by the client. Similarly, at the completion of the project, the measurement and valuation of all variations leading to the final account must be agreed. These variations will usually be calculated in relation to the itemized prices which have been given in the Bills of Quantities. These negotiations are considerd under *16.6 Fee Accounts and Certificates of Payment*.

It sometimes happens that the initial contract documents are not fully completed, but it is still desirable to negotiate a price for the intended work or obtain competitive tenders from a builder or contractor. This will enable an early start of site preparation and construction, as well as permitting close co-operation with a builder or contractor during the preparation of production information – of particular advantage when specific construction skills or resources have to be taken into account.

One method of involving specialized construction skills early in the design process is for the principal designer to prepare a 'short-list' of potential contractors, and then interview each to establish their methods of management, construction policy, the possible form and content of their pre-contract input, and their current contractual commitments. The contractor having the best potential suited to the particular project can then be appointed. However, the criteria to be used to calculate the price of the project must be agreed before the contractor's appointment so that a sum can be developed as the design process continues. This method of negotiation presumes a high level of trust between the parties involved, and also requires particular experienced cost expertise as well as a complete knowledge of current building costs.

Alternative methods include a *cost reimbursement scheme* where the selected contractor is paid for the work executed on the basis of audited accounts, to which is added an agreed percentage for overheads and profit. Sometimes an estimated lump sum price for a project is agreed prior to the commencement of construction work and any discrepancies – savings or extras – are shared equally between client and contractor. This is known as a *target contract*.

When it is necessary to obtain fully competitive tenders for project without completed tender documents one of the following techniques may be more appropriate:

(a) A list of the major items of work is prepared and a selection of suitable builders or contractors are invited to submit a *schedule of rates* for each. Once the work on site starts, the successful competitor will be paid for particular construction work at the agreed rates listed in the schedule, unless there are major changes. Work which the final design requires, and which has not been given a price rate, will be subject to negotiation.

(b) A *Notional Bill of Approximate Quantities* can be prepared by the quantity surveyor, and lists the approximate quantities measured from preliminary design drawings, or the quantities given in a full Bill of a similar project. This Notional Bill will be submitted for tenders and the successful contractor, once appointed, will be paid at the unit price rate agreed for the actual quantity of work measured on site. If there are any items not included in the Notional Bill, their cost must be negotiated separately.

(c) A *Two Stage Tendering* procedure may be adopted – the first stage based on a Notional Bill of Approximate Quantities as described previously. Once appointed, the successful contractor helps to develop the design and prepares the second stage tender based on contract documents which includes a full Bill of Quantities. Rather than appoint a contractor after the first stage tender, the client may prefer (or be recommended by the principal designer) to delay appointment until after the second stage tender in order to obtain more favourable competitive prices at this late stage. In this case, the pre-contract contractor may be paid a fee for his advice and services during the design process stage. Whichever stage of appointment is adopted, construction time can be shortened if the contractor is authorized to prepare the site, or order long term delivery of materials during the period between obtaining first and second tenders.

(d) A *Serial Tendering* procedure may be appropriate when a number of projects of a similar nature are to be constructed over a considerable period of time. The full Bill of Quantities for the project first constructed can be progressively updated for subsequent projects.

The above methods of tendering can also be used when negotiating a price with a preferred builder or contractor without adopting competitive techniques.

16.5 Contracts

It is preferable that building operations, like any other activity involving human and material resources, have a formalized set of rules based on a legal contract. This will ensure the rights and responsibilities of each party, and help in the successful production of a building. Various types of agreement can be drawn up, but it is usually advisable to adopt a *standard form of contract* because the individual provisions of non-standard forms will most likely not have been tested in the Courts of Law which means the accountability of each party is not always clear. Even though standard forms of contract result from the joint consultations of interested professional bodies and may have been well tested in practice, matters often arise which cause dissatisfaction between various parties as indicated by the legal sections of most current building trade journals. For this reason, a continuous process of re-assessment is necessary and revisions are issued from time to time, particularly in the face of the increasing complexity within the building industry.

Contracts between Client Team and Design Team

Standard Forms of contracts are available which stipulate the rights and responsibilities appertaining between a client and the principal designer, or specialist designer, or quantity surveyor. They are available from the professional body which represents each consultant, and state the role played by them during the design and construction periods of a project, the fees payable, and the liability taken. Although these formal agreements are, perhaps, the least used (most clients employ consultants through per-

sonal recommendation), the increased tendency towards costly litigation to solve disputes makes them almost mandatory in the interest of both parties.

A standard form of contract between client and building designer has been drawn up by the RIBA and is known as *Architects Appointment*. It is divided into four main parts:

Part 1 Preliminary and Basic Services – a description of the work normally carried out by an architect related to the RIBA Plan of Work.

Part 2 Other Services – a description of the wide range of additional services which can be carried out by an architect if required.

Part 3 Condition of Appointment – a description of the conditions which normally apply to an architect's appointment, including the responsibility and authority of both architect and client in relation to each other and to consultants or specialists, contractors, sub-contractors and suppliers.

Part 4 Fees and Expenses – a description of non-mandatory methods for calculating the fees for services and expenses incurred by an architect.

Contracts between Client Team and Construction Team

The most important standard form of contract for the building industry is that which exists between a client requiring construction activities and the builder or contractor who is prepared to execute them. There are not only several different forms of this contract, but also different contracts are prepared by different bodies. Currently the most common contract is that prepared by the *Joint Contracts Tribunal* (JCT) which has the following as its constituent body:

The Royal Institute of British Architects
The National Federation of Building Trades Employers*
The Royal Institution of Chartered Surveyors
The Association of County Councils
Association of Metropolitan Authorities
Association of District Councils

The Greater London Council

The Committee of Associations of Specialist Engineering Contractors

The Federation of Associations of Specialists and Sub-contractors

The Association of Consulting Engineers

The Scottish Building Contract Committee

Confederation of British Industry (observer only)

Contracts, including standard administration forms (Appendix C), are available for private sector clients and for local authority clients as follows:

Private Edition with Quantities
Private Edition without Quantities
Private Edition with Approximate Quantities
Local Authority Edition with Quantities
Local Authority Edition without Quantities
Local Authority Edition with Approximate Quantities

In addition to these, other forms of JCT Standard Form of Building Contract are:

– *Fixed Fee Form of Prime Cost Contract*, for use when the management contractor is to be paid a fixed fee to cover administrative overhead costs and profit, in addition to the (prime) cost for work executed by sub-contractors and materials/equipment provided by suppliers.

– *Agreement for Minor Works without Quantities*, for use when small amounts of work is to be carried out in accordance with the production drawings and specification only, and is paid for on the basis of the lump sum given in the tender prepared by the builder or contractor. This contract is not for use in Scotland.

A building contract known as *A Form of Building Agreement 1982* has been prepared by the Association of Consultant Architects (ACA) who represent over 400 private sector architectural practices. This Contract is shorter than the JCT Contract, and does not attempt to duplicate those areas of building legislation covered by common law. It should be used in conjunction with its own version of Architect's Instructions, Interim Certificates, Final Certificates and Taking-over Certificates.

The Society of Industrial Artists and Designers have similarly prepared their own

form of contract. This is known as the *Works Agreement* and is intended to be used with its own standardized documentation covering Instructions, Directions, Drawing Issues, Orders for Nominated Subcontractors, Interim Certificates, Final Certificates and Practical Completion Details. The Agreement is a specialist short form of contract which provides a suitable legalized basis for carrying out interior design projects and makes provision for the nomination of suitable sub-contractors.

Other forms of contract include one prepared by the National Federation of Building Trade Employers (NFBTE)* – for use where a contractor is to be responsible for both the design and construction of a project, and there is no reference to independent consultants. Another form of contract has been prepared by the Institute of Civil Engineers for use on civil engineering projects. The NFBTE* have also prepared contracts for use between a contractor and either nominated sub-contractors, or non-nominated sub-contractors; and for a contractor who prefers to use sub-contractors who supply labour only for the work involved.

16.6 Fee accounts and Certificates for Payment

One important area of concern in most standard forms of contract involves payments for the services connected with the design and construction processes.

Fee accounts

Fees paid by the client to various members of the Design Team are usually governed by the Conditions of Engagement issued by the professional body which represents a particular design discipline. For example, the RIBA's *Architect's appointment* described earlier lists those services normally offered by an architect in accordance with the Plan of Work, and sets out recommended fees and stages of payment (figure 115(a) and (b)). For the large projects, which occupy long periods of time, it is becoming usual for designers' fees to be paid by instalments in accordance with a scheduled programme agreed with the client at the outset of

* Now Building Employees' Confederation – see page 128.

Recommended percentage fee scales: New works

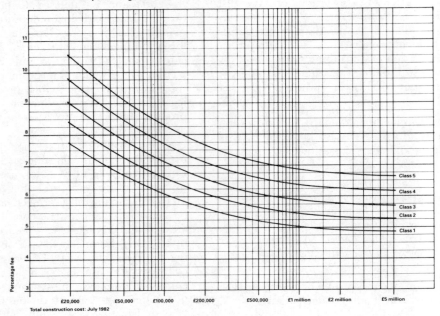

Recommended percentage fee scales: Works to existing buildings

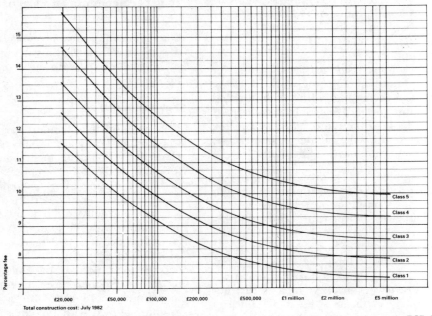

Figure 115(a)   Recommended fees and expenses related to class of work (from 'Architect's appointment', RIBA 1982)

Continued . . .

164

Classification of buildings, for guidance

Type	Class 1	Class 2	Class 3	Class 4	Class 5
Industrial	Storage sheds	Speculative factories and warehouses Assembly and machine workshops Transport garages	Purpose-built factories and warehouses Garages/ showrooms		
Agricultural	Barns and sheds Stables	Animal breeding units			
Commercial	Speculative shops		Supermarkets Banks	Department stores Shopping centres Food processing units Breweries	High risk research and production buildings
	Single-storey car parks	Speculative offices Multi-storey car parks	Purpose-built offices	Telecom and computer accommodation	Recording studios
Community		Communal halls	Community centres Branch libraries Ambulance and fire stations Bus stations Police stations Prisons Postal and broadcasting	Civic centres Churches and crematoria Concert halls Specialist libraries Museums Art galleries Magistrates/ county/ sheriff courts	Theatres Opera houses Crown/high courts
Residential		Dormitory hostels	Estate housing Sheltered housing	Parsonages/ manses Hotels	Houses for individual clients
Education			Primary/nursery/ first schools	Other schools including middle and secondary University complexes	University laboratories
Recreation			Sports centres Squash courts Swimming pools		
Medical social services			Clinics Homes for the elderly	Health centres Accommodation for the disabled General hospital complexes Surgeries	Teaching hospitals Hospital laboratories Dental surgeries

Figure 115(a)   ... continued

165

New works

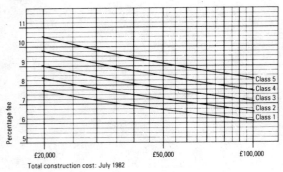

Works to existing buildings

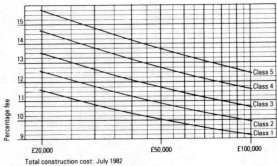

Figure 115(b)   Recommended fees and expenses related to class of work (From 'Architect's appointment' (Small Works), RIBA 1982)

negotiations. This programme can incorporate more frequent intervals between payments than provided by the work patterns given in the Plan of Work, and thereby enable the designer to receive more or less regular sums during a project's design and construction processes. (See also 15.3 *Design Team: Methods of operation*).

Certificates for Payment

Depending on which standard form of building contract governs the construction of a particular project, it is also usual for payments made by the client to the builder or contractor to be made at regular intervals. Normally the client's quantity surveyor prepares valuations (see page 117) at the periods stated in the contract (28 days or 14 days for large projects) in accordance with the given terms and conditions. Reports are made to the principal designer for guidance in preparing *Interim Certificates for Payment*. These Certificates give valuations of the work executed

on site; materials delivered to the site but not used; nominated sub-contractors' and suppliers' invoices; and other substantiating evidence of expenditure such as may result from 'daywork rates' (see page 160) to cover designers' instructions not already included in the contract documents. Thereafter, each interim certificate must contain an itemized statement of the principal sum due to the builder or contractor as well as the sums to be paid to sub-contractors for their work. With most forms of contract there is provision for the retention of a small percentage of the total sum given in each certificate. This is normally 5% (or 3% where estimated contract sum is £500,000 or more), and acts as a safety retainer in the event of subsequent default by the builder or contractor, sub-contractor or supplier.

Once the construction work has been completed a *Final Certificate of Payment* can be issued to cover the agreed costs of the project. At this stage, usually only half the retention sum previously withheld is paid – the remainder being kept for a period agreed in the contract during which it is assumed any defects will become apparent which are the sole liability of the builder or contractor, sub-contractor or supplier. After this agreed *defects liability period* has elapsed (generally three months for small works and six to twelve months for larger works), the residue retention sum can be released in total, or in part, according to the expenditure necessary to correct any faults. The retention sum does not cover faults in design, or any work executed by the builder or contractor in total accordance with the instructions from the Design Team (ie as given in the contract documents). Failures arising from the content of the contract documents are subject to separate negotiations between client and designer, although the experience of the builder or contractor may be used in the event of a dispute. Where no quantity surveyor is employed to prepare financial statements, the builder or contractor submits a claim direct to the principal designer who will then prepare the Certificates for Payment as already described. (See page 168.)

After preparation by the principal designer, Certificates of Payment are issued initially to the builder or contractor, who will then submit them to the client for payment. This procedure arises because the contract is between client and builder; members of the Design Team are only acting as agents who administer the rules of agreement. Nevertheless, when the principal designer issues the Certificate to the builder or contractor, the client and quantity surveyor (if involved) are automatically informed with a copy of the Certificate. Under the conditions of the building contract, the client must pay the contractor within a stipulated period, ie normally 14 days. Appendix C illustrates some of the various standard forms which are available from the RIBA and the CSD (formerly SIAD) to ease the administrative procedures briefly described above.

16.7 Programmes of work

Programmes of Work for the design and construction of a building project involve the co-ordination of many complex human and material resources to ensure economy and efficiency. The more detailed the programme, the less likely will unforeseen circumstances upset either the intentions or the time-tabling of a project, thereby reducing the chances of frustration and delays.

The three main areas requiring detailed programmes are as follows:

1 *Design Programmes* (figure 116) These can take the form of a *bar chart* and are often referred to as pre-contract programmes. They are prepared by the Design Team in order to establish how long it will take from the beginning of the design process to start of construction work on site. Apart from providing valuable information to the client – particulary when complicated negotiations are required for the release of capital for building – this type of programme allows the Design Team to accurately assess the demands on staff and the effects which the new project may have on other concurrently running work.

2 *Tendering Programmes* (figure 117) Sometimes, on behalf of a client, the principal designer will request builders or contractors who are tendering for a project to submit an outline, or tender programme. This can also

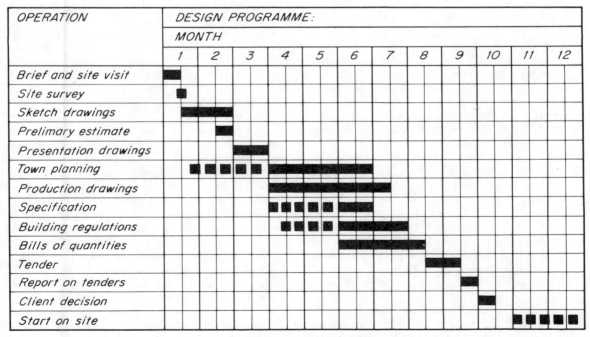

Figure 116   Design programme

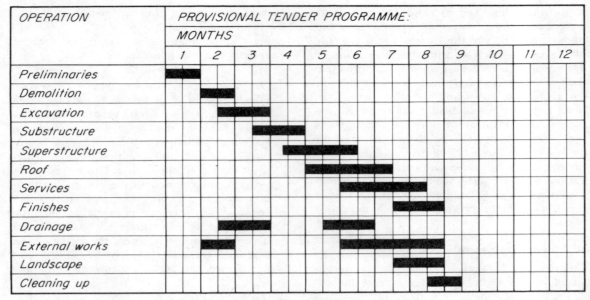

Figure 117   Tender programme

consist of a bar chart which indicates the approximate periods of time required for each major construction activity and, together with the total period indicated for the proposed project, provides a method of assessment concerning the organizational ability of the builder or contractor.

As mentioned earlier, the total period required for construction is very often the critical factor used when making a choice from a selection of tenders which has been priced competitively. Generally speaking, time delays are dearer for the client after work has commenced on site.

3 *Contract Programmes* (figure 118) Once the successful builder or contractor has signed the Building Contract, he must plan the project in great detail. Whereas the Design Team specifies the work to be done, the order in which it is executed is usually the responsibility of the contractor and even the smallest project demands a close examination of available resources, assessments of time, and the influences of sub-contracted work.

A *method statement* should be prepared from information given by the contract documents. This takes the form of a schedule breaking down the intended construction work into operations, and then labour and plant requirements (or availability) for each. The information can then be converted into a draft programme which should be issued prior to a general meeting between the management members of all the parties concerned with construction activities. Having had time to consider the proposals, this meeting will be able to confirm working methods and agree amendments as necessary. The contractor can then prepare his 'final' Contract Programme.

If the Contract stipulates a monetary penalty (*liquidated and ascertained damages*) for delays beyond the completion date required by the client, a prudent builder or contractor will ensure the work is finished earlier than the programmed finish. This potential time difference allows for the unforeseen eventualities which usually result in longer time periods for certain building operations. A contractor may find that the initial timetable determined by the programme soon becomes unrealistic because of delays caused by bad weather conditions, late delivery of materials, sub-contractors not being nominated quickly, or instructions not being given on time by the designers. If problems such as these develop, the contract programme

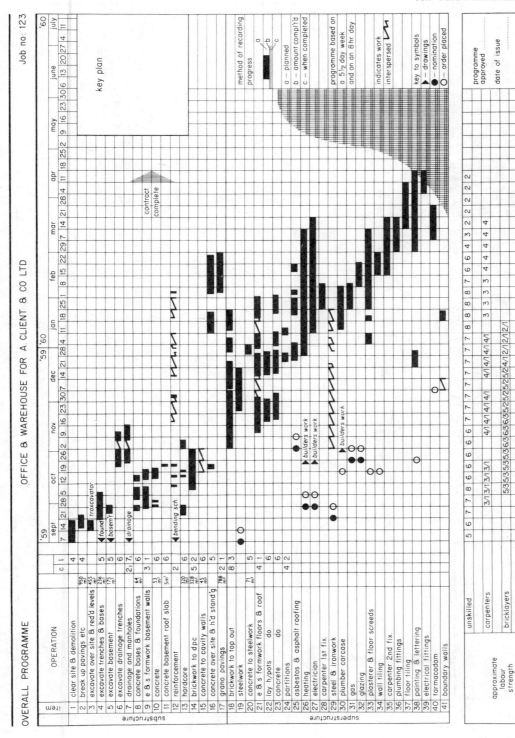

Figure 118   Contract programme

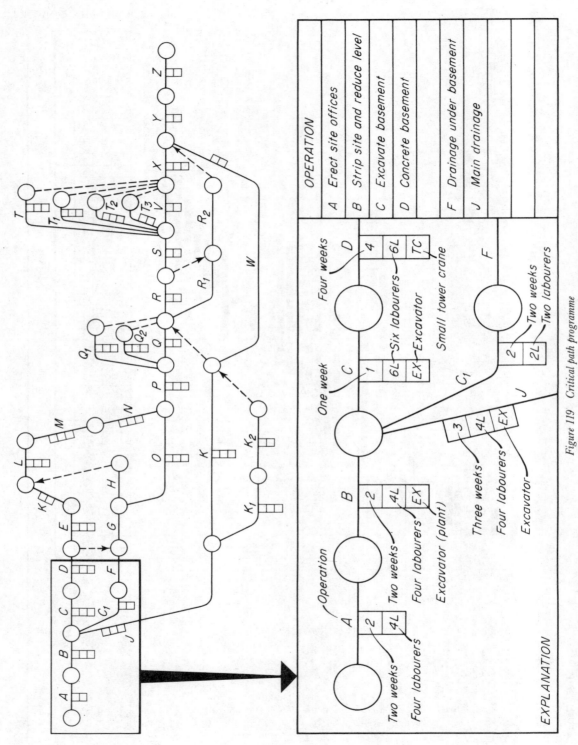

Figure 119 Critical path programme

should be up-dated to indicate a revised plan of action, and allowance made to cover the builder or contractor against unjust claims for delay.

Highly sophisticated programmes can be produced to provide a visual representation of the inter-relationship between materials and components deliveries, sub-contractor and trade work. These can provide a more accurate programme and are based on the *critical path method* (*CPM*) (figure 119). This method is sometimes considered too complicated for practical use in small to medium size projects because of the frequent reconstruction which may be necessary as a result of updating. Nevertheless, the critical path method more accurately reflects actual site techniques than is possible by bar charts as tasks performed on site do not divide naturally into 'design element' parts, or the sequential use of trades. Each task has its own breaks determined by the technology of production often peculiar to the particular site concerned, and a programme orientated towards reality should greatly improve the effective use of resources and progress of building work. No doubt the use of computer produced programmes with the associated ease of updating by continuous 'feeding' of data will make the CPM more desirable for future building work.

16.8 Maintenance manuals

A completed modern building is often complicated to operate and maintain. Therefore, it is to the advantage of all those directly concerned with its eventual use that a *maintenance manual* is produced. This manual should be prepared by members of the Design Team, and perform a similar function to the operation and maintenance book for a car or any other machine.

For building, the maintenance manual can be conveniently divided into three parts:

1 *Statement of the designers' intentions*, which includes key drawings, schedules and photographs of the newly completed building; a description of its function; details of specialist designers; contract information; consents, licences, and approvals obtained; details of Building Team; floor areas and permitted loadings; fire fighting methods, precautions taken and means of escape; and a list of those

maintenance contracts which are absolutely necessary as well as those considered desirable.

2 *Statement of the general maintenance required*, which includes regular cleaning instructions and methods when applicable; general maintenance log sheets; and a schedule of fittings and components which require regular attention and/or replacement.

3 *Statement of the mechanical services provided*, which includes a maintenance guide giving timing and frequency of required operations; fittings requiring periodic replacement; and drawings which are required for an understanding of facilities provided.

16.9 Feedback data

It is important that those concerned with the design and construction processes of a building also prepare data which will prove invaluable in the event that they will be concerned with a similar building in the future. This information should include a complete set of contract documents; an appraisal of the contract procedures between designer, client, builder or contractor, and sub-contractors; a detailed cost analysis; and an appraisal of design and technical criteria which incorporates details of operational procedures associated with the subsequent use of the building. This information may also prove invaluable in the event of legal claims arising as a result of disputes concerning responsibility for subsequent faults in the performance of a building. Claims of this nature could arise many years after the completion of a building when, perhaps, most of the original Design and Construction Team have dispersed.

Organizations requiring to retain this information for all or most of the projects for which they have been responsible will face a serious problem of storage, and even of retrieval when required for subsequent research. Large projects may have several thousands of large drawings and schedules, as well as other documentation. Storage problems can be overcome by one of the many micro-copying techniques ranging from simply reduction to *35 mm slide transparencies*, to *35 mm or 16 mm micro film or microfiche processes*. The latter process involves the photographic reduction of documents on transpar-

171

ent cards usually 150 mm × 100 mm in size. Each card carries up to 60 micro images set in rows which can be displayed on an enlarging screen for easy reading.

The use of micro-reproduction techniques is inseparable from the 'systematic' information storage method mentioned under 16.1 *Technical*

data, and can provide a total package of information on a particular project in a very small storage area. The CI/SfB system should be used, of course, for classification and ease of retrieval (see Table 0 – Physical Environments, page 140).

Further specific reading

Mitchell's Building Series

Structure and Fabric Part 1 Chapter 2 *The production of buildings*
Structure and Fabric Part 2 Chapter 1 *Contract planning and site organization*
Chapter 2 *Contractors' mechanical plant*
Components: Chapter 1 *Component design*
Chapter 3 *Industrialized system building*

Mitchell's Professional Library

Architectural Practice and Procedure, Philip H P Bennett
The Supervision of Construction, John W Watts
The Supervision of Installation, John W Watts

Building Research Establishment Digests

BRE Digest 53:   *Project network analysis*
BRE Digest 166:   *European product-approval procedures* Part 1
BRE Digest 167:   *European product-approval procedures* Part 2
BRE Digest 172:   *Working drawings*
BRE Digest 271:   *Project information for statutory authorities*
BRE Digest 289:   *Building management systems*

PART C

An analysis of a building in terms of one typical construction method

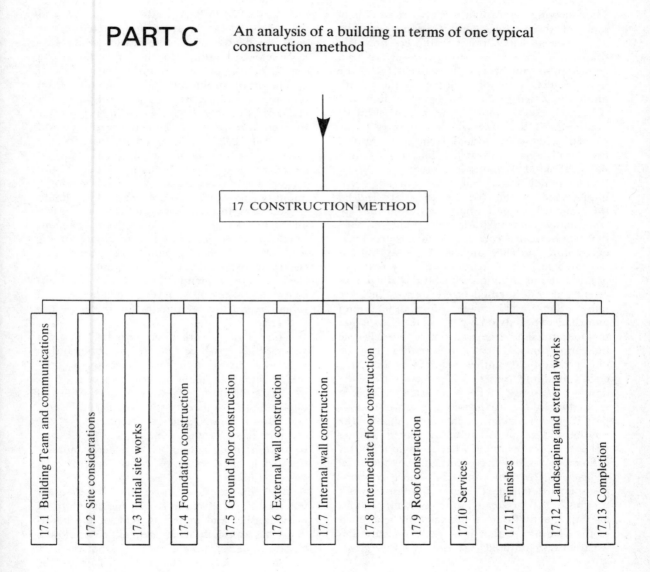

17 CONSTRUCTION METHOD

- 17.1 Building Team and communications
- 17.2 Site considerations
- 17.3 Initial site works
- 17.4 Foundation construction
- 17.5 Ground floor construction
- 17.6 External wall construction
- 17.7 Internal wall construction
- 17.8 Intermediate floor construction
- 17.9 Roof construction
- 17.10 Services
- 17.11 Finishes
- 17.12 Landscaping and external works
- 17.13 Completion

The following description will give an outline of some of the various considerations, activities and processes which result in the erection of a simple building. A two-storey house has been chosen primarily because it is likely to be the most familiar form of building to those with limited experience of the construction industry. A house is also the most common form of building in the UK today which continues to be erected using conventional techniques derived from traditional craft skills. This results in a particular kind of 'domestic' aesthetic.

This is not to say that house building techniques have not changed over the years. Indeed the standards of performance requirements are continually being upgraded to provide increasingly sophisticated enclosures which reflect ideas of comfort, stability and aesthetic appeal, as indicated in Part A. It should also be borne in mind that the techniques to be described represent only one method of erection – there are many others. Furthermore, this chapter is not an exhaustive study on a particular construction method and should be regarded as a brief basic indication only. Further information can be obtained from the other volumes of the Mitchell's Building Series. In order to help the designer to assess the effects of design decisions it is often useful for a sequence diagram to be prepared which diagrammatically shows the relationship of the activities involved. Figure 121 is a typical example where individual boxes define *activities* and circles contain the sequential *operation number*. One operation must be complete before the next begins to ensure work progresses at a steady rate without serious overlap of activities. A time scale can be applied to the diagram so that progress of construction is carefully monitored (see also 16.7 *Programmes of Work*).

17.1 Building Team and communications
CI/SfB (A1g)

The Building Team and method of communication adopted for the creation of a two-storey house could be less extensive in certain areas to those described in Part B. Also, the number and type of personnel involved and, therefore, the resulting degree of communication will obviously vary according to the problems to be solved. A large house with unusual, complicated or elaborate structural and/or environmental requirements may involve a correspondingly elaborate Building Team. The Client Team for the two-storey house to be considered here will most likely be a single individual or a couple who employ the services of a single principal designer (eg architect) from the Design Team. He or she may recommend the services of a quantity surveyor although his/her skills should be sufficient to control expenditure accurately for such a small project. The designer's skills should also encompass sufficient expert knowledge of the drainage, plumbing, heating and electrical installation required, although he or she can recommend the services of sub-contractors who will give recommendations and submit competitive estimates for these specialized areas.

The building work is best carried out by a general builder whose standard of workmanship is familiar to the designer. Alternatively, contract documents in the form of working drawings, specification of works (and Bills of Quantities, if a quantity surveyor is employed), and the contract can be prepared by the designer in order that competitive tenders can be submitted by a number of general builders situated in the locality of the site for the proposed building.

It is important, of course, that the designer maintains overall control over the standard of workmanship and materials used for the building work, especially when the builder employs a large number of sub-contractors rather than his own men. A form of contract which would be applicable for this project would be the *JCT Standard Form of Building Contract: Private Edition without Quantities (or Private Edition with Quantities)*.

The designer will be required to seek approval for the proposed scheme from the local authority's Town Planning Department and Building Byelaw Department. The house construction to be described will be governed by the Building Regulations 1985.

17.2 Site considerations CI/SfB (A3s)

(a) Selection (figure 120)

The characteristics of a site are likely to have an important influence on the way a building is designed and it is, therefore, important that the designer is employed as soon as the desire for a house has been established by the client. As many sites as possible should then be inspected, and the relative merits compared before final selection. A badly chosen site can be one which offers few amenities, or few aesthetic advantages; and also one which involves the client in considerable additional building expenses arising from physical properties requiring special constructional solutions. It is also important that the Town Planning authority is consulted regarding any problems which may arise in connection with suitable 'zoning' [see page 124].

The early involvement of the designer is particularly important today, since those sites which provide highly suitable facilities are becoming scarce and land is increasingly having to be used which was previously avoided. For example, it may be necessary to select more remote sites than normal and, as outlined in chapter 6 *Weather Exclusion*, it will be essential to investigate the full range of climatic conditions as these will affect the final form and appearance of the building fabric. Special consideration will influence the final form of building to be erected on a site which has poor subsoil or suffers from water-logging, or a site to be re-used following the demolition of existing structures where its physical properties may well have been altered during the life of the previous building.

(b) Survey

An Ordnance Survey Map (figure 102 – page 144) may give an initial guide about site features, but for more specific data it is necessary for a detailed accurate survey to be made. This should be done by a member of the Design Team, and will ascertain the precise shape and size of land involved, and the shape and position of any obstacles such as ponds, trees or buildings that exist within and around the boundaries of the site. It is also important to establish the exact path of the sun across the site, the prevailing direction of wind and rain, etc, and the way the site slopes.

In addition, details will be required of the public utility services (gas, electric, water, telephone, refuse collection, drainage system, transport routes, etc), and of private enterprises (cable television, refuse collection, maintenance contractors, etc). Greater details of these can probably be obtained from the local authority who may also be able to supply information about desirable access points to the site, future road widening schemes in the locality, preservation orders affecting buildings and/or views, rights of way and light, etc. Some planning authorities issue guidelines about the appearance of buildings to be erected in their locality. It may also be useful to make contact with the Building Inspector or Approved Inspector at this stage and discuss any general factors which may influence the construction methos applicable to a building to be erected on that particular site.

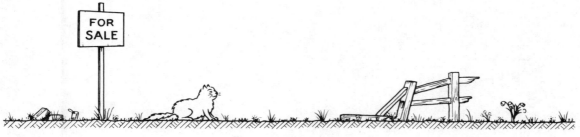

Figure 120   Site selection

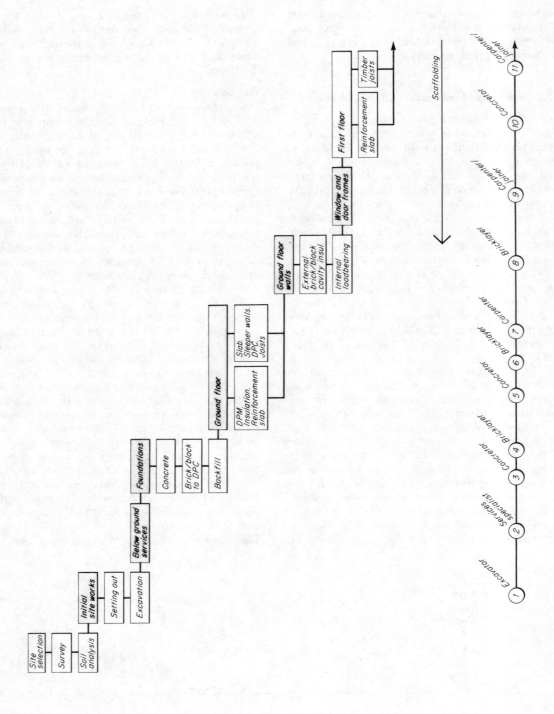

Site selection

Survey

Soil analysis

Initial site works

Setting out

Excavation

Below ground services

Foundations

Concrete

Brick/block to DPC

Backfill

Ground floor

DPM. Insulation. Reinforcement slab

Slab. Sleeper walls. DPC. Joists

Ground floor walls

External brick/block cavity insul.

Internal loadbearing

Window and door frames

First floor

Reinforcement slab

Timber joists

Scaffolding

1 Excavator

2 Services specialist

3 Concretor

4 Bricklayer

5 Concretor

6 Bricklayer

7 Carpenter

8 Bricklayer

9 Carpenter/Joiner

10 Concretor

11 Carpenter/Joiner

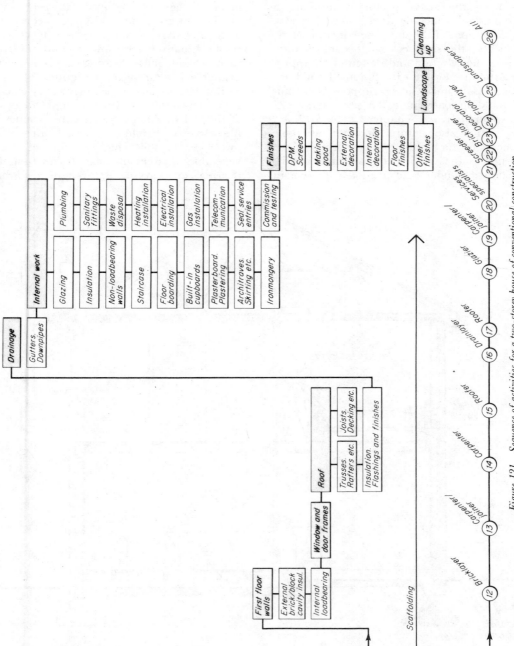

Figure 121   Sequence of activities for a two-storey house of conventional construction

177

(c) Soil investigation

Apart from the above ground conditions, a survey must also include data about the *below ground* conditions of the site. These include characteristics of the soil because this information influences the selection and design of an appropriate structural form for the proposed building.

The subsoil must safely support the combined building loads indicated as in figure 122 (for explanation see chapter 5 *Strength and Stability*) and also ensure that unreasonable movements of the building do not occur. If the supporting soils is sufficiently resistant and its characteristics under load are likely to remain satisfactory, the problems of support and movement will be easily resolved. However, few soils other than rock can resist these concentrated loads and it is usually necessary to collect the resolved loads at their lowest point and transfer them to adequate bearing soil known to be available on a *particular* site (figure 123).

Building loads can be fairly easily calculated from known data, but the strength or resistance of the soil to them requires expert understanding and very careful analysis. Underneath the top layer of soil which consists of decaying vegetable matter and has little strength, lie bearing soils varying in strength according to

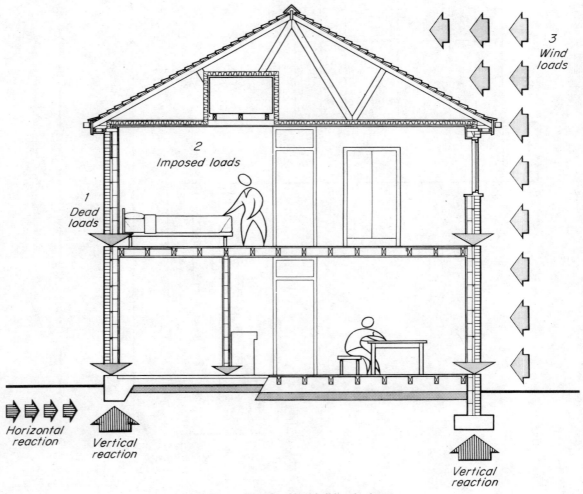

Figure 122   Combined building loads

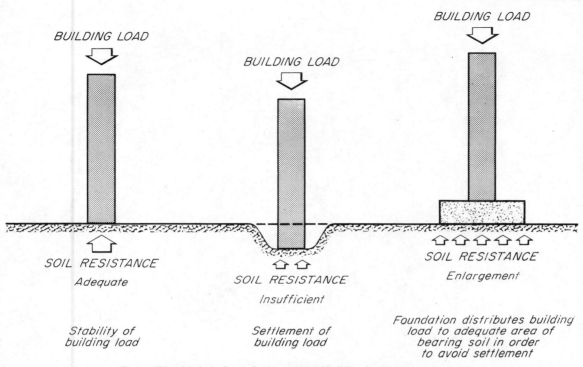

Figure 123   Method of transferring combined building loads to supporting soil

type and consistency. Consideration must be given to the soil, including the particle sizes from which it is formed; moisture and chemical content; behaviour under loads; the presence of inherent or induced peculiarities; and the likely effects of variations in its normal undisturbed condition.

There are a number of methods employed to assess soil conditions which can be used singly or in combination with one another depending upon the precise circumstances. Investigation can be done by reference to geological/topographical references or empirical knowledge from local residents, building inspectors, etc, and/or by physical exploration of the site. The latter may be accomplished by trial holes dug into the ground or by taking test cores of the soil. Both techniques can provide a reliable source of information which costs as little as 1% to 2% of the total building expenditure. It is important that *trial holes* are taken as near as possible to, but not on the actual line of, a proposed foundation in order to avoid the sub-

sequent excessive consolidation of the relatively loosely backfilled soil by the constructed foundation (figure 124). Trial holes need to be formed to sufficient depth to expose the likely soil which will support the foundation. Most of this lies in the volume of soil beneath the foundation known as the *bulb of pressure* (figure 125). For lightly loaded buildings with simple foundations this generally lies within 2 m below ground level. When necessary, investigations can be carried out beyond this depth by taking *sample cores* of soils, 150 mm–200 mm in diameter, using special equipment.

The distribution of soil types in the United Kingdom are indicated in figure 126; these include peat, clay, silt, sand and gravel and corresponding safe bearing pressures are also given.

The *steady* increase in loads resulting from the construction of a building causes initial consolidation of the soil beneath the foundation. The amount of this movement varies with the soil conditions and, in a correctly designed and formed foundation, causes little disturbance in

179

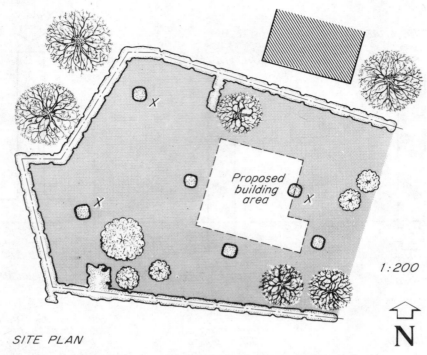

SITE PLAN

1:200

N

Figure 124   Suitable position for digging trial holes for soil investigation

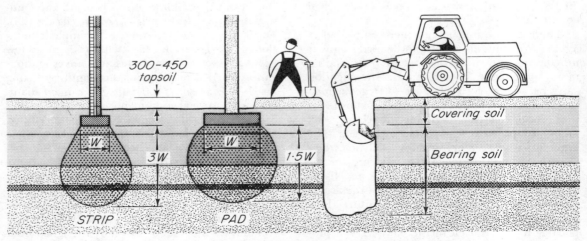

Figure 125   Influence of the 'bulb of pressure' on supporting soils

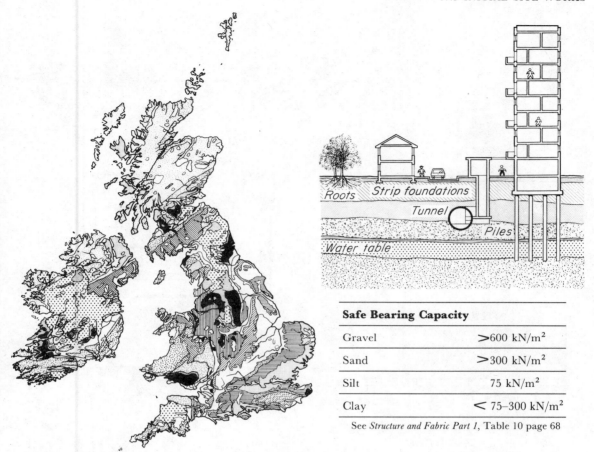

Safe Bearing Capacity

Gravel	$>600 \text{ kN/m}^2$
Sand	$>300 \text{ kN/m}^2$
Silt	75 kN/m^2
Clay	$< 75\text{–}300 \text{ kN/m}^2$

See *Structure and Fabric Part 1*, Table 10 page 68

Figure 126   Distribution of various types of supporting soils in the United Kingdom (see Geological Survey Maps for greater detail). The quality of bearing soils will also vary according to depth and previous uses

the building fabric above. Unreasonable movements of foundations (figure 127) can be caused by the use of poor materials (inadequate concrete mixes) and formation techniques, or commonly, they are a result of preceding soil movements. These movements derive from either excessive consolidation of the soil particles under load or by other factors independent of building loads such as:

– Swelling and shrinkage of the soil resulting from varying moisture content – seasonal fluctuations in rainfall and watertable, drought conditions, and the proximity of tree roots, boilers, kilns and furnaces are likely to affect clay soils.
– Frost heave – formation of ice lenses between

the soil particles of chalks and silts causing swelling.
– Swallow or sink holes – pockets of weaker soils in chalk and limestones.
– Slopes and landslips – movement of clay soils on sloping ground in excess of 1 : 10.
– Mining subsidence.
– Shock and vibration – soils near railways, motorways, or active industrial plant.

17.3 Initial site works (figure 128)
CI/SfB (11)

(a) Selection of plant
One of the first decisions the selected builder must make is whether it will be more economical to employ mechanical equipment

181

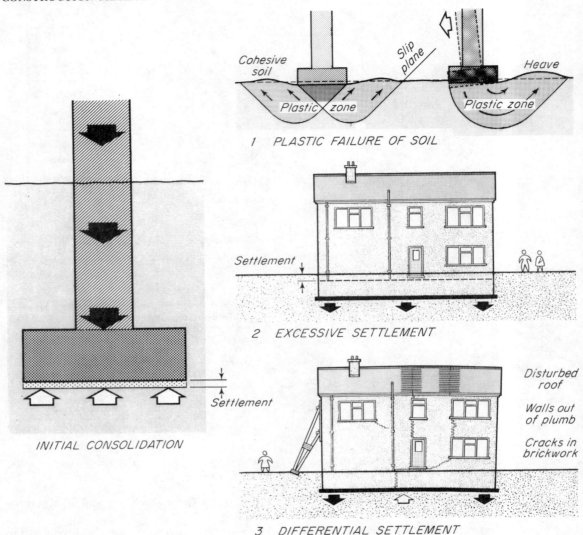

INITIAL CONSOLIDATION

1 PLASTIC FAILURE OF SOIL

Cohesive soil · Slip plane · Heave · Plastic zone · Plastic zone

2 EXCESSIVE SETTLEMENT

Settlement

3 DIFFERENTIAL SETTLEMENT

Disturbed roof

Walls out of plumb

Cracks in brickwork

Figure 127 Foundation movements. During construction gradual consolidation of the supporting soil will give rise to settlement, the effects of which are anticipated in the design. Unreasonable movements (1, 2 and 3) will cause damage and the foundation must be designed to avoid their occurrence

(plant) or hand excavation techniques for site work. Plant should normally be used where it can be kept fully occupied, the ground is not too boggy, the excavation is not too 'fussy', there is room to manoeuvre, and the transport costs of plant (mechanical equipment) to and from the site can be paid from the savings over the cost of hand excavation. Regardless of whether the builder owns certain items of plant or hires it for certain periods, the choice whether it should be used or not is a matter for expert consideration. It is possible, however, for a designer to take the use of plant into consideration (perhaps in consultation with builder) when designing a building so that maximum time saving can be obtained which could result in reduced costs.

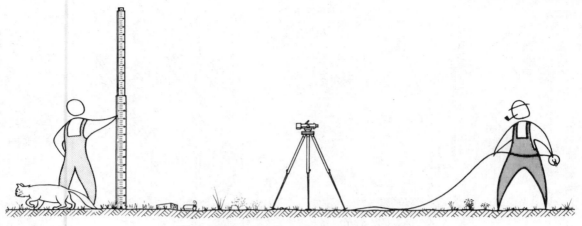

Figure 128   Initial site work: above ground survey

(b) Setting out

Having cleared the site of debris, unwanted shrubs, etc, and by using information from the designer's site plan, the builder can erect huts for workers (sanitary facilities, canteen, site office, etc) and the storage of materials and components. These should be located in a position which will be unaffected by access points, circulation about the site, and the subsequent building processes. The builder can then commence to set out the areas of the site which are required to be excavated. This is done by using a steel measuring tape to establish the exact position and perimeter dimension of the building or other excavated area, and then by providing *profiles* to give guidance during sub-structure operations. Profiles consist of timber pegs driven into the ground with timber cross bars fixed to them on which points are marked by nails to indicate lines of foundation and the walls above (figure 129).

The cross bar of the profile is fixed at a

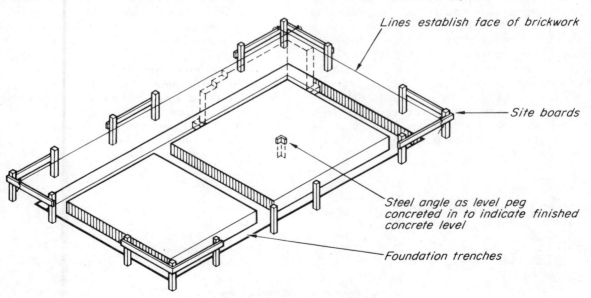

Lines establish face of brickwork

Site boards

Steel angle as level peg concreted in to indicate finished concrete level

Foundation trenches

Figure 129   Setting out trenches and brickwork

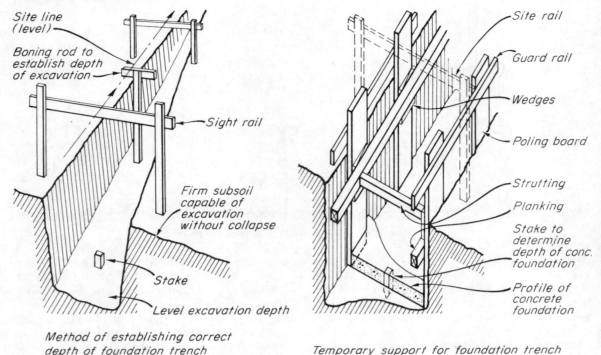

Method of establishing correct
depth of foundation trench
at consistent level

Temporary support for foundation trench

Figure 130   Setting out and casting foundations

known height or datum which is established by a measuring staff and an optical instrument known as a *level*. By periodically stretching a thin string or *line* between the top of the cross-bars during excavation and using a *boning rod* of known length, the amount of soil to be removed can be established (figure 130). It is always necessary to set up profiles away from the actual region of excavation to allow for working room. The setting out of trenches for drainage can be accomplished in a similar manner, but as the bottom of these excavations should be laid to a

fall to ensure the pipes they support are self-cleasing, measurements down from the temporary line stretched between profiles need to be taken at frequent intervals.

(c) Excavation (figure 131)

Excavation will consist of removing the top soil from the area of the proposed building and then forming the trenches for the foundations.

Topsoil exists on most sites for a depth of about 300 mm and is of considerable value for landscaping purposes. However, because it is

Figure 131   Clearance of top soil over area of building prior to commencement of foundation excavation

highly water retentive, propagates plant life and is highly compressible, it must be carefully removed from areas where it is not required and stored clear of building operations so that it can be re-used. If the quantity involved is not all wanted for its site of origin, arrangements will have to be made for it to be removed and, perhaps, sold.

The type of excavation necessary for the foundations obviously depends on the category of foundation chosen as suitable for particular subsoil type and characteristics. Further details about the influence of these factors are given in the next section 17.4(b) *Choice of foundation*.

17.4 Foundation construction CI/SfB (16)

(a) Function
The resolved dead, imposed, and wind loads may be transferred through the building by continuous walls, or through isolated piers, by columns or piers, or by combinations of these techniques. Because each of these structural systems results in foundation types which transfer the loads to the supporting soil in different manners, the selection of a foundation category (figure 132) should reflect the optimum relationship between building structure and the characteristics of the soil used as support. In other words, the aesthetic and technical factors resolving into a particular building form above ground are inter-dependent with the technical requirements of the foundation design below ground.

(b) Design
The following steps are necessary to arrive at a correctly designed foundation which takes into account the relationship between soil conditions and building loads.
1 Assess soil conditions by suitable investigation (see 17.2(c) *Soil investigation*).
2 From investigations decide permissible soil loading.
3 Determine combined building loads (dead, imposed and wind) and allow for self-weight of foundation. Typical loads for a single- and two-storey house are indicated in figure 133.
4 Establish form and size of foundation to suit loadings and correct for self-weight allowances.

5 Check soil resistances against permitted stresses and estimate permissible movements.
6 Allow for necessary movement joints for differentially loaded parts of the building or other structural effects on the foundation.

(c) Choice
For a two-storey house using continuous load-bearing masonry walls (brick, block or stone), a form of strip foundation will probably be most appropriate, and can consist of a horizontal strip *or* a vertical strip of mass concrete located beneath ground level. The choice between which of these two forms to adopt can be made during discussion between designer and builder as the structural significance of either is closely related to site conditions, plan shape of the building and the availability of appropriate resources. In both cases the depth of the foundation will depend upon the depth of adequate bearing soil, and the avoidance of certain phenomena which detrimentally affect the upper layers of the bearing soil, ie seasonal variations in moisture content, or tree roots etc. Generally a depth of 1 m to 1.2 m is adequate when a suitable subsoil is available. The width of excavation and hence the proposed strip foundation can be determined by calculation as stated earlier. The Building Regulations 1976 give 'deemed-to-satisfy' widths of foundation for certain building loads and soil types. For example, for a wall of a building weighing 50 kN/m to be supported on firm clay the recommended width is given as 600 mm. Nevertheless, in order to avoid eccentric stresses in the foundation, consideration must be given to allowing enough tolerance for the bricklayer to build a perfectly straight wall exactly on the centre line of the concrete mass of the foundation formed within a trench which has been relatively crudely excavated by machine (or hand).

The *vertical strip foundation* is generally used in firm soils where costs can be reduced because they are self-supporting and do not require the planking and strutting which the more loose soil types need. When safe soil bearing pressures and wall dimensions permit, the width of these foundations can be as little as 300 mm or the width of the smallest *mechanical* digging tool used for the excavation of the trench in which they are formed. (This is possible because the high verti-

185

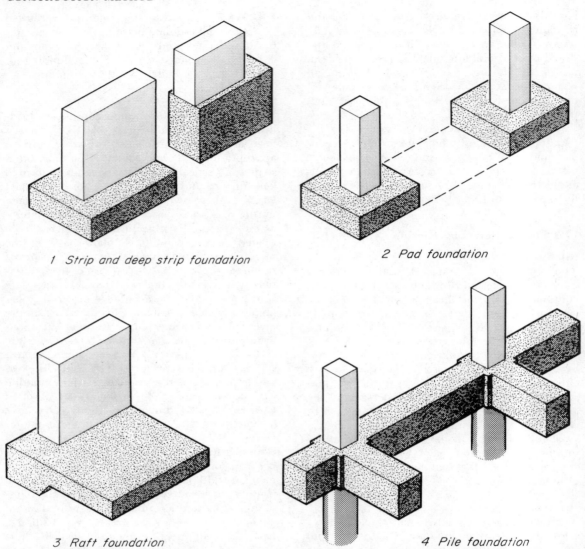

1 Strip and deep strip foundation

2 Pad foundation

3 Raft foundation

4 Pile foundation

Figure 132   Foundation categories – use dependent upon subsoil condition and form of building to be supported

cal 'thickness' or sides of the deep strip foundation are provided with lateral restraint by the
surrounding soil and offsets the likelihood of the
foundation overturning.) However, to avoid the
possibility of the development of the eccentric
stresses mentioned earlier, a tolerance dimension
of 75 mm minimum should be allowed either
side from the face of the wall above to the edge
of the foundation. With the *horizontal strip foundation* form, allowances must be made for a

bricklayer working *within* the trench (figure
134). He needs a minimum of 150 mm either
side of the proposed wall and this, together with
the structural consideration, may well determine the final width of foundation.

Generally, the chosen width of a vertical
strip foundation may relate very closely to the
width of the horizontal strip foundation, subject
to similar loading conditions. However, the
advantage of the deep strip foundation over the

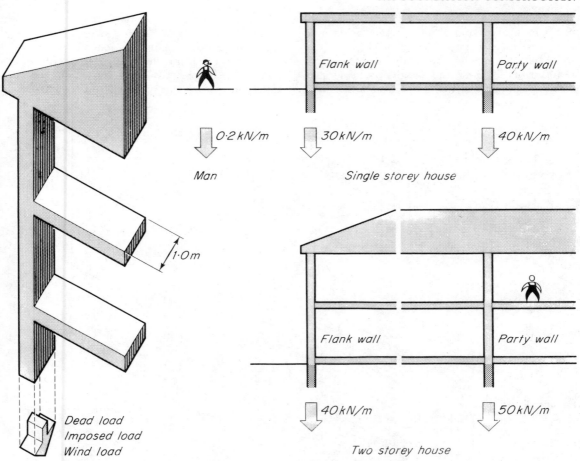

0.2 kN/m

Man

Flank wall   Party wall

30 kN/m   40 kN/m

Single storey house

1.0 m

Flank wall   Party wall

Dead load
Imposed load
Wind load

40 kN/m   50 kN/m

Two storey house

Figure 133 Typical loads on two-storey house

horizontal strip foundation is that it finishes just below ground level and does not create limiting working space for the bricklayer (figure 135). Nevertheless, their use should be approached with caution as it is difficult to inspect the bottom of a narrow trench to ensure consistency of soil before the concrete is placed, and also excessive soil movements can cause the narrow but high foundations to overturn unless adequate precautions are taken.

The thickness of the *horizontal strip* should never be less than 150 mm or the amount of projection from the face of the wall supported, whichever is greatest. This rule takes into account the general working tolerances applicable to the formation of an adequate amount of

structural concrete for practical purposes. Unreinforced or mass concrete strip foundations can fail under the compressive loads from the building above as a result of excessive bending or from excessive shear stresses developing within the cross section of the foundation. This latter form of failure generally occurs along the plane of maximum tensile shear which lies at an angle of about 45° to the horizontal. (In theory, the concrete lying outside this plane does little in distributing loads to the supporting soil but is, nevertheless, necessary for practical purposes in forming the foundation.) Therefore, if horizontal strip foundations are required to carry heavy loads or if the soil is weak, the corresponding increase in thickness of foundation (to maintain

187

the 45° angle of spread – see figure 134) could result in a large amount of concrete being used which in turn increases the dead load on the soil. In such cases, the foundations can be reduced in thickness and weight by using steel reinforcing bars to take up the tensile and bending stresses. These bars must be protected from corrosion by a covering of concrete.

The bottoms of foundation trenches should be protected from the weather to avoid temporary shrinkage or swelling by leaving the last 100 mm or so unexcavated until immediately

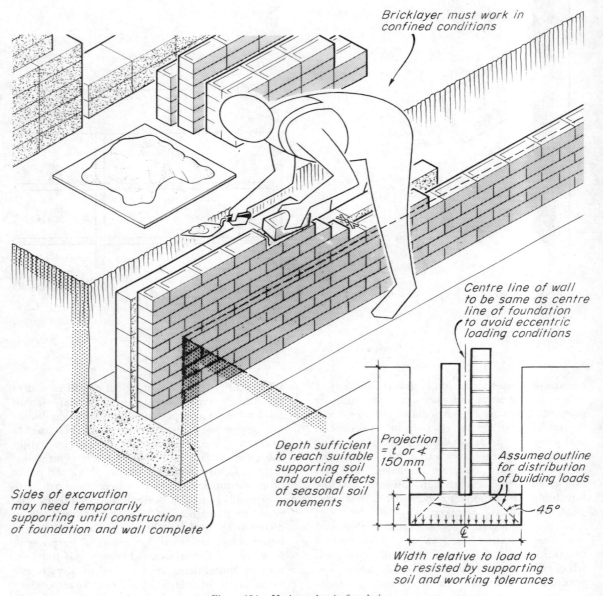

Bricklayer must work in confined conditions

Centre line of wall to be same as centre line of foundation to avoid eccentric loading conditions

Sides of excavation may need temporarily supporting until construction of foundation and wall complete

Depth sufficient to reach suitable supporting soil and avoid effects of seasonal soil movements

Projection = t or ≮ 150 mm

Assumed outline for distribution of building loads

45°

Width relative to load to be resisted by supporting soil and working tolerances

Figure 134   Horizontal strip foundation

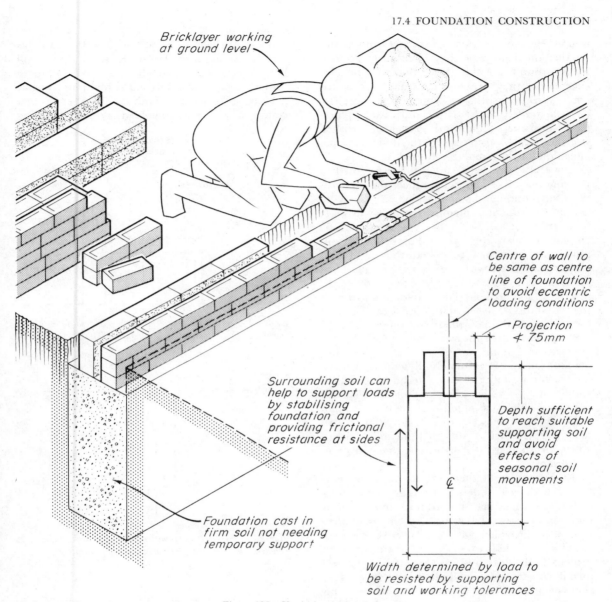

Bricklayer working at ground level

Centre of wall to be same as centre line of foundation to avoid eccentric loading conditions

Projection ≮ 75mm

Surrounding soil can help to support loads by stabilising foundation and providing frictional resistance at sides

Depth sufficient to reach suitable supporting soil and avoid effects of seasonal soil movements

Foundation cast in firm soil not needing temporary support

Width determined by load to be resisted by supporting soil and working tolerances

Figure 135   Vertical strip foundation

before pouring concrete. Alternatively, a blinding layer of concrete, about 50 mm thick can be placed at the base of the completed excavation. The accurate thickness of concrete can be established prior to pouring by inserting timber stakes at suitable intervals in the base of the excavation so that they project the required amount and the concrete is levelled to them. Immediately before the pouring of concrete

takes place, ie after excavation has finished, the bottom of the trench must be inspected and approved by a local authority building inspector, during his first formal site inspection.

(d) Materials
The materials employed to make foundations must be strong and durable – particularly as it

is not possible or practical to carry out maintenance on them once they have been installed in the ground. For this reason and with but few exceptions, they are formed from *good quality dense concrete*, consisting of cement, fine aggregate or well graded coarse sand, and coarse aggregate consisting of natural gravel or crushed stone all graded according to the designer's specification for the strength required.

Generally, ordinary *Portland cement* is used in the concrete mixes, but where the ground conditions surrounding the foundation are likely to contain detrimental chemicals which will react with this, special forms of cement should be employed as an alternative. For example, sulphate solutions which occur naturally in ground waters associated with clay soils, or the acidic solutions with peaty soils, attack and erode ordinary Portland cement. To avoid deterioration of the foundation *Sulphate Resisting Cement* could be used. High Alumina Cement, or Supersulphated Cement (currently not available in the UK) are alternatives but neither of these can be recommended since it is extremely difficult to achieve the essential quality of concrete under practical conditions. To avoid using any of these special cements (which are much more expensive than ordinary Portland cement) the foundation can be protected with an impermeable barrier wrapping such as rubber/bitumen coated polythene sheet.

In certain cases the concrete may be reinforced with steel bars. Depending on loading conditions, these may be ordinary mild steel or high-tensile steel bars and when used they should be provided with a protective concrete cover of not less than 75 mm. In order to protect both steel and concrete from the soil during construction, a blinding layer or excavation coverage of concrete can be provided. This could be 50 mm to 100 mm thick depending on site conditions and, if adopted, the protective concrete cover to reinforcing bars may be reduced to 40 mm.

(e) Construction

The sequence of the formation for the *vertical strip foundation* is indicated in figure 136. As strip foundations are used in conjunction with continuous walls of small bonded units (bricks, blocks, or stones), it is necessary that their top

surfaces are laid to a true level surface. In this way the erection process of the walling unit will be made quicker and the wall will be more stable as the necessity of providing mortar packing of varying depth beneath each first course of wall unit is avoided. When a building

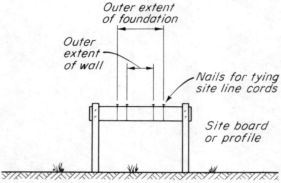

Figure 136(a)   Site board or profiles placed to establish position of future foundation (see figure 129)

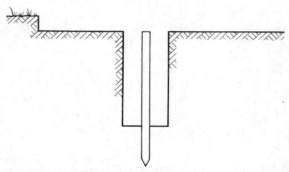

Figure 136(b)   After excavation foundation stakes inserted to establish consistent level for top surface of foundation

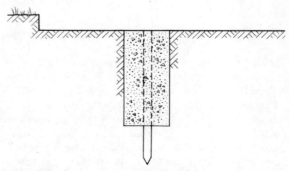

Figure 136(c)   Completed vertical strip foundation

190

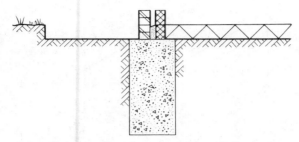

Figure 136(d)   Wall construction commenced just below natural ground level. Hardcore for solid floor construction laid and consolidated

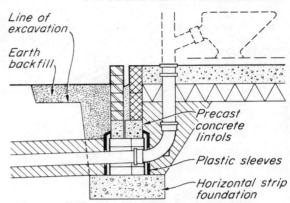

Figure 137(a)   Horizontal strip foundation: hole for service pipes formed by lintol construction

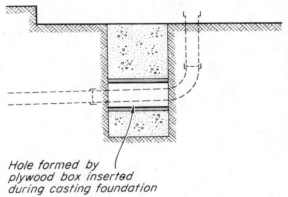

Hole formed by plywood box inserted during casting foundation

Figure 137(b)   Vertical strip foundation: hole for service pipes formed by shuttering

is to be erected on a sloping site and the required bearing strata follows the ground slope, excavation costs and the amount of walling below ground can be reduced by forming the foundation in a series of steps.

Services entering a building below ground (water, electricity, gas, telephone, etc) and those leaving a building (drainage, refuse systems, etc) usually pass through the wall below ground in the case of horizontal strip foundation, or through the foundation itself when vertical strip foundations are employed (figure 137). This presents little problem in the former case as apertures can be left in pre-determined positions by using precast reinforced concrete lintols or brick arches over openings, or by laying wall units in dry sand instead of mortar so that they can be easily removed prior to the service being installed. If the service pipe can be positioned as the below ground wall is being built, these processes can be simplified still further. However, when vertical strip foundations are used, the position for holes must be carefully predetermined and open ended timber boxes or plastic pipes placed in position during the concrete laying procedures. Whichever form of strip foundation is used, it is essential that the holes left are of adequate size to allow about 50 mm clearance all round the proposed services pipe in order to avoid the possibility of its subsequent deflection or fracture should any settlement of the foundation occur. It is always best to avoid passing services beneath foundations unless precautions are taken to ensure localised settlement of the wall above does not occur at these points through the removal of foundation support (bearing soil).

After the concrete has been poured and has hardened, the construction of the wall above can commence and is subsequently taken to a height of the *damp proof course* (see 17.6 *External wall construction*). At this stage the local authority building inspector will make his second formal visit to the site and inspect the suitability and quality of the workmanship and materials.

Once approval has been given, the remainder of *horizontal strip foundation trenches* can be backfilled: ordinary excavated soil generally being suitable for this purpose on the *outside* face of the foundation, with a denser material capable of greater consolidation being used for the backfill on the *inside* face of the foundation. This is because the backfill on the inside face of the foundation trench may be required to

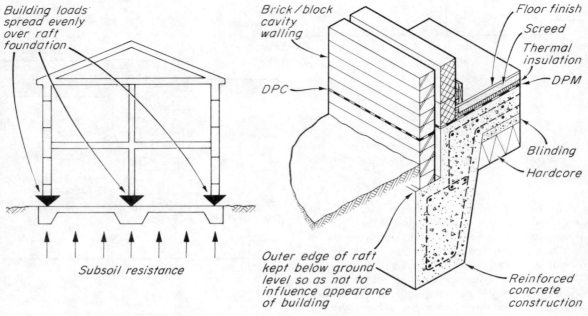

Figure 138   Typical details for raft foundation

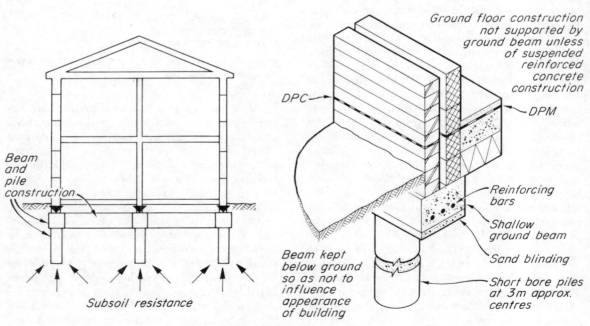

Figure 139   Typical short bore pile foundation

provide support for the ground floor construction.

Alternative forms of foundations to the horizontal or vertical strip may be necessary for a two-storey house when subsoil conditions are difficult. A *raft foundation* will be required when the soil is so weak that it is necessary to distribute the building loads within a wide volume of soil (figure 138); a *short bore pile foundation* will be required when seasonal variations in moisture or vegetation is likely to affect the stability of the bearing soil lying relatively near the surface (figure 139). Both these forms of foundations, together with the pad foundation applicable to framed buildings are described in detail in Mitchell's *Structure and Fabric Part 1*, chapter 4. (See page 220 for details of internal wall foundations.)

17.5 Ground floor construction
CI/SfB (16) + (43)

(a) Function
The performance requirements which the ground floor construction must fulfil include that of providing a level surface capable of supporting people, furniture, equipment, wheeled traffic, partitions and finishes, etc; and also of providing environmental control against the passage of water (liquid and vapour) and heat. Various factors may affect the method of construction as follows:

– existing ground levels
– desired finished floor level(s)
– required floor loading
– bearing capacity of supporting soil
– form of building foundation
– availability and cost of suitable fill material.

A ground floor construction which is continuously supported by the ground or through a fill material in contact with the supporting ground is called a *solid floor*: that which is supported by a foundation system at selected points is called a *suspended floor* (figure 140).

It is usual to commence some part (see page 119) of the floor construction as soon as the foundations and walls up to damp proof course level have been completed. This is because it is normally desirable to provide a firm and dry working surface for subsequent building operations, or for the storage of components and equipment requiring protection from moisture rising from the ground, eg windows, doors, frames, etc. This provision is particularly important when the area of the building site is small and leaves little space for storage. However, it is also necessary that any prematurely completed work is suitably protected against damage during subsequent building work and that no materials are used in the construction which will be adversely affected by the temporary exposure

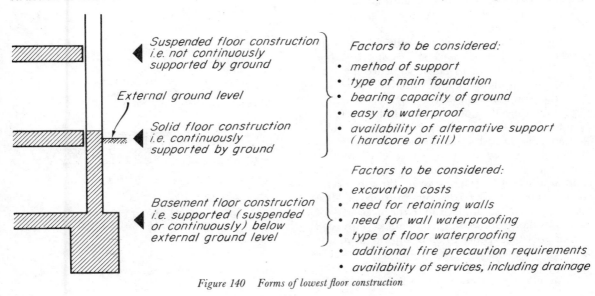

Figure 140   Forms of lowest floor construction

to the environment, eg timber joists, boards, screeds and finishes.

Before the work to the ground floor commences, trenches must be dug and the internal below ground service pipes for drainage, gas, water, telephone, television and electricity must be installed. These services may need special protection when they occur below the building (eg drainage pipes surrounded by concrete) and the trenches should be backfilled as solidly as practicable. It is important to supply the contractor with accurate information regarding the position of these services in order to avoid subsequent repositioning work.

(b) Solid ground floor construction

The concrete ground floor slab must be laid to the correct level, and on a firm base to avoid the possibility of future movements which could cause cracking resulting in an uneven top surface or moisture penetration from the subsoil below. As mentioned earlier, the top soil will have been removed from the formation area of the floor slab. The resulting reduced level should, therefore, be made good with a material or *fill* which is capable of full compaction so as to be strong enough to take the weight of the floor slab and the loads it carries, as well as restricting or reducing the upward capillary movement of water from the soil. A material which most conveniently conforms to these requirements and is easy to use is gravel, but less expensive materials include brick, tile or concrete rubble (*hardcore*) or other waste products such as blast furnace slag, pulverised fuel ash and colliery shale. However, care must be taken when using these latter products to ensure that their unit size allows thorough compaction, and that they do not contain sulphates which would attack the cement of the concrete slab as already mentioned under foundation materials. When completed, the top surface of the fill material or hardcore should be dry and even, and at a level which best ensures the top surface of the concrete floor construction butts against the perimeter walling without the necessity for vertical damp proofing techniques. This will avoid moisture penetration resulting from the lateral movement of water from outside (figure 141 (a)).

The fill or hardcore must be fully compacted

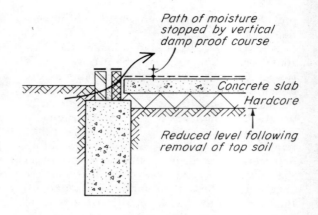

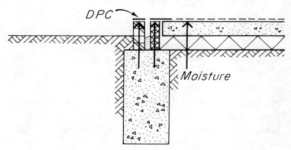

Figure 141(a)   Principles of waterproofing solid ground floor slab

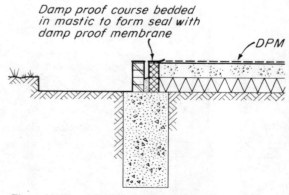

Figure 141(b)   Method of waterproofing solid ground floor slab

by mechanical hammers or rollers (depending on area involved). In practice it is usually difficult to ensure enough consolidation when fill or hardcore depths greater than about 600 mm are required – even if compacted in the usual 150/200 mm layers. Therefore, in order to avoid the possibility of settlement occurring in the floor, a suspended floor construction will be more appropriate when the desired finished floor level would require more than 600 mm fill or hardcore (see 17.5(c)).

The concrete for the slab is mixed and poured so as to obtain a *minimum* thickness of 100 mm, although it is preferable to use a thicker slab of 150 mm when any doubt exists about either the full compaction of the fill or hardcore, or about the subsoil when it is less than very stable. The thicker slab will also assist in spreading the point loads from loadbearing partitions when separate foundations for this purpose are not necessary; the greater mass of concrete will, of course, offer greater resistance to rising moisture. If the subsoil conditions are relatively weak, the 150 mm slab should be reinforced with a mild steel mesh. This should be placed in the slab after an initial layer of concrete has been poured and then be covered by the subsequent and final pour. As concrete shrinks on drying, the reinforcement will also help in avoiding shrinkage hair-cracking, although if large areas of uninterrupted concrete slab are required, movement joints should be incorporated by casting the slab in approx 3 m bays or to an appropriate modular dimension suitable to the plan area.

The surface of the concrete slab can be finished in various ways depending upon the position of the *damp proof membrane*. This membrane consists of a sheet of polythene (500/1000 gauge), or heavy reinforced building paper (bitumen impregnated), or an *in situ* coating of hot poured bitumen, or a minimum of three coats of cold-applied bitumen/rubber solution. The function of a damp proof membrane (dpm) is to resist moisture in liquid and vapour form coming into contact with the internal surfaces of the floor construction and thereby causing damage. The concrete slab itself may not be able to resist this moisture – particularly when the watertable is close to the ground surface – because of the fine hair-cracks which develop during drying and, unless it contains special admixtures, concrete is not moisture vapour proof.

If either a sheet or an *in situ* coating damp proof membrane is to be subsequently applied to the *top* of the concrete slab as indicated in figure 141(b), the top surface of the concrete must be trowelled smooth so that it is not punctured. The dpm must also be protected as soon as possible after it is laid, by a sand and cement mixture (in the proportion of 3 : 1) known as a *screed*. This screed should be a minimum thickness of 50 mm to avoid cracking and also trowelled smooth to a *perfectly level surface* in order to be satisfactory for the final floor finish – eg quarries, clay or pvc tile, carpet, etc. *Obviously, the dpm, screed and subsequent floor finish cannot be placed until most of the other building works have been completed because they are all susceptible to damage which will be costly to rectify (see 17.11 Finishes).*

Figure 142 indicates the alternative position for the dpm – *below* the concrete slab. In this case it is the fill or hardcore which must have the smooth surface to avoid sharp edges damaging the membrane and this is achieved by *blinding* consisting of a 30 mm to 50 mm thick layer of rolled ash or sand. After the dpm has been positioned, the concrete slab (and reinforcement if required) must be placed very carefully to avoid its puncture. The top surface of the slab should then be 'finished' with a rough surface by using a long plank of wood placed on edge and *tamping* across the wet surface. When dry, the resulting slightly irregularly raised surface texture will provide an excellent key for the screed. Because more positive bond is provided to the slab than when laid on the smooth dpm as indicated in figure 141(b), the finishing screed can be of reduced thickness, say 30 mm. If the screed is laid while the concrete slab is still green, and, therefore, providing an even better key, the screed thickness can be reduced to 12 mm. Indeed, as an alternative to providing a screed, the green concrete slab can be finished to a very smooth surface by a light mechanical buffing process using a special machine known as a *power float*.

The designer must consult with the contractor concerning which of the above two positions for the dpm is most suitable. Both have advantages as well as disadvantages. For example,

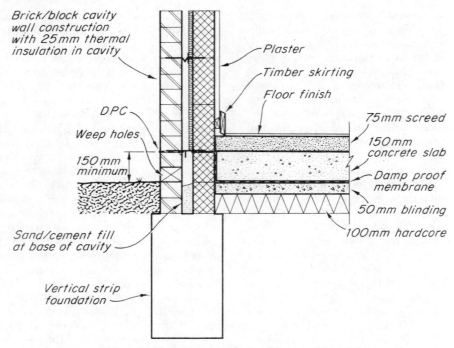

Figure 142   Damp proof membrane below ground floor slab

below the slab savings in both labour and materials can be made, and, if correctly executed, the slab remains dry throughout the life of the building and helps to minimise heat losses as well as providing a thermal reservoir (heat storage). The method of construction does, however, mean great care must be taken during construction to ensure that the dpm is not damaged. Unless a screed or some other finish (ie timber) is subsequently applied to the surface of the slab, it cannot be cast until quite late during building operations because the exposed surface will certainly be damaged. If a power float finish to the slab is required, it must be temporarily protected until most building operations are complete.

The amount of heat which may be lost through a solid ground floor construction depends on a number of factors. The greatest concern for heat conservation derives from the extent of external wall bounding the floor because this juxtaposition will provide a direct path for heat loss to the external air. This problem may be overcome by using perimeter insulation material between the junction of the

solid floor construction and the inside face of the external wall. This insulation should be waterproof and either extend downwards from the dpc/dpm position for 300 mm to 1000 mm depending upon climatic conditions, or may extend downwards for the thickness of the slab before running horizontally beneath it for about 600 mm to 900 mm.

Alternatively, insulating material can be placed directly on the dpm before the floor screed is applied as indicated in figure 143. This method of heat conservation is particularly useful when underfloor heating cables are to be inserted in the screed. However, care must be taken to ensure that the resulting *floating screed* (unattached) is of sufficient thickness to avoid cracking, and that it is separated from the insulation by a waterproof building paper to avoid saturation of the insulation and the loss of grout (cement/water) from the screed causing a 'cold bridge' link with the concrete slab below. The waterproof membrane will also act as a vapour barrier, and eliminate the possibility of interstitial condensation within the insulating material. A minimum thickness of 75 mm is required and this should be reinforced with a light mesh reinforcement (chicken wire).

(c) Suspended ground floor construction

As mentioned earlier, this form of ground floor construction should be adopted when the desired finish floor level is above the supporting ground to the extent that the difference cannot conveniently be made up with fill or hardcore (figure 140). The structural floor can be formed by timber joists (figure 144) or with reinforced concrete (cast *in situ* or by using precast concrete planks figure 148).

When *timber joists* are used, care must be taken to ensure that they are not adversely affected by the potentially moist condition existing in the void beneath the floor. Precautions include the provision of a *dpc* immediately beneath the timber wall plate which provides bearing for the timber joists through to the sleeper walls below; adequate *cross ventilation* of the void by air from external perforated bricks circulating around the timber joists and through gaps in the *sleeper walls* which are of 'honeycombed' construction; and the use of structural

timbers with not more than 18% moisture content *at the time of fixing*. All timber should be impregnated with a preservative solution to counteract the possibility of fungal attack (dry rot) and act as a precautionary measure against the failure of the construction techniques to keep the moisture content of the timber below 18% during its life in the building. Preservative impregnated timber also proves unpalatable to the grubs of wood eating insects (woodworm, house longhorn beetle, etc).

For further comments regarding timber joists, and their sizes and spacing see 17.8(b) *Materials for intermediate floors* and figure 164.

The floor *decking* can be tongued and grooved hardwood or softwood *boarding*, or chipboard *sheets* or plywood *sheets* (figure 145). The narrow face width of boarding provides greater flexibility when providing a floor in an irregular or *non-modular* width of room, and also permits easier access to the void below should it be required during the life of the building, eg for alterations or extensions of electric wiring contained within the void. However, sheet floor deckings are generally quicker to install – parti-

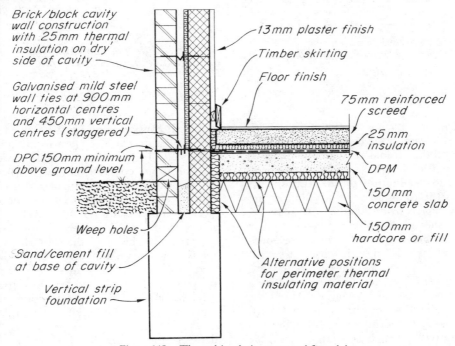

Figure 143   Thermal insulation to ground floor slab

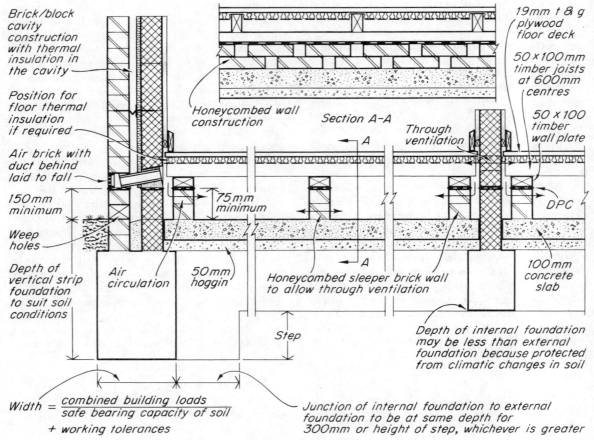

Brick/block cavity construction with thermal insulation in the cavity

Position for floor thermal insulation if required

Air brick with duct behind laid to fall

150mm minimum

Weep holes

Depth of vertical strip foundation to suit soil conditions

Honeycombed wall construction

Section A–A

A

75mm minimum

Air circulation

50mm hoggin

Honeycombed sleeper brick wall to allow through ventilation

Step

19mm t & g plywood floor deck

50 x 100mm timber joists at 600mm centres

50 x 100 timber wall plate

Through ventilation

DPC

100mm concrete slab

Depth of internal foundation may be less than external foundation because protected from climatic changes in soil

A

$$\text{Width} = \frac{\text{combined building loads}}{\text{safe bearing capacity of soil}} + \text{working tolerances}$$

Junction of internal foundation to external foundation to be at same depth for 300mm or height of step, whichever is greater

Figure 144   Suspended timber ground floor construction

cularly when the spacing of the timber joists and sizes of rooms have been closely co-ordinated with dimensions of available plywood or chipboard sheets. Also, provided the sheets are allowed to acclimatise to the atmospheric conditions of the room prior to installation and that they are correctly fixed, subsequent movements of the sheets are likely to have less effect on applied finishes than would the corresponding movements associated with softwood boarding. Boards as well as sheets should be impregnated with a preservative solution in a similar manner as the structural timbers, although, in the case of sheets, special hard wearing pre-treated flooring grades may be used as an alternative. When it is thought desirable to leave the top surfaces of the boards or sheets

exposed, consideration must be given to the selection of suitable wood grain patterns and the effects of subsequent seals, stains, or varnishes.

The *in situ* concrete slab used with the form of suspended floor construction indicated in figure 144, can be cast as soon as the walls have been completed up to dpc level in a similar manner to that described for solid floor construction. In this way, the slab serves as a relatively dry, clean and level surface for storage and work by other trades, ie preparation of purpose-made door and window frames. The remainder of the floor construction can be executed as more of the building is completed.

If it is considered desirable for the floor decking to be installed before 'heavy' trade work is complete a temporary layer of hardboard sheets must be installed to prevent damage to the surface of the boards or sheets. At any event, it is convenient if the services required to run beneath the floor decking can be installed before all the boards or sheets are finally fixed to the supporting joists. Alternatively, predetermined areas of decking are left open to allow services to be fed through the void below. When the services are installed early, they can be left projecting just above the decking, pending subsequent connections and completion of the above decking services.

Figure 146 illustrates a form of suspended ground floor construction where a concrete slab is not used. A slab cannot be considered as a water barrier since small cracks will develop through its thickness as a result of drying shrinkage. These gaps may be relatively large (2 mm to 3 mm) between two or more areas of slab cast next to each other on different days ('daywork' joints) and at the junction between a slab and

the external wall. Furthermore, concrete is not moisture vapour proof. By using a sheet membrane (polythene) lapped and folded at edges and sealed with the wall dpc, a more positive barrier against free moisture and moisture vapour can be provided. This form of construction can best be executed without the use of intervening sleeper walls, which means that the structural floor joists must be larger to span the resulting longer distances. Ventilation of the floor void must be provided as described earlier, and timbers should also be impregnated with a preservative treatment against fungal and insect attack.

This method of construction means that the dry working surface previously described is not available to the builder and that he must take special measures to ensure that the membrane barrier is not damaged during the installation of floor joists and decking. Therefore, it causes some inconvenience to the builder and his

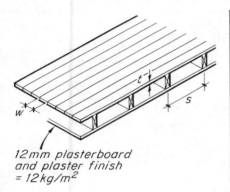

12mm plasterboard
and plaster finish
= 12 kg/m²

Boards	Finished hardness (t)	Nominal width (w)	Span (S)	Approx. weight, kg/m²
Softwood t and g	16	100 and 150	505	7
	19		600	9
	21		635	10
	28		790	13
	47	150	2560	23
	60	150	3230	28
	72	150	3830	32

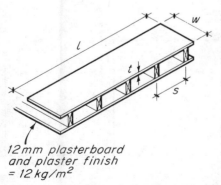

12mm plasterboard
and plaster finish
= 12 kg/m²

Sheets	Thickness (t)	Size (w × l)	Span (S)	Approx. weight, kg/m²
Plywood t and g (flooring grade)	12	600 and 1200 × 2440 and 2750	400	8
	15		600	9
	18		700	11
	25		1200	15
Chipboard t and g (flooring grade)	18	600 and 1200 × 2440 and 2750	400	12
	19		450-500	13
	22		600	15

Figure 145   Types of floor decking. Dead or self weight of flooring and ceiling can be used in conjunction with Table of Span/Depth for Intermediate Floors (figure 164) to obtain suitable timber joist sizes

199

opinion should be sought before it is specified by the designer even if considered essential to avoid excessive amounts of moisture occurring in the floor space beneath the decking. If the site is subject to water-logging it is probably best not to risk water standing freely on either concrete slab or membrane barrier. Figure 147 indicates a method of construction which does not employ either, but allows water to drain away during dry periods.

Generally speaking, a timber suspended floor provides better resistance to loss of heat to the surrounding ground by virtue of the better insulation properties of timber when compared with solid concrete. Additional thermal insulation can be provided, however, and this is usually draped over the timber joists prior to fixing the floor decking. Under these circumstances it is probably best not to provide a vapour barrier on the room side of the insulation (warmer) because it would only serve as a trap for water which may be spilled on the upper surface of the floor decking and seep through joints to the spaces below. Seeping moisture from this source will soon dry out (as will any interstitial

condensation) if the void is adequately ventilated.

Suspended ground floor construction using *reinforced in situ or precast reinforced concrete* units allow for 25% to 40% greater floor loadings than timber joists, and are capable of spans up to 6 m (figure 148). This may be particularly attractive when poor soil conditions require deep or complicated foundations and which should be formed as widely spaced as possible for reasons of economy. Also, concrete floor construction is not subject to dry rot or insect attack. It does, however, require formwork (permanent or temporary) for *in situ* construction, or lifting equipment (gantries or cranes) for precast concrete constructions. The latter also requires careful dimensional co-ordination to maximise the use of a minimum range of standard length and width components. As with solid floor construction, a suspended concrete floor construction can be provided with a timber floor finish (boards or sheets on preserved timber battens fixed to the concrete base), but does not provide the desirable springiness of a structural timber construction unless

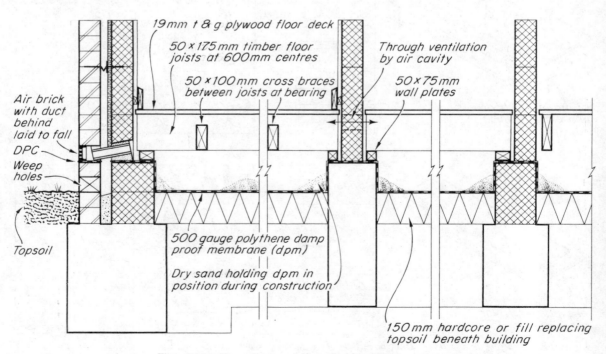

Figure 146   Alternative ground floor suspended timber construction 1

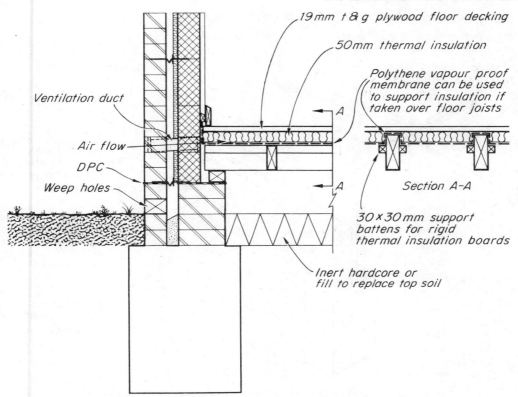

19 mm t & g plywood floor decking

50 mm thermal insulation

Polythene vapour proof membrane can be used to support insulation if taken over floor joists

Ventilation duct

Air flow

DPC

Weep holes

A

A

Section A–A

30 x 30 mm support battens for rigid thermal insulation boards

Inert hardcore or fill to replace top soil

Figure 147   Alternative ground floor suspended timber construction 2

special sprung fixing devices are installed with the timber battens.

17.6 External wall construction
CI/SfB (21) + (31) + (41)

(a) Function
The main function of an external wall involves providing the environmental control between the external and internal climates of a building. Also, depending upon the precise nature of the overall structural system, an external wall is usually required to support the combined dead, imposed, and wind loads of the roof and floor construction as well as its own combined loads and transfer them safely to a foundation. However, apart from the technical requirements, considerations of appearance must be a critical part of the design since to a very large extent an external wall determines the architectural character and quality of a building.

(b) Materials
Both technical and aesthetic criteria are influenced by the type and size of materials employed, workmanship, and the effects of weathering and general durability resulting from design and construction methods. Therefore, external walls can take many forms since they can be made from a very wide range of materials used either singly or in combination with others, *viz*:

– small blocks of brick and/or concrete laid in mortar
– dry jointed units of timber, metal, or concrete
– homogeneous matrix of reinforced concrete
– composite systems of timber, metal, or concrete preformed units
– rigid single sheets of glass, metal or plastics
– flexible single sheets of rubber or plastics.

When considering the construction of a dwelling in the UK, the first choice of materials for

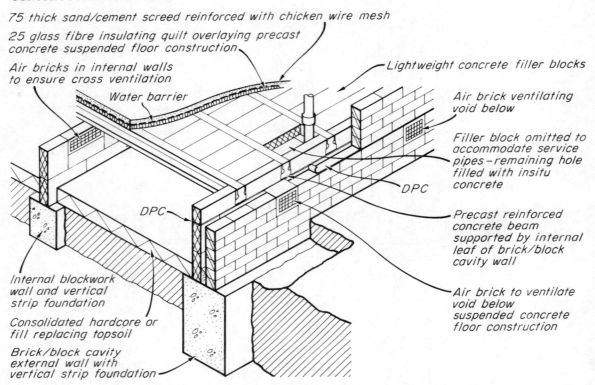

75 thick sand/cement screed reinforced with chicken wire mesh

25 glass fibre insulating quilt overlaying precast concrete suspended floor construction

Air bricks in internal walls to ensure cross ventilation

Water barrier

Lightweight concrete filler blocks

Air brick ventilating void below

Filler block omitted to accommodate service pipes – remaining hole filled with insitu concrete

DPC

DPC

Precast reinforced concrete beam supported by internal leaf of brick/block cavity wall

Internal blockwork wall and vertical strip foundation

Consolidated hardcore or fill replacing topsoil

Brick/block cavity external wall with vertical strip foundation

Air brick to ventilate void below suspended concrete floor construction

Figure 148   Suspended ground floor construction using precast concrete flooring units

an external wall is small blocks of brick and/or concrete laid in mortar (figure 149). This choice derives partly from traditional attitudes, but also because of availability of materials and craftsmanship as well as the inherent performance requirements known to be satisfactorily fulfilled by this form of construction. Bricks and blocks are unlikely to cause the spread of fire from either outside or inside because they are non-combustible and highly resistant to collapse as a result of being subjected to fire. When used correctly, bricks and blocks are also highly resistant to weather penetration and can provide satisfactory standards of sound control and thermal insulation. However, the selection of a most suitable form would vary according to the location of the external wall relative to the disposition and siting of the dwelling: this involves a thorough analysis of the performance requirements outlined in Part A of this book.

The majority of bricks used in the UK are manufactured from clay which, having been

moulded to suitable dimensions when in a plastic condition, is burnt to a hardened and vitrified state. There are three technical varieties according to use – *common bricks* for general and ultimately unseen work; *facing bricks* specially made or selected to give an attractive appearance; and, *engineering bricks* made to provide high levels of compressive strength *and* low levels of moisture absorption. The physical characteristics of the bricks are derived through the type of clay used and the manufacturing processes involved. Engineering bricks are of *special quality*, and rely on a special form of clay for their production, whereas common and facing bricks may be of *ordinary quality* (used in normal exposure condition) or of *internal quality* (used in protected position only) according to the combined strength and absorption characteristics of the clay used. Bricks can also be manufactured from shale (calcium silicate) or concrete.

Building blocks are larger than bricks and

are manufactured from clay, plaster, or concrete (figure 162). However, concrete is the most suitable material for incorporation in an external wall construction since its use avoids the special measures which have to be taken with the other materials to ensure adequate weathering and thermal insulating properties. Not unlike bricks, concrete blocks are manufactured in three basic

qualities: *Class A* for general purpose use including below ground; *Class B* also for general purpose use with better thermal insulating properties (less dense) but with less strength and less ability in withstanding the effects of excessive moisture conditions; and, *Class C* for internal protected and non-loadbearing use only. Also not unlike bricks, they are manufactured in solid, perforated or hollow horizontal section.

Walls constructed from bricks and blocks rely on their strength by being laid in horizontal courses so that their vertical joints are staggered or *bonded* across the face of the wall (see figure 21, page 30). In this way the compression loads which may initially affect individual or a series of bricks or blocks can be successfully distributed through a greater volume of the wall. The bricks and concrete blocks are held together by an adhesive mixture known as *mortar*, thereby completing the structural and environmental enclosure. This mortar also serves the function of taking up any dimensional variations in the bricks or blocks so that they can be laid in more or less horizontal and vertical alignment. Most mortars today usually consist of a water activated binding medium of cement and lime, and a fine aggregate filler such as sand, in the proportion of one part binder to three parts aggregate. The amount of water used to activate the binder in order for it to adhere to the fine particles of the filler and the face of the brick or concrete block must be kept to a minimum to avoid a 'sloppy' unworkable mix subject to excessive drying shrinkage. The use of lime as part of the binding medium serves to aid adhesion and avoids an otherwise very dry stiff mix which could be difficult to apply.

Mortar mixes on site should not be used when the air temperature is below 4°C since the possibility of freezing of the water content will

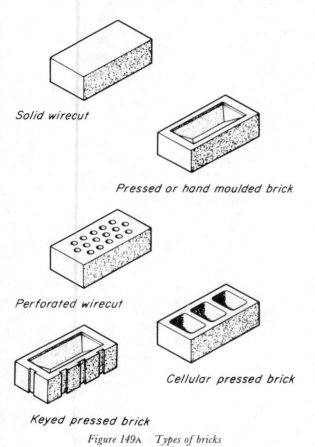

Solid wirecut

Pressed or hand moulded brick

Perforated wirecut

Cellular pressed brick

Keyed pressed brick

Figure 149A   Types of bricks

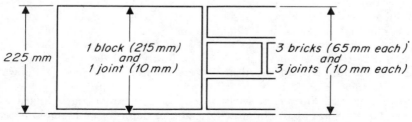

225 mm

1 block (215 mm) and 1 joint (10 mm)

3 bricks (65 mm each) and 3 joints (10 mm each)

Figure 149B   Relative course heights of brickwork and blockwork

adversely affect the strength of the mortar. When such conditions arise, the contractor will have to stop bricklaying, or provide protective shelters around the building or over that part to be immediately constructed, and probably also provide artificial space heating sources to ensure continuity of work for the bricklayers. These precautions must be applied to all on site 'wet' trade activities, including concrete laying and the application of plaster.

(c) Brick/block cavity walls

Bricks and concrete blocks can be used as a walling material in a number of ways as indicated in figure 150. Traditionally, solid brick walls were considered sufficiently weather resistant in many areas of the UK, but under severe exposure conditions quite thick solid brick walls may well permit water to penetrate to the inner face and would certainly not meet requirements of today for thermal insulation unless special precautions are taken. The most common form today derives from principles originated over 50 years ago, and uses the two materials together to create a *brick/block cavity wall construction*. This consists of an outer or external leaf of 102.5 mm thick brickwork (although this may be thicker to suit loading conditions), a separating air cavity ranging from 50 mm to 100 mm width, and an inner or internal leaf of 100 mm thick concrete block (although this may be increased to suit loading conditions or thermal insulating requirements) to which a finish may or may not be applied.

This method of construction accepts that water may permeate through the relatively thin outer leaf of brickwork, but is then prevented by the air cavity from reaching the inner leaf of concrete blockwork which remains dry and, therefore, provides good thermal insulating properties – particularly when lightweight or aerated blocks are employed. These properties can be further increased by inserting a rigid thermal insulating material on the cavity face of the blockwork, but this must be done with great care to ensure that, whenever possible, a minimum cavity of 40 mm is maintained to provide the water barrier. Alternatively, a 50 mm cavity can be used and filled completely with either an insulating board (batt) or insulating foam. Complete cavity filling techniques are easier for the bricklayer to implement, but can create problems of water penetration under conditions of severe exposure owing to the possibility of mortar which has squeezed out of the joints in the bricks or blocks forming a link across the cavity. Figure 151 indicates the former construction method which maintains a cavity barrier.

The relatively thin cross-section walling materials must be selected to withstand the effects of movements, mostly caused by dramatic temperature changes arising from winter to summer (seasonal) and, to a lesser amount, from night to day (diurnal). It is not uncommon for the annual temperature range of an external wall surface to be in the order of 90°C, although this will occur quite slowly over the four seasons, thereby giving time for corresponding movements to happen gradually. The continually alternating wetting and drying created by weather processes can also cause problems of movements as well as frost and sulphate attack of the outer brick leaf. Open pored bricks have a high rate of water absorption and may allow water to pass through their body. Although dense bricks have a low absorption rate and transfer little water through their body, large quantities of water running down the 'glass-like' wall face may be drawn in through capillary paths formed by fine hair cracks between the mortar used for the joints and the brick. (See figure 40). However, as mentioned earlier, the design of the cavity wall is such that once water has reached the inner face of the outer leaf, it can run down to a point of collection and then be redirected to the outside through strategically placed *weepholes*. (See figure 36.) These can be provided by vertical joints (perpends) left open without mortar filling. The cavity must be kept clear of obstructions and damp proof courses must be inserted to stop and redirect water at points where the cavity is unavoidably bridged, eg around door and window openings, air bricks, etc.

Water which may come up from the ground or down from the exposed tops of a wall is also barred from those parts of the wall where it will cause damage or loss of environmental control by a damp proof course linked, when appropriate, to damp proof membranes as described for floor and roof construction.

215   solid brick wall
 13   plaster
'U' value = 2·1 W/m²°C

102·5 brick outer leaf
  50   cavity
102·5 brick inner leaf
  13   plaster
'U' value = 1·7 W/m²°C

102·5 brick outer leaf
  50   cavity
100   lw block inner leaf
  13   plaster
'U' value = 1·0 W/m²°C

These forms of construction no longer comply with The Building Regulations, 1976 and The Building (second amendment) Regulations, 1981 for domestic construction

---

  25 external skin
  50 insulation
215 solid brick
  13 plaster
'U' value = ·55 W/m²°C

  25 external skin
  50 insulation
200 solid lw block
  13 plaster
'U' value = ·4 W/m²°C

102·5 brick outer leaf
  50   cavity
150   lw block inner leaf
  13   plaster
'U' value = ·6 W/m²°C

102·5 brick outer leaf
  45   cavity
  25   insulation
100   lw block inner leaf
  13   plaster
'U' value = ·6 W/m²°C

  25 external skin
100   lw block outer leaf
  50   cavity
100   lw block inner leaf
  13   plaster
'U' value = ·57 W/m²°C

102·5 brick outer leaf
  50   insulation
100   lw block inner leaf
  13   plaster
'U' value = ·45 W/m²°C

  25 external skin
100   lw block outer leaf
  50   insulation
100   lw block inner leaf
  13   plaster
'U' value = ·35 W/m²°C

Figure 150   Use of bricks and blocks to form external walls

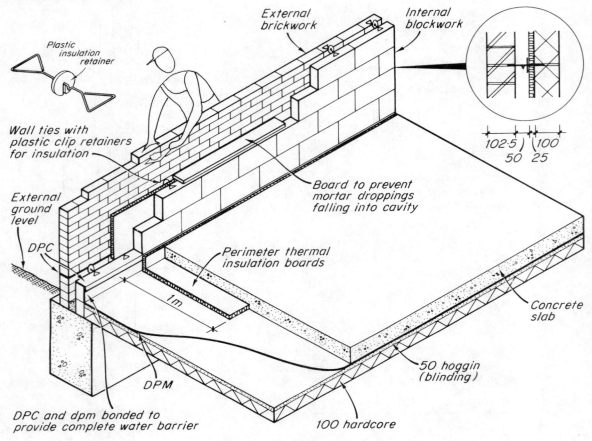

Figure 151   Construction of brick/block cavity wall

(d) Construction

The brickwork and blockwork are normally laid in stretcher bond, although half bats or 'snap headers' may be used to achieve the appearance of Flemish or other bonds. The brick and block leaves must be held together by *wall ties* placed at intervals which will ensure that the two leaves develop a mutual stiffness and act in unison in resisting the combined building loads. The design of wall ties also ensures that water entering the outer leaf is stopped before reaching the inner leaf by a *drip*, which occurs centrally over the cavity. The ties should be embedded within the horizontal mortar joints to a minimum penetration of 50 mm in order to provide adequate restraint to each leaf of the wall. They must be completely surrounded by

the mortar of the joint, and ideally laid to a slight fall from the inner leaf down to the outer leaf of wall construction. A range of overall lengths is available to suit different cavity widths, and the ties are produced in non-corrosive metals and plastics to varying shapes. It is essential that ties are kept clear of mortar droppings which could accumulate during the course of construction as these may negate the drip mechanism of the ties and thereby provide a bridge for water to reach the inner leaf. Any mortar droppings, therefore, must be removed daily as the work proceeds – temporary openings called 'coring holes' or 'clean outs' are provided at 1 m centres to allow their clearance from the bottom of the cavity.

Where the type of cavity wall is that indicated by figure 151, the construction method is briefly as follows:

- The dpc material is *bedded* on a layer of mortar on each leaf of the brick/block cavity foundation wall. It is important that the dpc is the same width of the walling below and the lengths are adequately lapped at the ends in order to prevent rising moisture travelling past to the walling above.
- A layer of mortar is placed on the dpc to provide a level bedding for the cavity wall construction. The load from each leaf is then evenly distributed to the foundation wall, the dpc is protected from damage which may be caused by the rough brick or block underface (bed face), and the bond between dpc mortar and walling material ensures minimal horizontal water penetration by capillary attraction.
- The first row of wall ties is also positioned at 600 mm centres within the layer of mortar placed on the dpc. Polythene retaining discs on the ties should be spaced at a distance from the blockwork inner leaf to suit the thickness of thermal insulation material.
- The blockwork inner leaf is built up to one course above the next row of wall ties which should not have retaining discs fitted at this stage.
- Excess mortar oozing from joints between newly placed blocks should be removed from surfaces, and all mortar droppings must be cleaned from cavity and wall ties.
- Rigid glass fibre insulating boards or batts are placed between polythene discs on the lower wall ties and against the blockwork inner leaf. Care should be taken to ensure edges of the insulating boards are butted closely together before the polythene retaining discs of the upper row of wall ties are fitted to hold the insulation firmly against the inner leaf.
- The brickwork outer leaf is built up to the top level of the insulation batt, ensuring that mortar does not drop into the remaining cavity. In practice this is difficult to achieve unless a special cavity batten is used which is suspended on cords and which collects the droppings as they fall. The batten is subsequently drawn up at each stage of wall construction to be cleaned.

- The blockwork inner leaf is built up and followed by the brickwork outer leaf and, as work progresses the wall ties and insulating batts are positioned, ensuring at all times that the cavity, the wall ties and joints between batts are kept clean of mortar.

The usual procedure is to build the corners or the extremities of the wall to a height of about eight or twelve brick courses (600 mm to 900 mm), care being taken to ensure all edges are vertically plumb. The base of the corner is then extended along the wall and raked back as the work is carried up. The intermediate portion of the wall is built between the two corners, the bricks and blocks being kept level and straight by building their upper edges to a string line stretched between the corners of the building. Notwithstanding this building sequence, the whole of the walling should be carried up simultaneously: no part should be built higher than 900 mm to avoid the risk of unequal settlement on the foundations before the mortar has sufficiently set. The outside face of the mortar joints can be finished in a number of ways as illustrated in figure 152.

At the end of a day's work, or during rain, exposed areas of insulation, including edges, must be protected with polythene sheets. As mentioned earlier, building work involving wet mixes of materials should not be carried out when the temperature is at, or likely to approach, 4°C unless special techniques are employed. Whenever necessary, newly completed work should be adequately protected overnight or during non-working days to avoid the detrimental effects of freezing.

Figure 153 illustrates an alternative construction method for an external wall using a timber frame technique incorporating thermal insulation, and clad both sides to provide a complete environmental enclosure. This construction technique is described fully in Mitchell's Building Series: *Structure and Fabric Part 1*, chapter 5.

(e) Doors and windows

Doors and windows can provide the weak link in those performance requirements for an external wall concerned with weather exclusion, sound control, thermal comfort, fire protection and security (see appropriate text in Part A). In addition, their location, size, shape, proportion

207

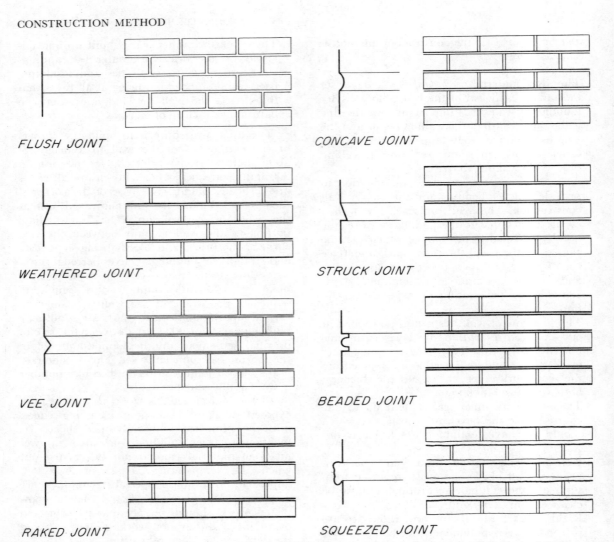

FLUSH JOINT

CONCAVE JOINT

WEATHERED JOINT

STRUCK JOINT

VEE JOINT

BEADED JOINT

RAKED JOINT

SQUEEZED JOINT

Figure 152 Types of mortar joints. Concave joint is better than flush joint when bricks are uneven; struck joint, beaded joint, and raked joints, should not be used in exposed condition because water will lay on exposed edge of bricks. The squeezed joint will also allow trapped water to penetrate brickwork. Ordinary mortar joints can be raked back and re-pointed in coloured mortar to any of the above profiles: coloured joints can have a dramatic effect on the overall appearance of the brickwork

and material profiles profoundly influence the overall and detail appearance as well as aesthetic aim of a building. The designer must carefully select the sizes of doors and windows relative to their location to ensure all these factors have equal value.

The frames for doors and windows can be 'built-in' openings left in the brick/block cavity wall construction, or can be positioned earlier so that the wall is constructed around them. The latter has many advantages when it comes to forming a weather proof joint between wall construction and frame. However, the contractor must ensure that the frames are available on site at an early stage of the contract, and that they are adequately protected from physical damage as well as the weather during storage and after their positioning. It is important that the designer has ensured that the horizontal lengths and vertical heights of wall construction

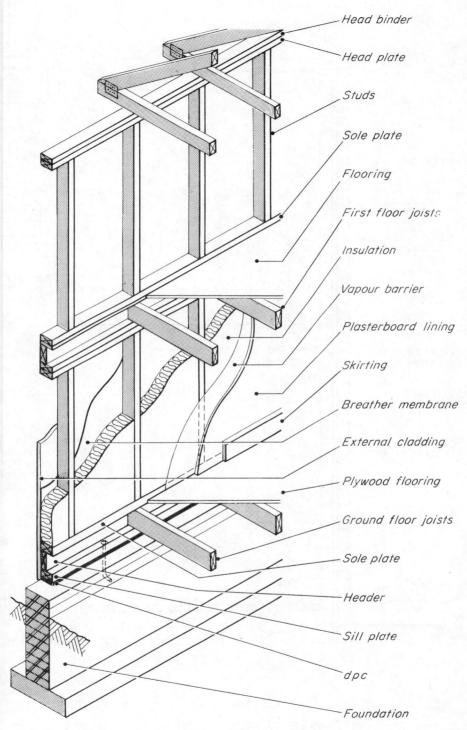

Head binder

Head plate

Studs

Sole plate

Flooring

First floor joists

Insulation

Vapour barrier

Plasterboard lining

Skirting

Breather membrane

External cladding

Plywood flooring

Ground floor joists

Sole plate

Header

Sill plate

dpc

Foundation

Figure 153   Typical timber wall construction using platform frame

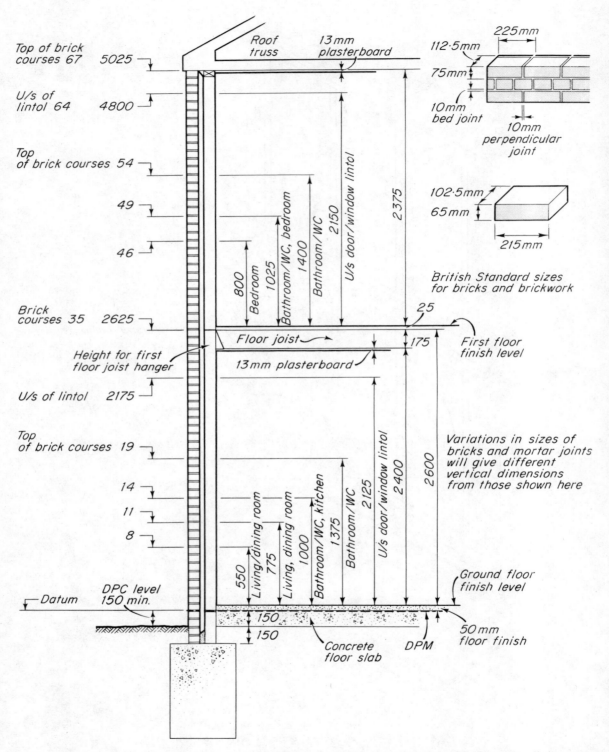

Figure 154 Co-ordination of vertical brick courses and positioning of openings as well as horizontal structure can eliminate waste of materials and speed construction sequences — see also figure 15 page 21 and figure 149 page 203

conform with the overall dimensions and the distances between individual door and window frames (figure 154). Materials used for door and window frames include timber (softwood and hardwood), metals and plastics. Timber window and door frames should be treated in the factory with a preservative solution and, ideally, a primer so that all hidden surfaces are protected after building-in. Both preservative and primer must be chemically compatible with subsequent paint systems. Alternatively, timber frames can be delivered to site having been treated with a decorative preservative stain which only requires re-coating after installation.

Before any wall construction is built around the frames they should be checked with a level to ensure perfect verticality and then they can be securely propped (figure 155). As the wall is constructed around the frames, they will become permanently fixed by building-in right angled galvanised mild steel lugs screwed to their side edges as the work proceeds (figure 156(a)). The cavity of the wall construction is closed by cutting the blockwork and making a 90° return to the inside face of the brick outer leaf. As this work is carried out, a vertical dpc is *bedded* between the two wall materials to form an effective water barrier. To ensure that the sides or jambs of the cavity wall construction are sufficiently robust to withstand the forces exerted by the door or window frames (wind pressures, slamming, etc) the cavity wall ties around the opening are positioned at increased vertical centres of 300 mm or every three horizontal brick courses.

Once the cavity wall construction has reached the top of the door or window frame, a lintol construction must be formed to carry the walling above the opening. Together with the number and proportion of openings, the method of forming a lintol has a critical influence on the appearance of the building. Furthermore, if not formed correctly, the lintol construction will also give rise to damp penetration problems as well as condensation due to occurrence of cold bridges. Lintols can be formed with reinforced

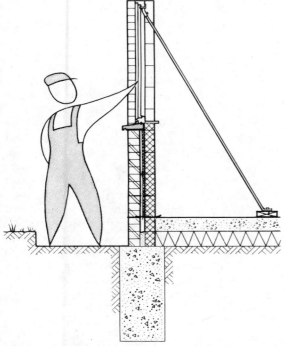

Figure 155   Timber window frame temporarily propped until lintol construction has been completed

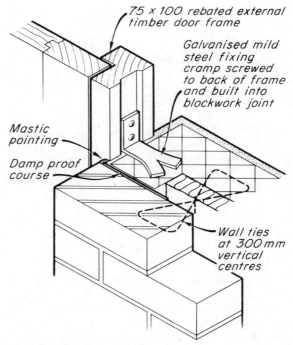

75 x 100 rebated external timber door frame

Galvanised mild steel fixing cramp screwed to back of frame and built into blockwork joint

Mastic pointing

Damp proof course

Wall ties at 300 mm vertical centres

Figure 156(a)   Fixing method for external door (or window) frame to brick/block cavity wall

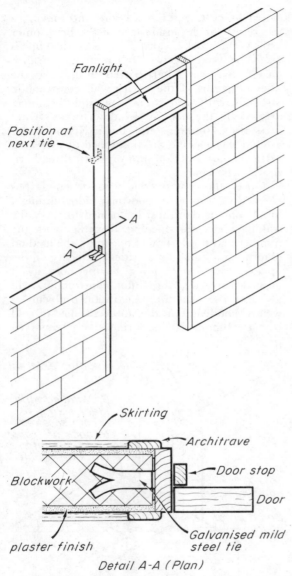

Figure 156(b) *Fixing method of door lining to internal blockwork wall*

concrete; with preformed steel section; or with a combination of these two which may or may not include a structural contribution of the walling material itself. When calculating the thermal insulation value given by a door or a window opening and its effect on the total value provided by the whole external wall enclosure, it is possible to make *either* the jamb and lintol construction have a U value corresponding to that of the surrounding wall, *or*, if it provides a lesser value, include it as part of the value applicable to the door or wall (see 8.3 *Thermal insulation values*).

Reinforced concrete lintols can be formed *in situ* when the openings to be spanned are sufficiently large to make the bulk of the lintol too heavy to lift into position. Otherwise, they are generally preformed or precast by setting up a precast unit mould into which the steel reinforcement is placed at the bottom on spacers to ensure adequate concrete cover. Concrete can then be poured into the mould. When a large quantity of similar shape and size lintols are required a *batch precast unit* mould can be employed. This and the other methods of forming reinforced concrete lintols are indicated in figure 157.

The various forms of lintol construction are shown in figure 158: they follow similar principles of environmental control, but each would affect the overall appearance of the building differently:

1 *Rectangular section precast or in situ reinforced concrete lintol*
   The full depth of the concrete lintol shows on the face of the building. Because of its fully rectangular section, a separate cavity tray must be provided in order that water percolating in the cavity wall is diverted through the weepholes to the outside. The effects of heat losses (cold bridge and condensation) should be lessened by providing insulation material of the internal faces of the lintol although this goes only a small way towards matching the total thermal insulation value of the wall supported and 'pattern staining' is likely to occur (see 8.3 *Thermal insulation values*). The provision of a pelmet board for curtain hanging helps to hide the resulting dark staining. Depth of lintol depends upon the span.

2A *'Boot' section precast or in situ reinforced concrete lintol*
   Similar to above, except lintol is shaped to reduce amount of exposed concrete on face of wall. The sloping shape in the cavity provide support for the cavity tray which

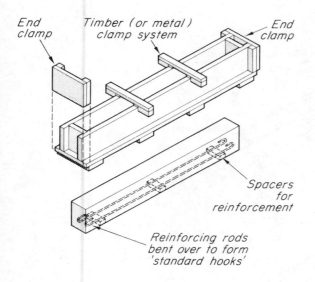

End clamp

Timber (or metal) clamp system

End clamp

Spacers for reinforcement

Reinforcing rods bent over to form 'standard hooks'

PRECAST UNIT

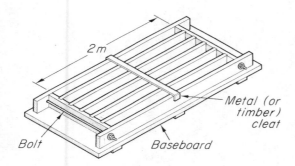

2m

Bolt

Metal (or timber) cleat

Baseboard

BATCH PRECAST UNITS

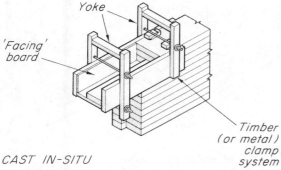

Yoke

'Facing' board

Timber (or metal) clamp system

CAST IN-SITU

Figure 157 Methods used for casting reinforced concrete lintols

would otherwise be liable to damage during construction – especially when cavity is being raked clean of mortar droppings.

2B *'Boot' section precast or in situ reinforced concrete lintol*

Similar to 2A except face depth of externally exposed concrete is further reduced. To add to the visual effect of this lintol, the front edge or toe of the lintol can be slightly recessed back from the brick face to create a shadow.

3A *Composite metal and precast reinforced concrete lintol.* (See also figure 39 page 46.)

This uses a precast concrete lintol to support the blockwork inner leaf and a preformed metal section to support the brickwork outer leaf. The metal lintol also acts as a cavity tray but it is recommended that it is heavily galvanized or painted with bituminous paint prior to placing in position in order to prevent corrosion. This composite lintol system maintains the cavity wall construction right down to the head of the door/window frame, and with the internally insulated concrete lintol goes some way in maintaining an accepted degree of thermal insulation at this point in the wall construction. Except for the narrow edge of the metal lintol, the supporting mechanism over openings is not visible and gives a different aesthetic than the exposed concrete lintols previously described.

3B and 3C *Composite metal and precast reinforced concrete lintol with exposed brick arch construction*

Both show a brick lintol construction although 3C would not actually provide structural support to the outer leaf unless reinforced internally. In these examples both brick arch forms are, in fact, supported by the preformed non-ferrous or galvanized metal section.

4A *Preformed metal lintol*

The full advantage of preformed lightweight metal sections are taken in this example and there is no heavy precast concrete employed. The lintols can be supplied with the cavity filled with insulating foam which give thermal insulation properties corresponding to that of the wall supported. Again, the lintol is virtually unseen after construction is complete.

213

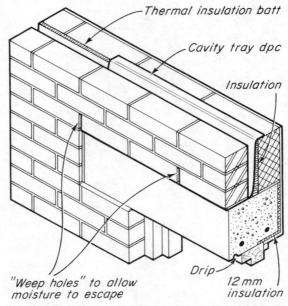

Figure 158(1)   *Rectangular section. Precast or* in situ *reinforced concrete lintol*

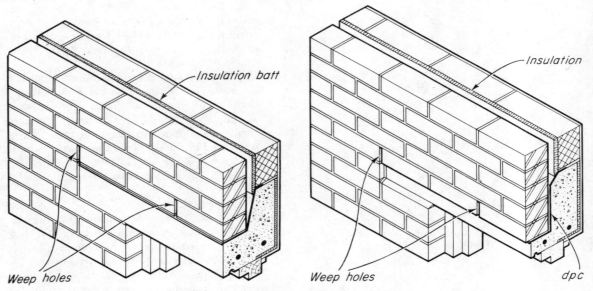

Figure 158(2A *and* B)   *Boot section. Precast or* in situ *reinforced concrete lintol*

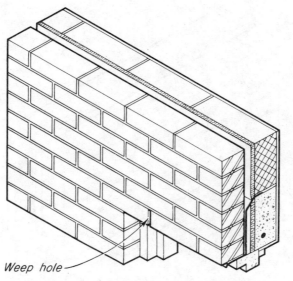

Weep hole

Figure 158(3A) Composite metal and precast reinforced concrete lintol

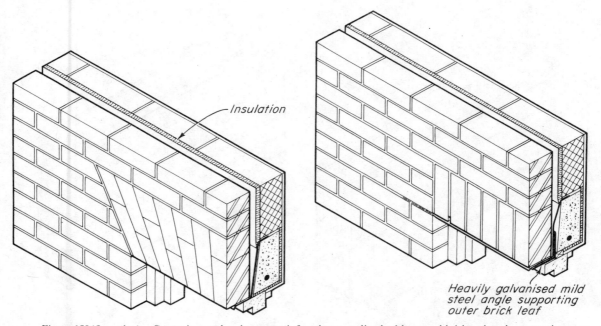

Insulation

Heavily galvanised mild
steel angle supporting
outer brick leaf

Figure 158(3B and C) Composite metal and precast reinforced concrete lintol with exposed brickwork arch construction

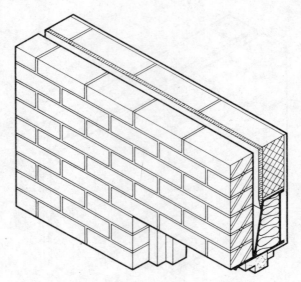

Figure 158(4A)   Preformed metal lintol

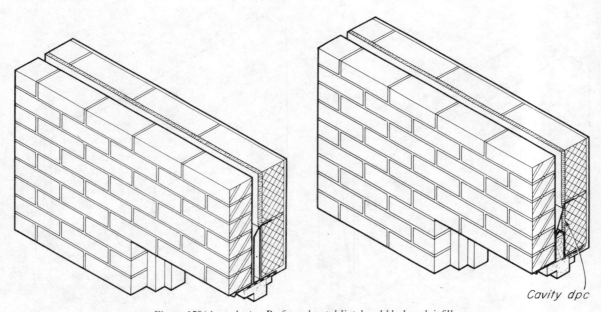

Cavity dpc

Figure 158(4B and C)   Preformed metal lintol and blockwork infill

4B and 4C *Preformed metal lintol with block infill*

Both these maintain the inner blockwork leaf as part of the lintol construction and although giving a slightly reduced insulation value than the wall construction they provide a consistency of internal surfaces for fixings and finishes.

These forms of lintol construction would be suitable for doors and windows: alternative forms of cill construction for each of the two components are indicated in figure 159.

(f) Scaffolding

As the work progresses beyond a height where it is unreasonable for the bricklayer to lift his materials from ground level, it will be necessary to erect scaffolding to support raised working platforms. The scaffolding generally consists of tubular steel or aluminium alloy connected by special fittings or couplings; the platform consists of scaffold boards of softwood. Figure 160 and some subsequent construction sequence figures indicate *putlog scaffolding* which consists of a single row of uprights set away from the wall for a sufficient distance to accommodate a working platform. The uprights (standards) are connected or coupled together with horizontal members (ledgers) which are tied back to the wall under construction by means of transoms and bridle fixings at openings. Scaffolding platforms are provided at the required height by coupling putlogs to the ledgers, and their flat-

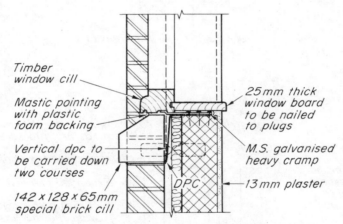

Figure 159A   *Combined brick and timber window cill*

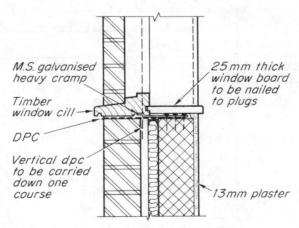

Figure 159B   *Timber window cill*

217

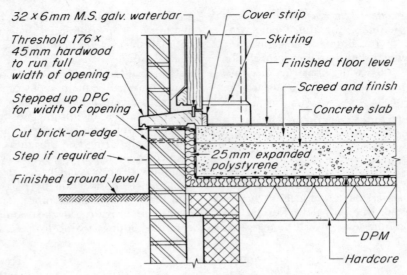

32 × 6 mm M.S. galv. waterbar — Cover strip

Threshold 176 × 45mm hardwood to run full width of opening — Skirting

Stepped up DPC for width of opening — Finished floor level

Cut brick-on-edge — Screed and finish

Step if required — Concrete slab

Finished ground level — 25 mm expanded polystyrene

DPM

Hardcore

Figure 159c   Timber door threshold 1

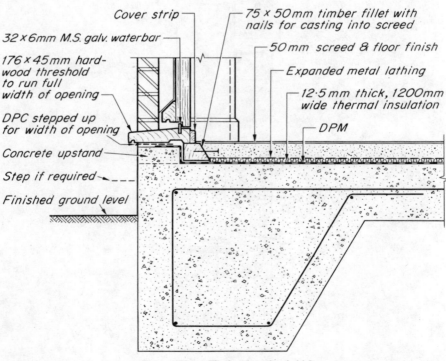

Cover strip — 75 × 50 mm timber fillet with nails for casting into screed

32 × 6 mm M.S. galv. waterbar — 50 mm screed & floor finish

176 × 45 mm hard-wood threshold to run full width of opening — Expanded metal lathing

DPC stepped up for width of opening — 12·5 mm thick, 1200mm wide thermal insulation

Concrete upstand — DPM

Step if required

Finished ground level

Figure 159d   Timber door threshold 2

218

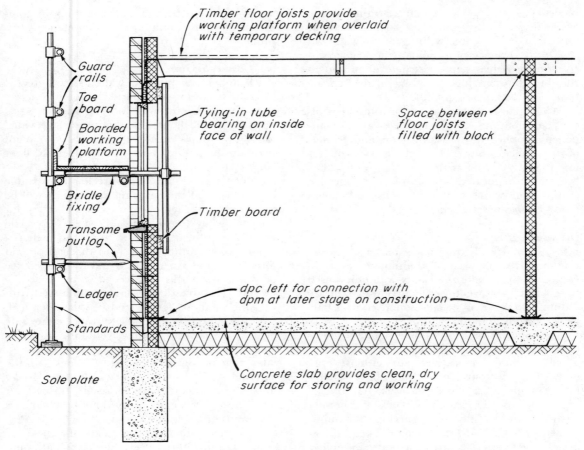

Figure 160   Construction sequence: first lift

tened ends are supported on the outer leaf of the cavity wall construction. After the scaffolding is no longer required, it will be dismantled and putlog holes filled or made good.

17.7 Internal wall construction
CI/SfB (22) + (32) + (42)

(a) Function
Internal walls, or partitions, provide the physical space separation within a building which is necessary to isolate certain activities in order to provide privacy and security. They may be required to provide fire protection between the spaces they enclose as well as between the spaces and general circulation routes. A sufficient amount of thermal insulation and sound control may be necessary to permit some check on overall performances. Internal walls must be robust enough to take various fixings for furniture and equipment, and be sufficiently durable to withstand the wear and tear associated with the activities of the building in which they are located. They should be of pleasing appearance, colour and texture which together should be compatible with the overall aesthetic character of the building of which they form a part.

Internal walls can be non-loadbearing or loadbearing. In the latter case they will assist in distributing the combined loads (dead, live and wind) of the building down to the foundation system and then to the supporting soil. Also, they may sometimes be required to act as

219

butresses and provide lateral restraint to the external walls. By acting as sides of cellular forms (in conjunction with external wall and rigidly fixed floor/ceiling joists), the whole constructional system acts in structural unison in resisting the combined building loads (see 5.3 *Continuous structures*, figure 20).

When planning the layout of internal walls consideration must be given to the various components applied to the walls, floors and ceilings of the spaces they define. It is particularly important that, as with external walls, door openings are located so as to avoid excessive cutting and wastage of the bricks and blocks (see 4.7 *Dimensional co-ordination*).

(b) Materials

Like external walls, internal walls may also be built from bricks or blocks, although rarely in combination or to form a cavity wall unless used to separate two dwellings (ie party walls), or are required to serve as specific sound control barriers, eg walls to music rooms.

The internal walls of a house separate rooms in which the activities are relatively quiet. Also, at most, they are required to support only domestic loads from floors or roof above. Therefore, a single 102.5 mm thick brick (half brick), or 75 mm to 100 mm thick block is generally adequate and satisfies slenderness ratio requirements (see 5.7 *Slenderness ratio*). These thicknesses and materials are adequate in giving support for internal doors and frames, as well as fixings for cupboards, sanitary appliances, kitchen units, etc. They will also provide an adequate domestic standard of sound control, thermal insulation, and fire resistance.

As an alternative to brick or blocks, internal partitions can be constructed from preserved timber using 50 mm × 100 mm vertical studs at 400 cc (to suit 1200 mm wide plasterboard cladding) braced at mid point by staggered timber noggins of the same cross sections (figure 161). The timber sections are planed prior to erection to ensure that the wall is of constant thickness with parallel faces to facilitate easy fixing of the plasterboard. Stud partitions are generally non-loadbearing and are easy to construct, lightweight, adaptable and can be clad as well as infilled with various materials to give different finishes and properties. Stud partitions

are of 'dry' construction and should not be commenced until the building has been made waterproof.

(c) Construction

Brick and block internal walls can be erected directly from the ground floor solid concrete slab; from a sleeper wall forming part of a suspended ground floor construction; or, provided loads are not excessive, from some point between the main supports of a suspended floor (figure 162).

When subsoil conditions are less than stable or the loads on the partition are high, it will be necessary to provide separate foundations for internal walls (see figures 142 and 143). The wall should incorporate a dpc, and when a separate foundation is not used this can be laid directly on the concrete slab and the first course of bricks or blocks bonded to it by a bed of mortar. This dpc should have been lapped with the dpc of the external cavity wall inner leaf, and must also extend a short distance either side of the internal wall so that eventually it can be stuck to the dpm laid on the concrete slab, at a later stage of construction. At junctions with the external wall, the partition will be *block bonded* into the inner leaf: that is to say it will be built into this leaf vertically every three brick courses (225 mm) or every block course.

First floor internal walls can also be constructed using bricks or blocks. Continuing upwards from the ground floor internal wall present no particular problems, although when floor joists penetrate the partition care must be taken to ensure the brick or blockwork between is continued *solidly* to the underside of the wall above. If there is no wall below, a first floor non-loadbearing wall can be carried on double floor joists which have been bolted together at about 600 mm centres to ensure they act in unison. However, as the partition load may cause supporting joists to deflect slightly and, therefore, damage subsequently applied finishes, it is probably best to insert a steel beam at this point (figure 163).

A non-loadbearing internal wall running *across* the timber joists can be carried by providing a timber sole plate at its base. Loadbearing partitions not continued from the floor below should always incorporate a Rolled Steel Beam

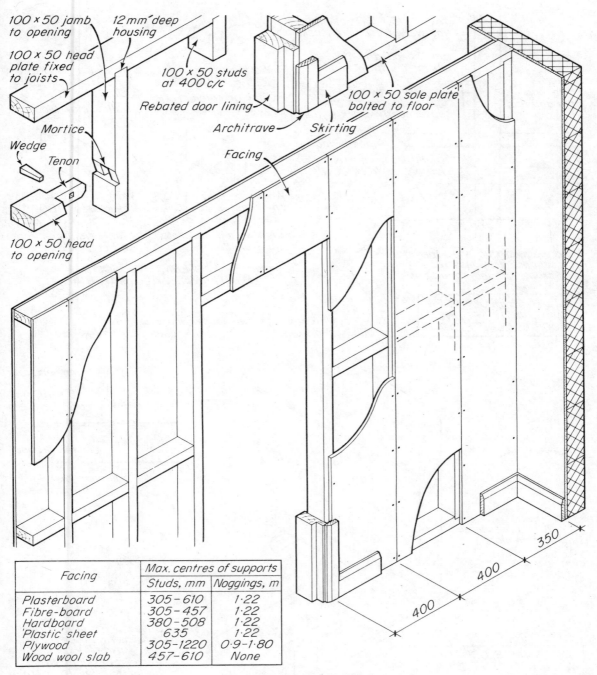

100 × 50 jamb to opening

12 mm deep housing

100 × 50 head plate fixed to joists

100 × 50 studs at 400 c/c

Mortice

Rebated door lining

100 × 50 sole plate bolted to floor

Wedge

Tenon

Architrave

Skirting

Facing

100 × 50 head to opening

Facing	Max. centres of supports	
	Studs, mm	Noggings, m
Plasterboard	305 – 610	1·22
Fibre-board	305 – 457	1·22
Hardboard	380 – 508	1·22
'Plastic' sheet	635	1·22
Plywood	305 – 1220	0·9 – 1·80
Wood wool slab	457 – 610	None

350

400

400

400

Figure 161   Stud partition details

221

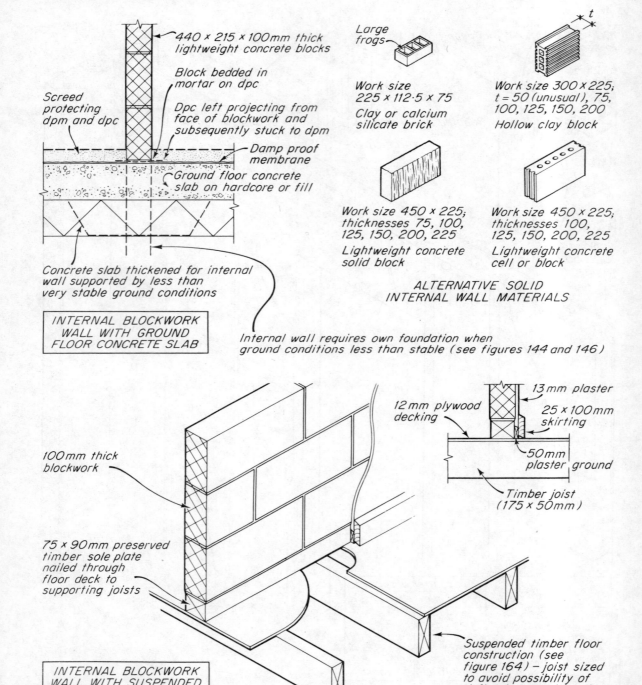

440 × 215 × 100mm thick lightweight concrete blocks

Block bedded in mortar on dpc

Dpc left projecting from face of blockwork and subsequently stuck to dpm

Screed protecting dpm and dpc

Damp proof membrane

Ground floor concrete slab on hardcore or fill

Concrete slab thickened for internal wall supported by less than very stable ground conditions

INTERNAL BLOCKWORK WALL WITH GROUND FLOOR CONCRETE SLAB

Large frogs

Work size 225 × 112·5 × 75
Clay or calcium silicate brick

Work size 300 × 225;
t = 50 (unusual), 75, 100, 125, 150, 200
Hollow clay block

Work size 450 × 225;
thicknesses 75, 100, 125, 150, 200, 225
Lightweight concrete solid block

Work size 450 × 225;
thicknesses 100, 125, 150, 200, 225
Lightweight concrete cell or block

ALTERNATIVE SOLID INTERNAL WALL MATERIALS

Internal wall requires own foundation when ground conditions less than stable (see figures 144 and 146)

100mm thick blockwork

13 mm plaster

25 × 100mm skirting

12 mm plywood decking

50mm plaster ground

Timber joist (175 × 50mm)

75 × 90mm preserved timber sole plate nailed through floor deck to supporting joists

Suspended timber floor construction (see figure 164) – joist sized to avoid possibility of deflection resulting from weight of internal wall

INTERNAL BLOCKWORK WALL WITH SUSPENDED GROUND FLOOR

Figure 162   Methods of supporting internal walls

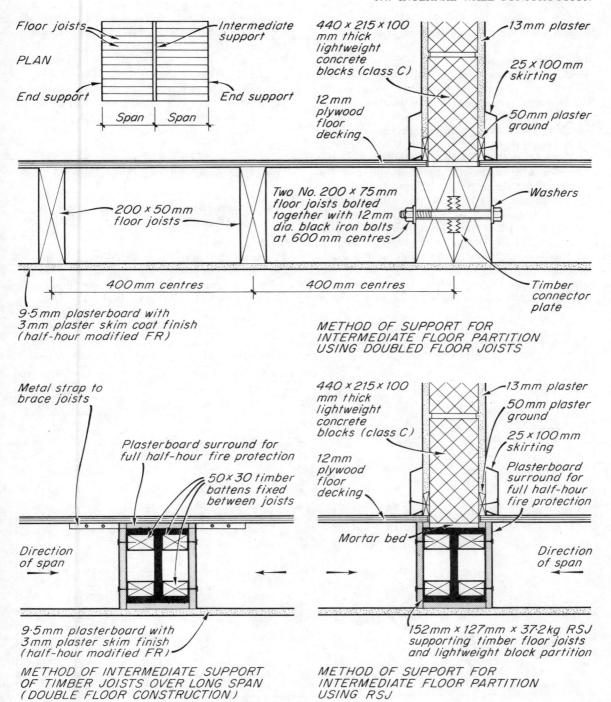

Figure 163   Methods of support for first floor partitions

[RSJ] or a Universal Beam [UB] support in the intermediate timber floor construction.

(d) Door openings

Doors may be incorporated in the internal wall by building-in standard *door linings* using galvanized metal ties similar to those used for the external door and window frames. (See figure 156(b).) Often the space over the door is filled with a window or *fanlight* rather than solid brick or blockwork. The use of a fanlight avoids the need for a lintol in this position and also for cutting the brick or blocks to bond between the walls either side of the opening. Although the use of fanlights may be economical, care must be taken to ensure that the interrupted lengths of internal walls are adequately stabilized at door openings. When an internal wall is load-bearing, the floor joists over will provide this stability – particularly when they are securely anchored to the top of the wall. If the partition is non-loadbearing and does not support floor or ceiling joists it is necessary to firmly connect the top of the wall to the underside of joists with timber wedges. Often the top of the brick or block partition is capped by a 50 mm deep timber wall plate which facilitate this process by providing a level top surface. When a loadbearing partition is not required to continue up to the next floor, this timber plate also provides a level fixing for the supported joists. (See figure 160.)

Brief comments regarding the fixing of door linings are given under 17.11(b) *Joinery*.

17.8 Intermediate floor construction
CI/SfB (23) + (33) + (43)

(a) Function

The construction of the external cavity wall and the internal walls of the house continue until the correct height is reached for the first floor timber joists to be put in position. The function

Size of joist in millimetres graded GS or MGS	Dead load in kilogrammes per square metre supported by joist, excluding the mass of the joist								
	Not more than 25 kg/m²			More than 25 but not more than 50			More than 50 but not more than 125		
	Spacing of joists in millimetres								
	400	450	600	400	450	600	400	450	600
	Maximum span of joist in metres								
38 × 75	1.05	0.95	0.72	0.99	0.90	0.69	0.87	0.79	0.62
38 × 100	1.77	1.60	1.23	1.63	1.48	1.16	1.36	1.24	1.00
38 × 125	2.53	2.35	1.84	2.33	2.12	1.69	1.88	1.73	1.40
38 × 150	3.02	2.85	2.48	2.83	2.67	2.26	2.41	2.23	1.83
38 × 175	3.51	3.32	2.89	3.29	3.11	2.71	2.82	2.66	2.27
38 × 200	4.00	3.78	3.30	3.75	3.55	3.09	3.21	3.03	2.64
38 × 225	4.49	4.24	3.70	4.21	3.98	3.47	3.61	3.41	2.96
44 × 75	1.20	1.08	0.83	1.13	1.02	0.79	0.98	0.89	0.70
44 × 100	2.01	1.82	1.41	1.83	1.67	1.31	1.51	1.39	1.12
44 × 125	2.71	2.56	2.09	2.54	2.38	1.90	2.08	1.92	1.56
44 × 150	3.24	3.06	2.67	3.04	2.87	2.50	2.60	2.45	2.03
44 × 175	3.77	3.56	3.10	3.53	3.34	2.91	3.02	2.86	2.48
44 × 200	4.29	4.06	3.54	4.02	3.80	3.31	3.45	3.26	2.83
44 × 225	4.81	4.55	3.97	4.51	4.27	3.72	3.87	3.66	3.18
50 × 75	1.35	1.22	0.93	1.26	1.14	0.89	1.08	0.99	0.78
50 × 100	2.22	2.03	1.58	2.03	1.85	1.46	1.66	1.53	1.23
50 × 125	2.84	2.72	2.33	2.70	2.55	2.10	2.27	2.09	1.71
50 × 150	3.40	3.26	2.84	3.23	3.05	2.66	2.76	2.61	2.21
50 × 175	3.95	3.78	3.30	3.75	3.55	3.09	3.22	3.04	2.64
50 × 200	4.51	4.31	3.76	4.27	4.04	3.52	3.67	3.46	3.01
50 × 225	5.06	4.83	4.22	4.79	4.53	3.95	4.11	3.89	3.39
63 × 150	3.66	3.52	3.17	3.50	3.38	2.97	3.09	2.92	2.54
63 × 175	4.25	4.10	3.68	4.07	3.93	3.45	3.59	3.40	2.96
63 × 200	4.84	4.67	4.20	4.64	4.48	3.93	4.09	3.87	3.37
63 × 225	5.43	5.24	4.70	5.21	5.02	4.41	4.59	4.34	3.78
75 × 200	5.10	4.93	4.51	4.90	4.72	4.27	4.43	4.20	3.67
75 × 225	5.72	5.52	5.06	5.49	5.30	4.79	4.97	4.71	4.11

Figure 164   Span/depth for intermediate timber floor joists. See figure 145 page 199 for dead loads of finishes (Swedish Timber Council)

of the *intermediate floor*, of which these joists form a part, is to provide a level surface with sufficient strength to support the dead loads of flooring and ceiling plus imposed loads of people and furniture. In addition, it must provide a degree of sound control and fire protection between the two levels. Often an intermediate floor must also assist in providing lateral restraint to heights of external and internal walls.

(b) Materials

The materials used for the structure of domestic intermediate floors are reinforced concrete, or pressed metal joists, or timber joists. Although reinforced concrete construction (*in situ* or precast units) have many advantages associated with sound control and fire resistance, the materials are heavy and can involve complicated methods of assembly. Whereas pressed steel joists do not have these advantages but are very lightweight, timber joists construction remains most adaptable.

In working out the layout of timber joists consideration must be given to their spacing, the clear spans and openings to be formed or 'trimmed' for the staircase, etc. Figure 164 indicates typical spacings and spans. Variations in spans can be achieved for a given cross sectional size of timber according to the actual stresses it can withstand. This capacity depends upon characteristics of the timber (structure, grain, knots, etc) and it is now possible to obtain joists *stress graded* to provide known limits of loading which make the most economical use of timber.

The sizing of joists can be arrived at by calculation, or by reference to Tables contained within the clauses of the Building Regulations or supplied by timber-interested organizations such as the Timber Research and Development Association (TRADA), the Swedish Timber Council, etc. An approximate, although uneconomical, guide can be given by the formula $D = (\text{span in mm}/24) + 50$, where $D = $ depth of joist in mm. This formula assumes that joists have a breadth of 50 mm and are used at 400 mm centre spacing.

As mentioned under suspended ground floor construction, timber joists should be treated with a preservative against fungal and insect attack prior to final placing in the building. Joist lengths can be treated before delivery to site, but on site 'touching up' will always be necessary as a result of cutting. Particular care needs to be taken with the ends of joists because these are most susceptible and are placed in vulnerable positions, such as adjoining the external wall construction.

The decking for timber intermediate floors can be tongued and grooved hardwood or softwood boarding, or chipboard or plywood sheets. These are indicated in figure 145. Tongued and grooved profiles are preferred rather than butt jointed boards because they reduce the effects of curling caused by thermal/moisture movement. In addition, the reduction of the amount of clear gaps is essential in obtaining a satisfactory resistance to the travel of flames and smoke in the event of the outbreak of a fire. A timber tongued and grooved board decking to floor joists acts in conjunction with a 9.5 mm plasterboard finish to form the ceiling to give the '*modified' half hour fire resistance* required by the Building Regulations. This means that for at least half an hour the floor must not collapse or allow the passage of flame or smoke, but the insulation requirement is reduced to 15 minutes (see 9.2 *Fire resistance*). Brief comments regarding the fixing of floor decking are given under 17.11(b) *Joinery*.

(c) Construction

For the cavity wall construction already described, the level of the top of 175 mm × 50 mm joists should coincide with the level of 35 courses of brickwork (each brick 65 mm + 10 mm for bed joint) from the top of the dpc as shown in figure 154. This will provide a joist top level of 2625 mm and an underside joist level of 2450 mm: an allowance of 50 mm for screed will permit a 2400 mm clear storey height (excluding plasterboard ceiling), which is ideal for the storey height door frames as mentioned earlier (figure 153). Traditionally the ends of the joists are built into the inner blockwork leaf of the cavity wall as this prevents the end of the joists twisting as well as providing lateral restraint to the external wall by preloading (see 5.3 *Continuous structures*). However, with the tendency of today to use an increased amount of sap wood for structural timber, and also to avoid the risk of cold bridge or condensation which might occur if the inner leaf is inter-

225

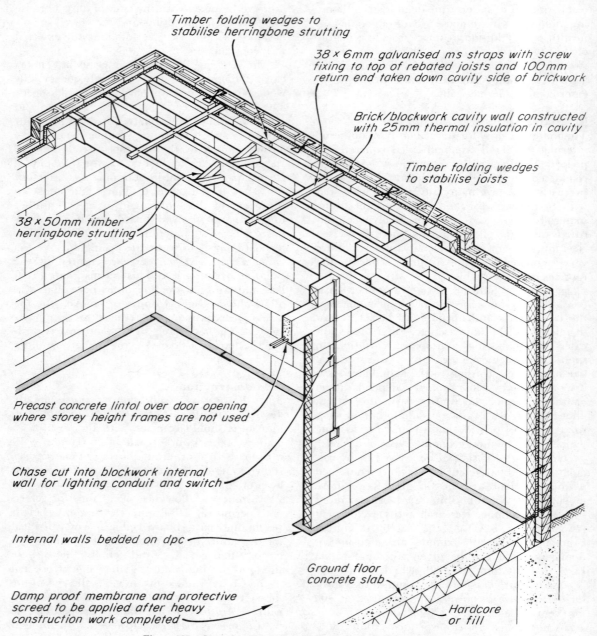

Timber folding wedges to
stabilise herringbone strutting

38 × 6mm galvanised ms straps with screw
fixing to top of rebated joists and 100mm
return end taken down cavity side of brickwork

Brick/blockwork cavity wall constructed
with 25mm thermal insulation in cavity

Timber folding wedges
to stabilise joists

38 × 50mm timber
herringbone strutting

Precast concrete lintol over door opening
where storey height frames are not used

Chase cut into blockwork internal
wall for lighting conduit and switch

Internal walls bedded on dpc

Ground floor
concrete slab

Hardcore
or fill

Damp proof membrane and protective
screed to be applied after heavy
construction work completed

Figure 165 Lateral bracing of walls by intermediate floor construction

rupted, it is probably wiser to use a *metal hanger* to support the joists. These are made of galvanized mild steel and can be supplied with a 'hooked' tail which provides the necessary lateral restraint to the blockwork leaf of the cavity wall. The joists will be prevented from twisting by the shape of the hanger (figure 165). In order to avoid cutting the blocks of the inner leaf so that the tail of the joist hangers align perfectly with the top of the joists, it will be necessary to provide one course of 140 mm high blocks (150 mm including joint) instead of the normal 215 mm high blocks (255 mm including joint). At the other end to the external wall, timber joists can be lapped over partitions where the bedded *wall plate* mentioned earlier assists in spreading the loads taken by the joists, and provides fixing for them by cross or 'skew' nailing. Another method for accurate levelling of joists over partitions is to use a bearer bar cogged to the joists.

Having fixed the structure of the floor, a temporary decking of plywood can be laid over the joists in order to provide a working platform for subsequent building processes. The final finished decking should not be fixed until the building is weather proof and all necessary work in the void of the floor has been complete. Similarly, a ceiling of plasterboard should not be fixed until electric conduits, etc, have been installed and there is little risk of subsequent damage occurring. The plasterboard for the ceiling is usually fixed at the same time as any other plasterboard, eg for stud partitions, ductwork, etc (see 17.11 *Finishes*).

(d) Lateral bracing of joists

When the timber joists span more than 2.4 m it is necessary to provide lateral bracing or strutting in order to restrict the movements due to twisting, and vibrations which would damage ceiling finishes (figure 165). The strutting can consist of 38 mm × 50 mm softwood battens or galvanized steel sections fixed diagonally between joists in a herringbone pattern. They should occur at mid span or at not less centres than 50 times the width of one joist. Alternatively, blocks of timber can be fixed between the joists at alternative centres for each space – these are easier to fix but provide less restraint. It is important that the restraint provided to the

joists by strutting is continued through to the flanking walls. This is done by ensuring the first joist next to the wall is positioned so as to allow a 40 mm gap into which timber *folding wedges* are driven on the line of the strutting.

(e) Lateral bracing of walls

Hooked tail joist hangers provide restraint to walls once in position and the joists have been securely nailed in position. Lateral bracing may also be required for walls which run parallel with the direction of span of the timber joists – particularly when the flanking wall is an external cavity wall. Bracing in this direction can be provided by galvanized mild steel straps fixed at not more than 2 m centres, each strap being screwed to at least two joists and has a 90° return end which is built into the cavity wall construction (figure 165).

(f) Notches and holes in joists

As mentioned earlier, structural timbers used today can be stress-graded to give maximum economical depth/span ratios. It is, therefore, necessary that certain rules are observed should it be necessary to notch or drill holes in timber joists so that services (electric cables, central heating pipes, etc) can be accommodated within the floor space. For example, it is recommended that *notches* should occur within the middle half of the span of any joist, should be in the upper cross section and never be more than one-eighth the depth of joist. Holes should be situated within the middle two thirds of the span of joist and their diameter should not exceed one sixth of the depth of the joist. In practice, it may be very difficult to control tradesmen on site from placing notches or holes anywhere in the joists but in the most suitable position for their particular services. To overcome this problem, a prudent designer will oversize the joist sizes when services are required to be concealed in an intermediate floor space.

(g) Openings in floor

Openings must be left in the intermediate floor so that staircases can be fitted and, perhaps, to allow service pipes which have been grouped together or central heating air ducts to pass from one floor to another. This means that the timber joists which would normally span from one wall to another would be stopped short and

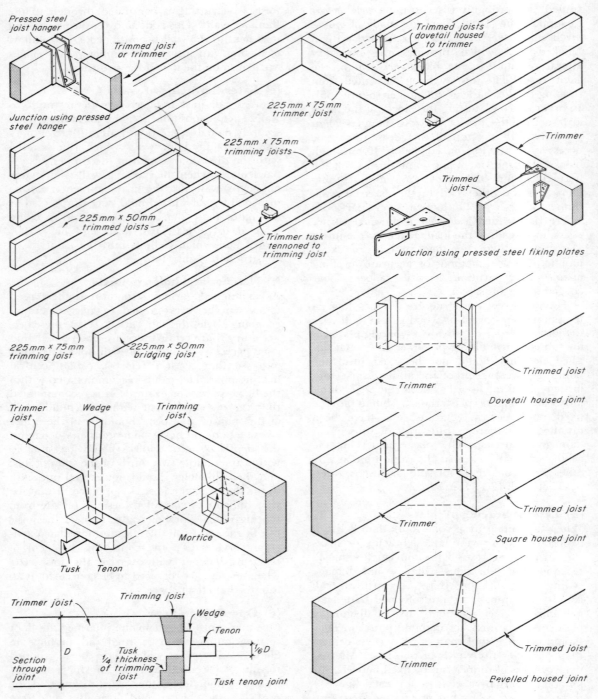

Pressed steel joist hanger

Trimmed joist or trimmer

Junction using pressed steel hanger

Trimmed joists dovetail housed to trimmer

225 mm × 75 mm trimmer joist

225 mm × 75 mm trimming joists

Trimmer

Trimmed joist

225 mm × 50 mm trimmed joists

Trimmer tusk tennoned to trimming joist

Junction using pressed steel fixing plates

225 mm × 75 mm trimming joist

225 mm × 50 mm bridging joist

Trimmer

Trimmed joist

Dovetail housed joint

Trimmer joist

Wedge

Trimming joist

Mortice

Tusk   Tenon

Trimmer

Trimmed joist

Square housed joint

Trimmer joist

Trimming joist

Wedge

Tenon

Section through joint

D

Tusk ¼ thickness of trimming joist

⅙D

Tusk tenon joint

Trimmer

Trimmed joist

Bevelled housed joint

Figure 166   Trimming to timber floors

228

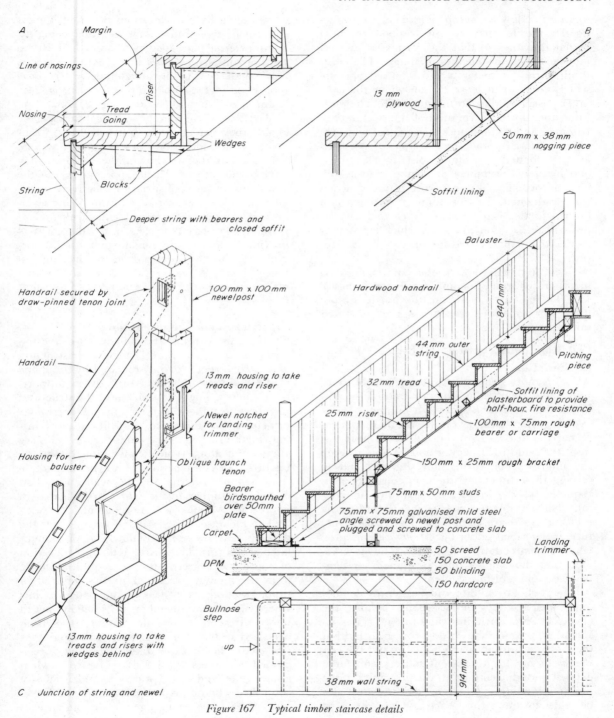

A

Margin

Line of nosings

Riser

Tread
Going

Nosing

String

Wedges

Blocks

Deeper string with bearers and
closed soffit

B

13 mm
plywood

50 mm x 38 mm
nogging piece

Soffit lining

Handrail secured by
draw-pinned tenon joint

100 mm x 100 mm
newelpost

13 mm housing to take
treads and riser

Handrail

Newel notched
for landing
trimmer

Oblique haunch
tenon

Housing for
baluster

Bearer
birdsmouthed
over 50 mm
plate

Carpet

Baluster

Hardwood handrail

840 mm

44 mm outer
string

Pitching
piece

32 mm tread

Soffit lining of
plasterboard to provide
half-hour fire resistance

25 mm riser

100 mm x 75 mm rough
bearer or carriage

150 mm x 25 mm rough bracket

75 mm x 50 mm studs

75 mm x 75 mm galvanised mild steel
angle screwed to newel post and
plugged and screwed to concrete slab

50 screed
150 concrete slab
50 blinding
150 hardcore

DPM

Bullnose
step

Landing
trimmer

up

914 mm

13 mm housing to take
treads and risers with
wedges behind

38 mm wall string

C   Junction of string and newel

Figure 167   Typical timber staircase details

229

trimmed to allow for an opening. It is good practice for the designer to provide floor trimming layout drawings so that the carpenters can set out the joists for maximum usage. The joists forming the perimeter of the required opening will have to be at least 25 mm wider than the other joists as indicated in figure 166. The joists interrupted by the formation of the opening are trimmed and secured to a *trimmer joist* which in turn is carried by *trimming joists*. Methods adopted for fixing trimmed joists to a trimmer, and for fixing trimmers to trimming joists will vary according to the expertise of the contractor's tradesmen. Timber joints (tusk tenon and housed joints) can be skilfully carried out, or pressed steel section employed which are similar to the joist hangers used for the main joist supports.

Typical details of a timber staircase to be incorporated in an opening are shown in figure 167. These can be made on site, but it is more usual for prefabricated staircases to be fitted into a prepared opening.

17.9 Roof construction   CI/SfB (27) + (47)

(a) Function
A roof is more vulnerable to the effects of wind, rain, snow, solar radiation and atmospheric pollution than any other part of a building. While avoiding the harmful results of these phenomena, it must also prevent excessive heat loss from the building in winter and be able to keep the interior cool during hot seasons. And in serving these functions, the roof construction is required to defy gravity by spanning horizontally between loadbearing elements (unless the roof and walls are combined in a tensile structure as described under 5.6 *Membrane structures*), while accommodating all stresses encountered, including movements due to changes in temperature and moisture content.

The roof of a building must also provide adequate defence for the occupants against the effects by fire, including limiting its potential spreading powers. This is done by ensuring that the *external* cladding is able to withstand the *penetration* of fire from adjacent sources and for a defined period according to the proximity of the roof to a boundary. The roof cladding must also be able to limit the *spread of flame* over its surface.

Although hybrid variations occur, it is most convenient to consider roofs as either being *flat* if the external face is at not more than 10° to the horizontal, or *pitched* if the external face is more than 10° to the horizontal. A *short span* roof is considered to be up to 7.6 m, *medium span* 7.6 m to 24.4 m, and *long span* over 24.4 m.

Apart from the purely technical factors, the choice between flat or pitched roof forms involves consideration of aesthetic appeal. When a flat roof is chosen, the building may have a 'cut off', or 'block' appearance, but the semblance of the roof surface is relatively unimportant unless it is overlooked from higher levels or used as a terrace or for vehicular traffic. Conversely, the selection of a pitched roof will provide a dominant visual feature, the importance of which will depend on the viewing distance. For this reason the shape, colour and texture of the roof surface are very important factors.

(b) Pitched roof construction
Of the many considerations involved in the selection of an appropriate form, the construction of a pitched roof is mainly dictated by the shape of the building and the span between supporting elements. The arrangement of the structural members forming the roof may follow similar principles, but their size, disposition and jointing methods will vary considerably. Figure 168(a) indicates a typical conventional domestic pitched roof construction. However, for simple spans without internal support walls, a method employing timber *trussed rafters* as indicated in figure 168(b) could be more economical.

The modern trussed rafter roof indicates the development in empirical knowledge of triangulated roof forms. They involve the incorporation of more recent scientific techniques resulting from the analysis of actual stresses induced in timber sections as well as at the critical points where two or more members are joined to act in unison. Instead of forming joints on site using oversized sections for structural safety and convenience, the trussed rafter employs stress-graded timbers (17.8(b)) joined when appropriate and with precision under factory controlled conditions by galvanised mild steel *truss plates*. The resulting prefabricated, economical and lightweight truss rafters are delivered to the site at an appropriate time in the building programme. They are then hoisted into position, placed on the supporting wall and relatively rapidly fixed. However, it is particularly important that the light trusses are adequately protected from the elements during the period from when they leave the factory until they are finally covered by the external roof cladding. Should these trusses be allowed to become exposed to the weather for any length of time, the relatively thin sections of timber will become saturated and cause them to twist and loosen the truss plates. Twisting is still likely to occur after the roof construction is complete, particularly if the central heating of the building causes rapid drying.

(c) Pitched roof bracing

Rafters are spaced at close centres, usually 600 mm, to provide direct support for roof cladding and ceiling, which in turn helps in giving the necessary *lateral bracing* to the individual trussed rafter. Each rafter is securely nailed to the wall plate already placed on a mortar bed on top of the cavity closure block as indicated in figure 171. Because of the light weight of the trussed rafters and the tendency for wind to lift the whole roof construction, it is important that this wall plate is anchored to the supporting wall by galvanized mild steel straps at about 1.5 m centres. When a flank wall of the house continues up above the lowest part of the rafter to form a gable wall, it is important that lateral bracing is provided for this otherwise free-standing wall by similar steel straps as indicated in figure 169. Finally, the whole roof system should be provided with wind braces running diagonally from a bottom corner to a top corner. This brace will prevent the roof timber from collapsing sideways as a result of laterally applied wind forces. Other timber braces are generally required to provide stability when trussed rafters are used in order to prevent twisting and deflection from vertical alignment.

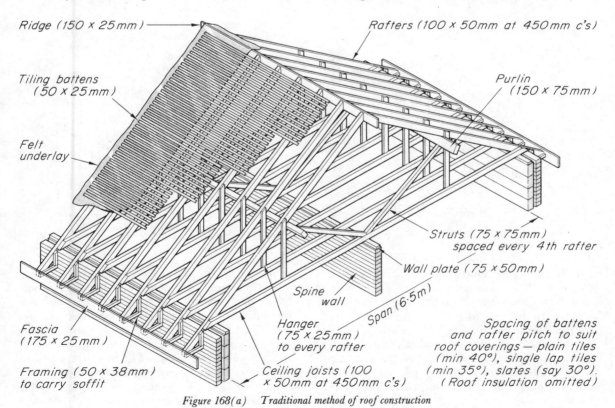

Figure 168(a)   Traditional method of roof construction

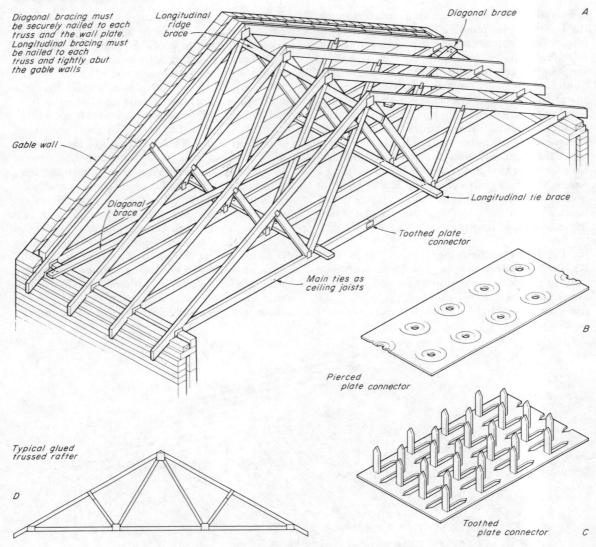

Diagonal bracing must be securely nailed to each truss and the wall plate. Longitudinal bracing must be nailed to each truss and tightly abut the gable walls

Longitudinal ridge brace

Diagonal brace

A

Gable wall

Diagonal brace

Longitudinal tie brace

Toothed plate connector

Main ties as ceiling joists

Pierced plate connector

B

Typical glued trussed rafter

D

Toothed plate connector

C

Figure 168(b)   Trussed rafter roof construction

(d) Pitched roof cladding

On a pitched roof, gravitational forces move rainwater down the slope and, aggravated by wind currents, try to move it inwards (figure 37). Because the roof has a relatively rapid degree of run-off, the external covering usually consists of relatively small *lapped* units such as tiles or slates, which are dry jointed and supported independently at intervals by timber battens. As the slope of the roof decreases, the resistance to the inward flow of water becomes less. Therefore, the amount of water penetration likely depends on the angle of slope and the amount of horizontal overlap provided by the slates or tiles. The movement of water around the tiles or slates depends on the *angle of creep*: the bigger the overlap (which corresponds to the size of cladding unit), the less the pitch can be (figure 170). However, slates and tiles should only be regarded as a *water check*, and not as a

232

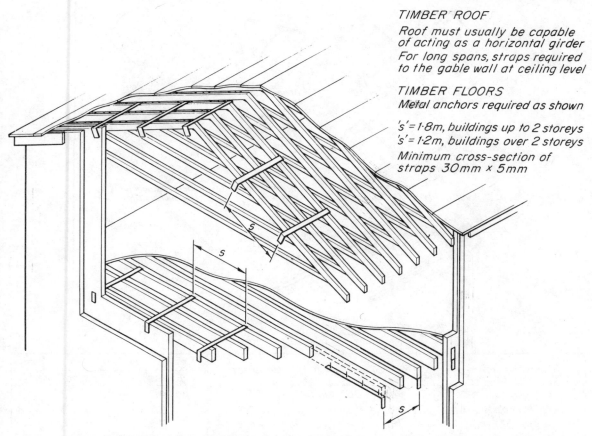

TIMBER ROOF
*Roof must usually be capable
of acting as a horizontal girder
For long spans, straps required
to the gable wall at ceiling level*

TIMBER FLOORS
Metal anchors required as shown

's'= 1·8m, buildings up to 2 storeys
's'= 1·2m, buildings over 2 storeys
*Minimum cross-section of
straps 30mm × 5mm*

Figure 169   Means of providing laterial restraint to external brick/block cavity walls

water barrier which is usually formed by the independent waterproof membrane known as *sarking* and indicated in figure 37.

Working from the scaffolding platform at roof level, the sarking felt is unrolled and laid across the rafters starting at the lowest point (eaves). The sarking is placed over a tilting fillet and lapped over the fasçia board to form a small projection or drip. It is important that the sarking is allowed to sag slightly between rafters in order to form the space which permits the water permeating around the slates or tiles to be caught by the sarking and transferred by gravity to the eaves drip and rainwater gutter. The sarking is nailed into position before the next horizontal length is placed so as to permit an overlap to be formed of at least 150 mm. All sides of the roof are continued upwards until the sarking is taken over the top of the roof or *ridge*. Vertically sloping edges of the roof or *hips* are overlayed with strips of sarking to ensure a complete enclosure. Slate or tile battens are then fixed at horizontal centres to suit the type and size of cladding unit.

When cladding to the pitched roof is to be in slates, each horizontal row is laid starting again from the eaves. Each slate is butt jointed at the side and overlapped at the head so as to form *two thicknesses* of slate over each nail hole as protection making, in all, three thicknesses of material at the overlaps. The side butt joint should be left slightly open so that water will drain quickly. Special length slates are required at eaves and sloping edges or *verges*. Each slate is fixed twice by yellow metal, aluminium alloy, copper or zinc nails. The slate should be holed

233

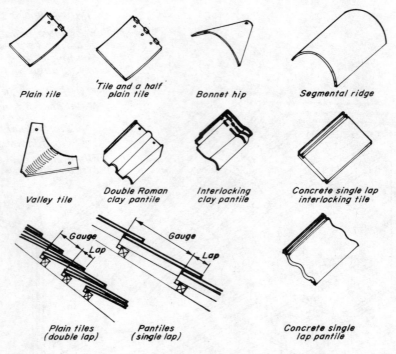

Plain tile   'Tile and a half plain tile   Bonnet hip   Segmental ridge

Valley tile   Double Roman clay pantile   Interlocking clay pantile   Concrete single lap interlocking tile

Gauge   Lap   Gauge   Lap

Plain tiles (double lap)   Pantiles (single lap)   Concrete single lap pantile

Description	Size	Head lap	Pitch	
			Exposed	Sheltered
Plain tile	165 wide × 265 long	65	40	35
Double Roman pantile	406 wide × 419 long	76	35	30
Single pantile	430 wide × 380 long	76	30	$17\frac{1}{2}$

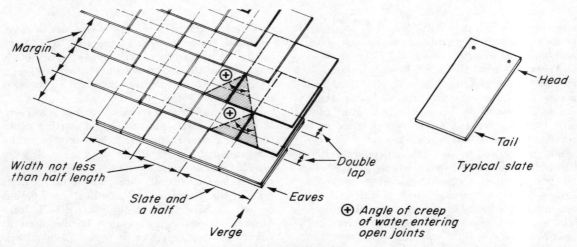

Margin

Width not less than half length

Slate and a half

Verge

Eaves

Double lap

⊕ Angle of creep of water entering open joints

Head

Tail

Typical slate

Figure 170   'Angle of creep' depends on pitch of roof. The steeper the pitch the narrower the angle will be and this can be used to determine minimum acceptable size of tile or slate

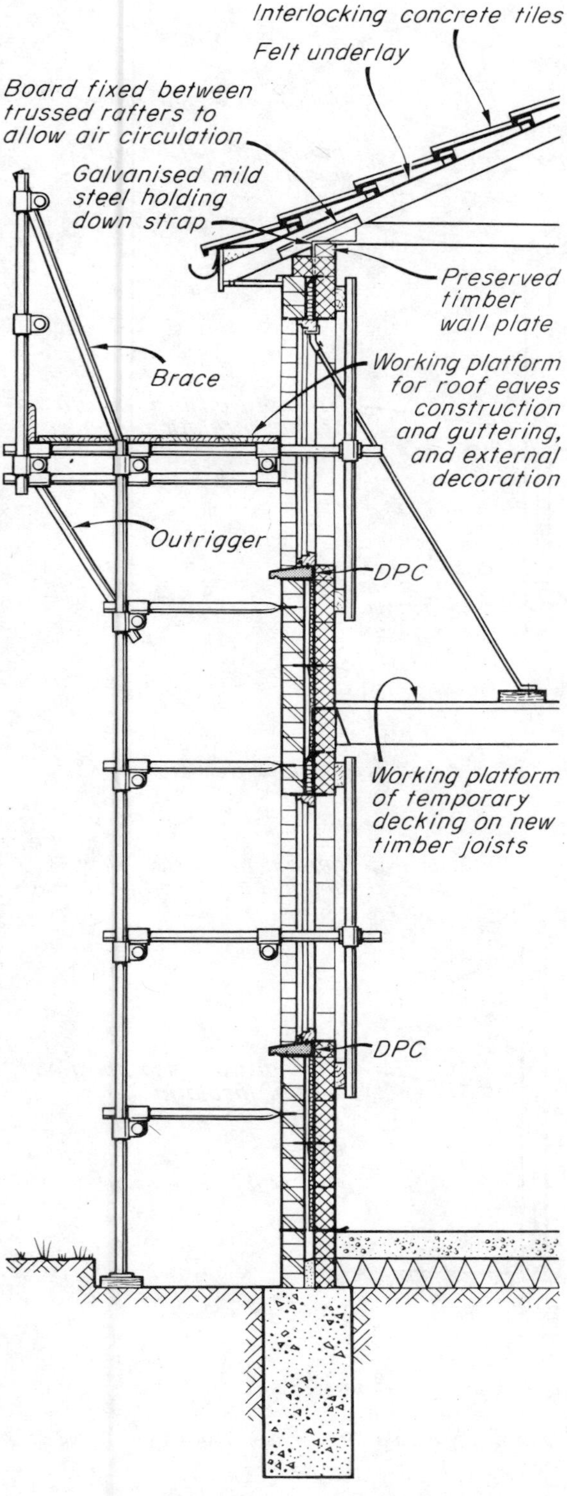

Interlocking concrete tiles

Felt underlay

Board fixed between trussed rafters to allow air circulation

Galvanised mild steel holding down strap

Brace

Outrigger

Preserved timber wall plate

Working platform for roof eaves construction and guttering, and external decoration

DPC

Working platform of temporary decking on new timber joists

DPC

Figure 171   Construction sequence: final lift

so that breaking away of the edges of the hole (spalling) will form a counter sinking for the nail heads. This is best done by machine on site so that the holes can be correctly positioned by the fixers.

Clay plain tiles are fixed in a similar fashion except each tile is retained on the battens by nibs, and unless the roof is very exposed, they need only be nail-fixed every third course.

On completion of the external waterproofing system to the roof (figure 171) the rainwater gutters and drain pipes should be installed to prevent the water collected by the roof surface from cascading down the face of the walls below and causing later problems. Work can also commence on work to the interior of the building which has hitherto been left until a dry interior 'climate' could be ensured.

(e) **Thermal insulation and ventilation**

Pitched roofs using slate or tile external cladding can be either *warm roofs* or *cold roofs* according to the positioning of the insulation (figure 172). An effective vapour barrier must always be incorporated on the warm side of the insulation.

Warm roof construction allows the roof space to catch the rising warm air from the rooms below, or even be heated when it is to form a room itself. Also, the roof timbers are protected from external solar heat gains which could cause excessive thermal movement and crack ceiling finishes, etc. The thermal insulating material is placed over the rafter prior to fixing the sarking. As the sag in the sarking necessary to allow permeating water from the roof cladding to run to the eaves may be difficult to form, an alternative constructional system is required. This involves the incorporation of vertically fixed *counter battens* over the sarking and insulation on the lines of the rafters so that the space is created between the underside of the horizontal slate or tile battens (figure 173).

A similar system of counter-battening is also necessary when it is considered necessary to fix boarding to the top face of the rafter to provide additional resistance to wind penetration. This boarding also gives the roof greater internal stability as well as greater defence against illegal entry. (Easy access for a burglar into a dwelling

235

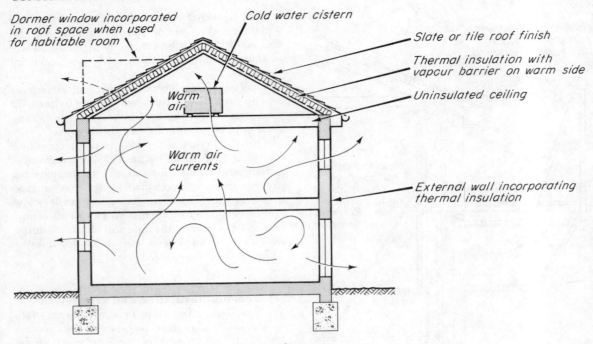

Dormer window incorporated in roof space when used for habitable room

Cold water cistern

Slate or tile roof finish

Thermal insulation with vapcur barrier on warm side

Uninsulated ceiling

Warm air

Warm air currents

External wall incorporating thermal insulation

'WARM' pitched roof

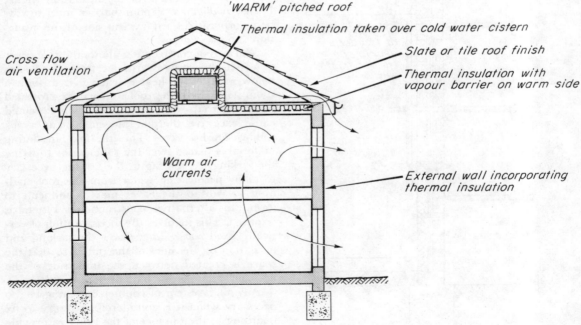

Thermal insulation taken over cold water cistern

Cross flow air ventilation

Slate or tile roof finish

Thermal insulation with vapour barrier on warm side

Warm air currents

External wall incorporating thermal insulation

'COLD' pitched roof

Figure 172 'Warm' and 'cold' pitched roofs: the National Building Regulations 1976 require a maximum U value of 0.6 W/m²°C. (See figure 52)

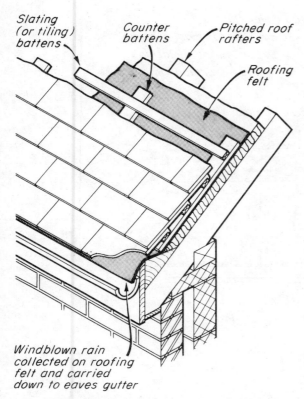

Slating (or tiling) battens

Counter battens

Pitched roof rafters

Roofing felt

Windblown rain collected on roofing felt and carried down to eaves gutter

Figure 173   Use of counter battens to provide gap between roofing felt and slating battens (see also figure 37)

is through the roof – tiles can be lifted and the sarking felt cut away!).

The advantage of cold pitched roof construction lies in *energy conservation* – the roof space is unheated, and most of the heat in the rooms below is used solely for those spaces. However, thermal movements caused by solar gain within the roof space are likely to provide problems of movement and special cover strips will be required at junctions between ceiling and the walls below. Having completed the outer cladding of the roof and made it waterproof, the builder is more able to accommodate insulating material over and/or between the ceiling joist later in the building programme, just prior to the finishes. However, care must be taken to ensure that the thermal insulation is placed *over* water cisterns so that the lost heat from the rooms below is at least used to prevent them freezing. Water pipes, including overflow pipes,

must be separately lagged if they are located within the cold roof space.

As no vapour barrier system can be guaranteed to be entirely effective, it is very important that the space above the thermal insulation is adequately ventilated to remove the last vestige of moist air which could cause condensation. This is particularly important when 'cold roof' construction is employed and the Building Regulations require the provision of cross ventilation by a continuos gap along two opposite sides of a rectangular roof. However, it is possible a for this gap to be blocked by insulation during laying, and unless protected by a galvanized mesh screen, it could allow insects, etc, to enter the roof space. Proprietary roof ventilating devices, or boards fixed between rafter (figure 169), overcome these problems. Moreover, the effectiveness of ventilation can be more easily achieved by establishing the method most suitable relative to the air flow required, determined by wind speed, the availability of a free flow of air and the plan shape of the roof space. Nevertheless, the Building Research Establishment (Digest 270) has concluded that 'air speeds required to avoid condensation (0.1 m/s) are much lower than those which cause deterioration in performance of insulation (1.2 m/s)'.

The use of tiles or slates provides a roof cladding which gives good protection against the effects of an external fire: they will not allow flames to penetrate into the roof space or allow flames to travel along their surface.

(d) Flat roof construction

The structure of a flat roof is similar to intermediate floor construction except it must be waterproofed externally, thermally insulated, and, as lighter loads are usually required to be carried, the timber joist sections can be reduced in size. Because gravity cannot carry the rainwater away, a *water barrier* must be provided consisting of continuous impervious membrane, such as jointed metal sheets, asphalt, multi-layer roofing felt, or single layer plastics. These have to be laid on a *firm decking to falls* to ensure that water drains away to the water collection points. On a timber joist flat roof the falls are generally provided by *firring pieces*. Figure 174 illustrates typical details: particular care is required by the

CONSTRUCTION METHOD

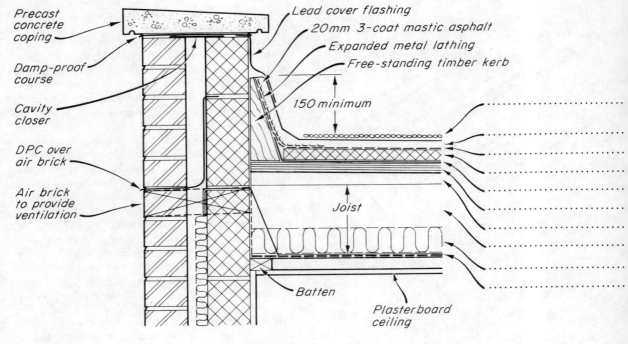

Precast concrete coping

Damp-proof course

Cavity closer

DPC over air brick

Air brick to provide ventilation

Lead cover flashing

20mm 3-coat mastic asphalt

Expanded metal lathing

Free-standing timber kerb

150 minimum

Joist

Batten

Plasterboard ceiling

PARAPET

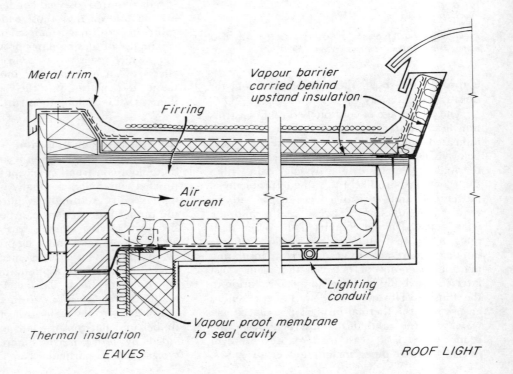

Metal trim

Firring

Vapour barrier carried behind upstand insulation

Air current

Lighting conduit

Thermal insulation

Vapour proof membrane to seal cavity

EAVES

ROOF LIGHT

Figure 174   Typical flat roof construction. Details are for a 'cold' roof construction where thermal insulation and vapour barrier are held against the underside of structural timber joists by battens placed at right angles. In order to ensure the vapour barrier remains as imperforate as possible, electrical lighting conduits are not installed in the structural zone of the roof

······Solar reflective chippings or
      reflective paint finish as required

······20mm 2-coat mastic asphalt

······Sheathing felt separating membrane

······Board insulation

······Underlay or vapour barrier

······Plywood decking

······Firrings on joists to provide fall

······Cold void

······Thermal insulation

······Vapour barrier

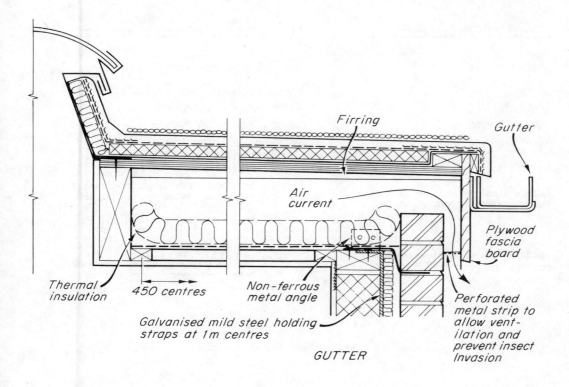

Firring

Gutter

Air current

Plywood fascia board

Thermal insulation

450 centres

Non-ferrous metal angle

Galvanised mild steel holding straps at 1m centres

Perforated metal strip to allow vent-ilation and prevent insect Invasion

GUTTER

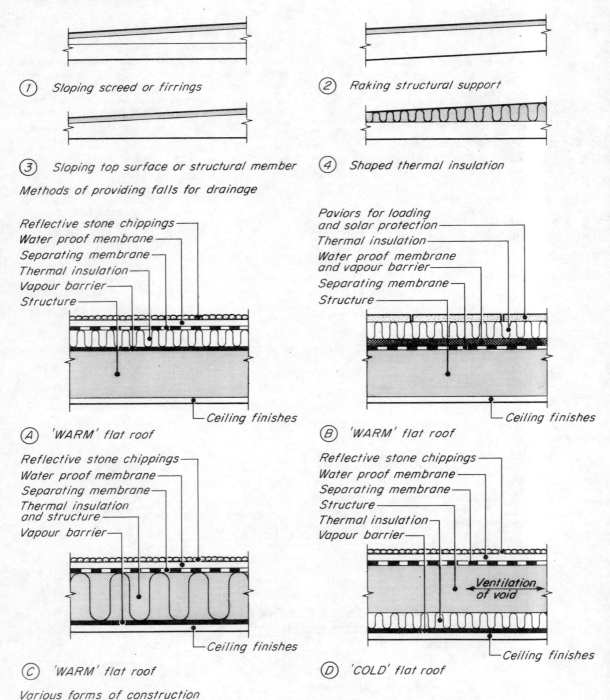

① Sloping screed or firrings

② Raking structural support

③ Sloping top surface or structural member

④ Shaped thermal insulation

Methods of providing falls for drainage

Reflective stone chippings
Water proof membrane
Separating membrane
Thermal insulation
Vapour barrier
Structure

Ceiling finishes

Ⓐ 'WARM' flat roof

Paviors for loading
and solar protection
Thermal insulation
Water proof membrane
and vapour barrier
Separating membrane
Structure

Ceiling finishes

Ⓑ 'WARM' flat roof

Reflective stone chippings
Water proof membrane
Separating membrane
Thermal insulation
and structure
Vapour barrier

Ceiling finishes

Ⓒ 'WARM' flat roof

Reflective stone chippings
Water proof membrane
Separating membrane
Structure
Thermal insulation
Vapour barrier

Ventilation
of void

Ceiling finishes

Ⓓ 'COLD' flat roof

Various forms of construction

Figure 175   'Warm' and 'cold' flat roof construction

designer when detailing, and by the builder when constructing the junctions between a 'flat' roof and any upstand, such as walls or projecting service pipes. When inadequately formed, these points usually provide sources of water penetration and, therefore, failure of the roof system. It should also be borne in mind that flat roof construction generally forms the major source of building failures in the UK – faults being attributed to poor design, inadequate materials, and bad workmanship.

Flat roof construction can also provide a 'warm' roof or a 'cold' roof, and various systems are shown in figure 175. The jointed metal sheets and asphalt provide adequate protection against the effects of an external fire, but multi-layer roofing felts and single layer plastics systems must be finished with a coating of stone chippings to provide the equivalent protection. Stone chippings also give protection against solar radiation and mechanical damage and may, therefore, be incorporated for this reason on asphalt membranes.

17.10 Services   CI/SfB (5−) + (6−)

(a) Below ground drainage (figure 176)

Below ground drains for a house are installed in trenches formed at appropriate falls by excavation techniques similar to those used for foundation trenches. If the builder needs to hire an excavator or a specialist sub-contractor for this purpose, it will be more economical to form all trenches – foundation and drainage – during the same period of building work. Lengths of pipes can be installed and connected together up to convenient points where the above ground drainage commences, such as gullies and wc outlets. The trenches can then be backfilled so as not to cause inconvenience to subsequent site and building work. Alternatively, the builder may prefer to carry out the majority of excavations for drainage (other than that affecting service pipes through foundation walls and ground floor slabs) once the bulk of the work on the building itself has been completed. In this way the installation of the drains, and subsequent backfilling processes, becomes part of the general work relating to external landscaping. Similarly, the connection to the public sewer can be delayed until most of the site drainage work is complete; or can be done at a very early stage in the building programme by forming the link drain between the sewer and the nearest manhole on site. This is known as an *interceptor manhole* and is specially constructed to divide 'private' drainage from 'public' drainage, as well as accommodating any difference in levels which may exist between the two drains. Local authority permission must be sought when connections to a public sewer are required and the work should be carefully planned to cause minimum inconvenience to the public. Normally the local authority form the connection using their own direct labour.

Drainage 'tails', left for future connections with above ground drainage, must be temporarily sealed against the leakage of smells and gases from the main sewer, and/or the possibility of them becoming blocked by debris. Backfilling drainage trenches must be done with great care and the exact method employed for casing the pipes will depend to a large extent on local conditions.

It will be necessary for the local authority to inspect and approve all drainage work and the subsequently installed plumbing work. Although they are looked at during their installation for general approval, the final tests on drainage by the Inspector generally takes place after the trenches have been backfilled. Consequently, it is desirable for the builder to give the drains a preliminary water test *before* backfilling commences in order to avoid the trouble and expense of opening up again at a later date if leaks become evident as a result of the local authority test.

(b) Above ground drainage and plumbing (see figure 73)

The drainage above ground can be installed as soon as practicable. This generally means once the basic structural work has been completed (walls, floors, and roof): care having been taken to ensure that all necessary holes have been formed in the correct position. Once the interior of the building has been made 'dry' by completion of the external enclosure, plumbing work can also commence.

Sanitary fittings connecting to the internal drainage system are not usually installed until after the basic finishes of the building have been

241

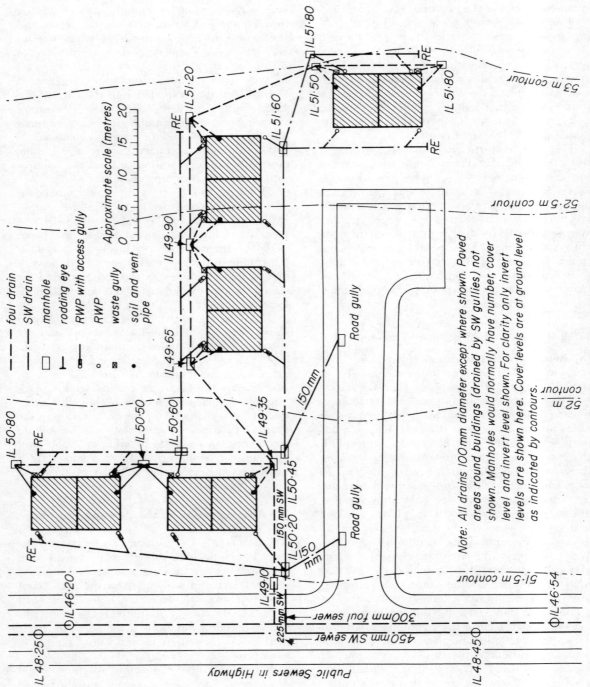

Figure 176 Below ground drainage: this layout should be read in conjunction with figure 73 page 86

Note: All drains 100mm diameter except where shown. Paved areas round buildings (drained by SW gullies) not shown. Manholes would normally have number, cover level and invert level shown. For clarity only invert levels are shown here. Cover levels are at ground level as indicated by contours.

foul drain
SW drain
manhole
rodding eye
RWP with access gully
RWP
waste gully
soil and vent pipe

Approximate scale (metres)
0   5   10   15   20

IL 51·80
IL 51·20
RE
IL 51·60
IL 51·50
RE
IL 51·80
RE
53 m contour

IL 49·90
52·5 m contour

IL 49·65
IL 49·35

150 mm

Road gully

52 m contour

IL 50·80
RE
IL 50·50
IL 50·60
IL 50·45
IL 50·20

IL 49·35

150 mm SW

150 mm

Road gully

IL 49·10
IL 46·20
IL 50·50

RE

51·5 m contour

IL 46·54

150 mm SW
225 mm SW

300mm foul sewer
450mm SW sewer

IL 48·25
IL 48·45

Public Sewers in Highway

Road gully

completed. This is because it is often easier for tradesmen, such as plasterers, to carry out their work on walls and ceilings unhampered by pipes and fittings, and save on the 'cleaning down' processes after they have finished. Plumbing and heating pipes are generally left until after plastering for similar reasons. However, when the fittings and pipes are to be surrounded with duct enclosures or other built-in features, or where finishes of a less 'messy' nature than wet plastering processes are used, it is often more convenient to install them as soon as the internal drainage is complete. These matters should have been considered during the design stage of the project, and normally form a subject of dis-cussion between designer and builder during the progress of the work on site.

(c) Water (figures 71 and 177)
The water supply pipe is taken from the Water Board stop valve located on the perimeter of the site to a stop valve situated at a convenient point within the building. From this point the rising main serves drinking water facilities and a cold water storage tank installed at high level.

(d) Gas and electricity (figure 178)
The gas pipe and electric service cable are brought independently from their mains posi-tion outside the site to a point of termination by a meter. (The entry of the gas connecting pipe

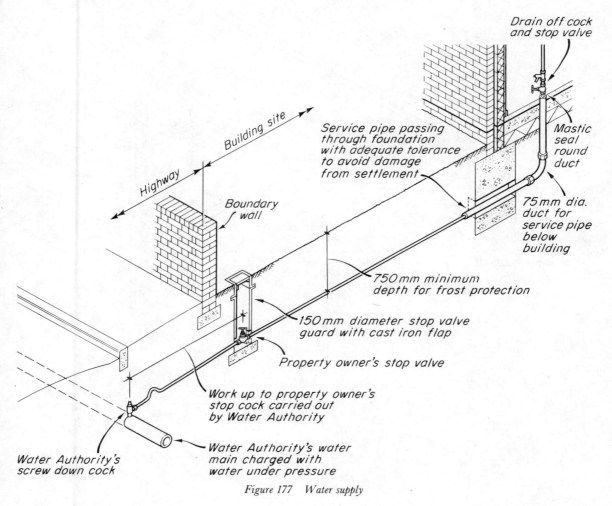

Drain off cock and stop valve

Service pipe passing through foundation with adequate tolerance to avoid damage from settlement

Mastic seal round duct

75 mm dia. duct for service pipe below building

Building site

Highway

Boundary wall

750 mm minimum depth for frost protection

150 mm diameter stop valve guard with cast iron flap

Property owner's stop valve

Work up to property owner's stop cock carried out by Water Authority

Water Authority's screw down cock

Water Authority's water main charged with water under pressure

Figure 177   Water supply

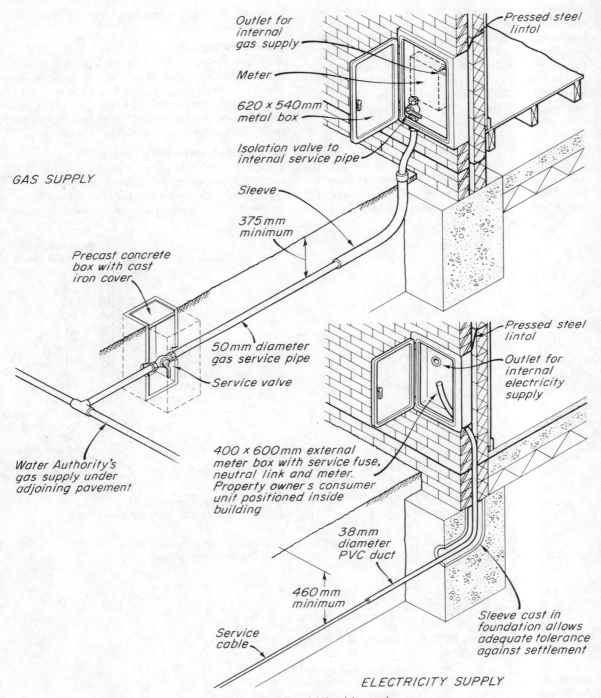

GAS SUPPLY

Outlet for internal gas supply

Meter

620 x 540mm metal box

Isolation valve to internal service pipe

Pressed steel lintol

Sleeve

375mm minimum

Precast concrete box with cast iron cover

50mm diameter gas service pipe

Service valve

Water Authority's gas supply under adjoining pavement

Pressed steel lintol

Outlet for internal electricity supply

400 x 600mm external meter box with service fuse, neutral link and meter. Property owners consumer unit positioned inside building

38mm diameter PVC duct

460mm minimum

Service cable

Sleeve cast in foundation allows adequate tolerance against settlement

ELECTRICITY SUPPLY

Figure 178   Gas and electricity supply

244

should not be through the same duct as that which contains the electrical cable.) This work is carried out by the Gas and Electricity Boards to a programme agreed with the builder.

Unless separately ducted, the work of installing internal gas pipes and electricity cables from the meter positions should be carried out before basic finishes to the building are commenced. This work may also be carried out by the relevant Boards, but, for electric wiring, it is more usual that this is carried out by an electrical sub-contractor – a satisfactory price for the work having been given through competitive tendering procedures. Once the gas and electrical services have been installed to the layout requirements stipulated by the drawings and specification of the designer, the work must be inspected by the Inspectors of the two Boards for compliance with safety regulations.

(e) Telecommunication systems
(figure 179)

Like the other services described, the telephone, computer, and cable television terminals may have localized common supply points from which connections are made to a building. Alternatively, they can be supplied from remotely located exchanges. These installations are carried out externally and internally by specialist engineers after most of the building work, including finishes, have been completed.

17.11 Finishes   CI/SfB (4 −)

As soon as the constructional processes reach the stage where the structural elements of the house (walls, floors and roof) have more or less been finished, and the weatherproofing completed by window/door glazing and roof claddings, etc, the work on the interior can commence in earnest. Initially, this will involve the erection of dry construction components such as stud partitions, and internal door frames and linings as described earlier. At about the same time, the service trades – gas, water, plumbing, heating and electrical – will be continuing their installations from the previously incorporated terminal points. In close association with all these activities, the building contractor will wish to make a start on the coverings for the internal surfaces of walls, floors and ceilings, as well as building the staircase, cupboard fitments, and

ducting work. Therefore, whereas previous trades were working relatively independent of each other and progress relied on consecutive working programmes, the processes now required are much more interdependent and the degree of organization required increases. Builder and sub-contractor tradesmen must work closely together to ensure that the correct labour skills and adequate material resources are available so as not to cause unreasonable delays which invariably accumulate to result in frustrations and, of course, the implementation of financial penalties.

(a) Plasterboard and plastering

Stud partitions and joisted ceilings are usually faced with *plasterboard* as it is reasonably strong, flexible in use, and contributes towards fire protection. Plasterboard consists of a plaster core encased in and firmly bonded to specially prepared durable lining paper. A grey lining paper indicates that it is meant to receive a coating of wet plaster, and when ivory coloured, so that it can be decorated direct. Plasterboard used without wet plaster (except in joints) is referred to as *dry lining*.

Plasterboards are available in a number of standard sizes to conform with the centres of structural studs or joists, and is supplied to various thicknesses. When finished with a 5 mm coat of wet plaster, a 9.5 mm thickness applied both sides of a stud partition will provide the required half hour fire resistance. Alternatively, a 12.5 mm thickness plasterboard on its own to each side will give the same degree of fire resistance. The 'modified half hour' fire resistance required for the floor (see 17.8(b) *Materials*) can be achieved with only a 9.5 mm thickness plasterboard providing tongued and grooved floor decking is used and joists are not less than 38 mm wide.

Plasterboards are secured by galvanized or sheradized nails. Inevitably, some cutting will be necessary, particularly around door openings, electric lightswitches and power points, and when splays (to form bulkheads round staircase) are incorporated in the design of the stud wall and/or ceiling. It is essential that all cross joints in the plasterboards are staggered and that they are covered with a *jute scrim* which forms a seal, and helps to prevent subsequent

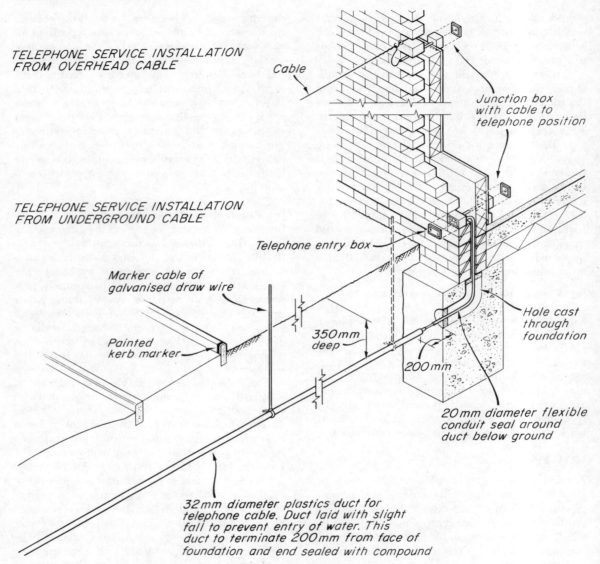

TELEPHONE SERVICE INSTALLATION
FROM OVERHEAD CABLE

Cable

Junction box
with cable to
telephone position

TELEPHONE SERVICE INSTALLATION
FROM UNDERGROUND CABLE

Telephone entry box

Marker cable of
galvanised draw wire

Painted
kerb marker

350 mm
deep

Hole cast
through
foundation

200 mm

20 mm diameter flexible
conduit seal around
duct below ground

32 mm diameter plastics duct for
telephone cable. Duct laid with slight
fall to prevent entry of water. This
duct to terminate 200 mm from face of
foundation and end sealed with compound

Figure 179   Telephone cable supply

cracking of future finishes caused by thermal movements. It is also good practice to form the joints at right angles to the support of the sheets, and for each joint to be backed by a timber batten or *noggin*. When all the plasterboard is in place, the plastering can commence to the plasterboard as well as the internal blockwork surfaces. A plaster finish fulfils the function of camouflaging irregularities in the backing wall

as well as exposed conduits on the surface. It also provides a sufficiently hard surface to resist damage by impact which is also smooth enough to be suitable for direct decoration. Gypsum plasters are suitable for this and the choice of mix, type and number of coats will depend upon the characteristics of the background and its surface texture. Generally, the lightweight concrete blocks used for the internal leaf of the external cavity wall and for the internal partitions, and the surface of plasterboard, provide a suitable key for the direct application of plasters without further preparations. Although one-coat plasters are also available for blockwork, it is more usual to use a two-coat process. The undercoat is applied by means of a wooden float or rule which are worked between 'dots' or 'runs' of plaster to give a true level surface. This undercoat is about 10 mm thick and is scratched before drying to provide a suitable key for the finishing coat. The thin finishing coat is applied to a depth of about 3 mm and finished with a steel float to provide a glass-like surface.

Junctions between backing walls of different materials (plasterboard adjoining block walls) require special treatment if cracking is to be avoided as a result of subsequent differential movements. Jute scrim reinforcement can be used at these points although a much better detail can be adopted by using alloy edge reinforcing strips, as indicated in figure 180. These can also be used on the corners of walls to prevent damage by impact. Plaster reinforcing details between roof ceiling joists and wall below is also indicated in figure 180.

When a 'warm roof' construction is used it will be necessary to install the vapour barrier prior to the fixing of the plasterboard lining for the ceiling. Although plasterboards are available with a thin aluminium foil bonded to their back face for this purpose, with the high degree of insulation required it is wise to incorporate an additional independent vapour barrier. This is because the aluminium foil will probably be damaged during erection and will be pierced for electric light fittings. The independent vapour barrier is best incorporated as part of the insulation which can be laid once the ceiling and electrical work has been completed. As stated earlier, it is important that the warm roof insulation is taken *over* any water cisterns in the roof space and that any water pipes are suitably lagged to prevent the possibility of freezing.

(b) Joinery

As soon as practicable, the upper surfaces of the horizontal structural elements should be made available by fixing the decking in an order which allows other work to be carried out from them. The staircase, which links these surfaces, should also be constructed in parallel, and door linings as well as window cills, etc, must be fixed so that plasterboard linings and plastering can continue without hindrance.

The intermediate floor decking can be *tongued and grooved* hardwood or softwood boarding, or chipboard or plywood sheets (see figure 145). When narrow face width boarding is used it will be necessary for the boards to be cramped securely together before nailing to the joists in order to close the joints between each, and lessen the effect of shrinkage once the central heating has been switched on. Areas requiring future access for maintenance and alterations to electric wiring, junction boxes, etc, can be accommodated by removing the tongues from the perimeter board affected and using screws to form a removable panel. The use of 't & g' chipboard, or plywood sheets (say 1.2 m × 2.4 mm), will speed the laying of floor decking provided their standardized dimensions are dimensionally co-ordinated with the centres of the structural joists and the overall sizes of individual rooms. Access areas can be provided by removable screw down panels in a similar manner to hardwood or softwood boarding. Flooring grades of chipboard sheets need particular attention if subsequent bowing of the floor deck is to be avoided as a result of moisture absorption. It is recommended that the boards are allowed to acclimatize to the moisture levels of the room in which they are to be used before being fixed, and that movement gaps are left around the perimeter of the floor at junctions with walls.

Door linings and window boards are normally included in the 'first fixing' programme of the joinery. Linings are carefully set up to be plumb and then screwed to the plugs already built into blockwork or to the double stud and head timbers of stud partitioning. They should be of a width which allows for the thickness of plasterboard and/or plaster. Internal window boards are levelled and fixed direct to the blocks using cut nails or screws or, depending on the precise detail, they can be nailed to timber packing pieces already fixed to the top course of the blockwork opening. Sometimes, tiled cills are required and in this case the tiles are bedded in mortar direct to the blockwork. 'Second

247

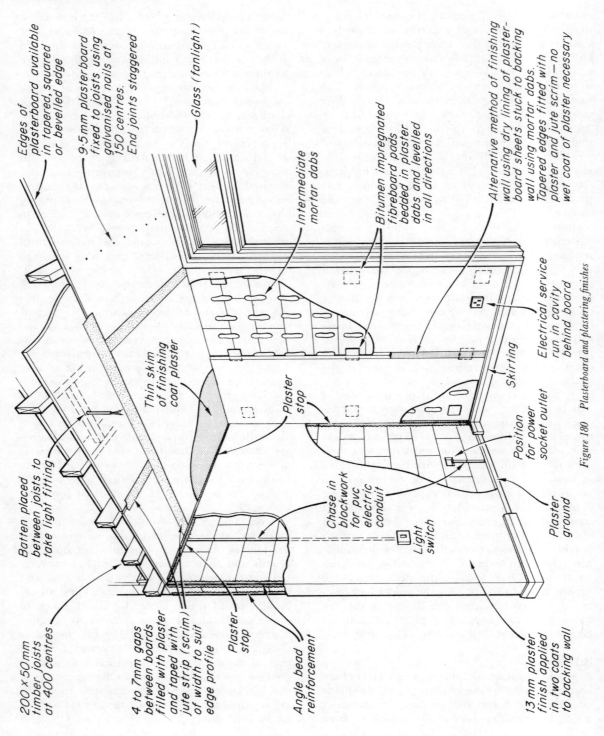

Edges of plasterboard available in tapered, squared or bevelled edge

9·5mm plasterboard fixed to joists using galvanised nails at 150 centres. End joints staggered

Glass (fanlight)

Intermediate mortar dabs

Bitumen impregnated fibreboard pads bedded in plaster dabs and levelled in all directions

Alternative method of finishing wall using dry lining of plaster-board sheets stuck to backing wall using mortar dabs. Tapered edges fitted with plaster and jute scrim — no wet coat of plaster necessary

Electrical service run in cavity behind board

Thin skim of finishing coat plaster

Skirting

Plaster stop

Position for power socket outlet

Batten placed between joists to take light fitting

Chase in blockwork for pvc electric conduit

Light switch

Plaster ground

200 x 50mm timber joists at 400 centres

4 to 7mm gaps between boards filled with plaster and taped with jute strip (scrim) of width to suit edge profile

Plaster stop

Angle bead reinforcement

13mm plaster finish applied in two coats to backing wall

Figure 180 Plasterboard and plastering finishes

fixing' joinery, which includes items such as doors, architraves, skirtings, shelves, cupboard units and ironmongery (locks, handles, etc), should not be installed until just prior to final finishing. When considering the design of joinery items, it is important to select appropriate sizes for timber sections to allow for the reduction in size of the original rough or sawn (*unwrot*) standard sections caused by their being smooth (*wrot*). The amount of reduction will vary according to species of timber and the original sawn size – the range for softwoods being from 1.5 mm of *each* face to be planed for up to 22 mm sections, to 6.5 mm of each face for sections over 150 mm. Joinery sections are, therefore, specified by using the prefix '*ex:*' (out of) against the unwrot size of timber to be used, eg an ex: 100 mm × 25 mm softwood skirting resulting in a wrot size of 94 mm × 19 mm. For further clarification, see Mitchell's Building Series: *Components:* Figure 4.3 (page 66) *Reduction from basic sizes to finished sizes to accurate sizes by processing of two opposed faces of softwoods.*

After the main construction work has been completed, but before the 'second fixing' of the joinery, the ground floor damp proof membrane can be installed if it has not already been incorporated under the solid concrete slab, or if a suspended floor construction has been used. After the dpm has been laid on the top surface of a concrete floor slab, the protective floor screed should be placed. For obvious reasons, if the final floor finish is to be applied by the builder (plastics tiles, wood strip or parquet, etc), the dpm and screed should not be commenced until the very last period of the construction programme.

(c) Decorations

Once the construction work has been completed those parts of the house which require decoration can be appropriately prepared (figure 181). The more care that is taken with preparatory work, the better the finished result will be.

Sometimes the externally decorated areas will have been prepared and finished during an earlier stage of the building programme. This involves bringing in the painting and decorating tradesmen out of sequence relative to the work to be done inside the building. This procedure may still be economical since it allows the scaf-folding to be removed much earlier in the contract, and may mean that the external landscaping can proceed earlier and unhampered by construction activities. However, it is important that allowance is made for protecting the newly finished work while any construction work is still being carried out in the interior of the building. Protection is particularly important for window and door openings, especially as these are often supplied to site in some pre-finished form (primed or preservative decorative staining to timber window and door frames, or anodized aluminium sections). They will require protection against mechanical damage arising from loading material to upper levels from scaffolding platforms.

Before any of the woodwork is to be painted on site, it should be carefully rubbed down, all nails punched below the surface and any necessary filling done. When choosing a particular manufacturer's priming, undercoat and finishing coat paint it is essential to check that they are chemically compatable with any preservative treatment which has been used on the joinery. The outside junction between door and window frames with the brickwork should be sealed with a mastic.

Internally the walls to be decorated should receive their final preparation. This process includes the removal of temporary fixing nails, the repair of damaged plasterwork, and the smoothing of rough surfaces generally. In a new building, there may be a small amount of movement due to initial structural settlement, and to the drying out of structural timbers, blockwork, etc. Drying out could persist for some time, and is likely to cause some cracking in plaster finishes, particularly during the first heating season. Plastered blockwork walls need a 'drying out' period after their completion because of the amount of water used in their construction. For this reason, their initial decoration should be regarded as temporary. Certain emulsion paints allow the walls to 'breath' and residual moisture escape without causing excessive or speedy breakdown of the surface film. It is normal to choose a neutral colour for the temporary finishing emulsion. Blemishes will be less obvious, and the occupants of the dwelling will be able to more easily appreciate the quality of space and light

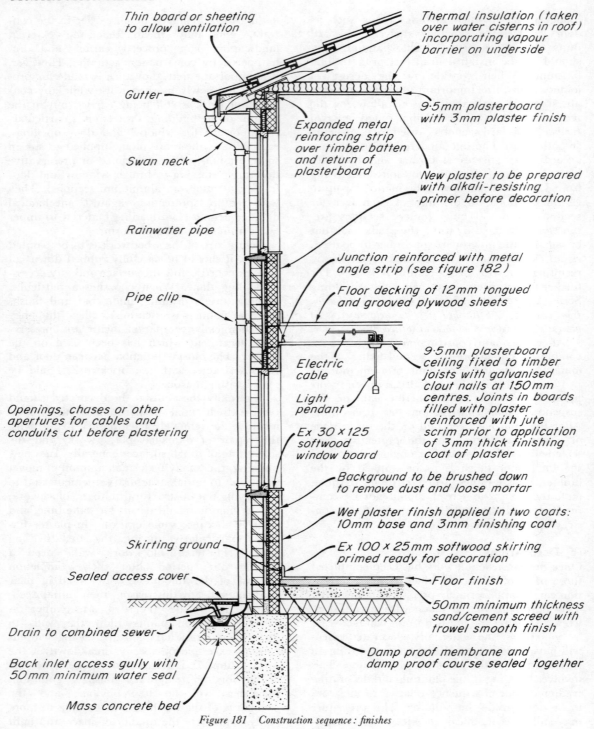

Thin board or sheeting to allow ventilation

Thermal insulation (taken over water cisterns in roof) incorporating vapour barrier on underside

Gutter

9·5mm plasterboard with 3mm plaster finish

Expanded metal reinforcing strip over timber batten and return of plasterboard

Swan neck

New plaster to be prepared with alkali-resisting primer before decoration

Rainwater pipe

Junction reinforced with metal angle strip (see figure 182)

Floor decking of 12mm tongued and grooved plywood sheets

Pipe clip

9·5mm plasterboard ceiling fixed to timber joists with galvanised clout nails at 150mm centres. Joints in boards filled with plaster reinforced with jute scrim prior to application of 3mm thick finishing coat of plaster

Electric cable

Light pendant

Openings, chases or other apertures for cables and conduits cut before plastering

Ex 30 x 125 softwood window board

Background to be brushed down to remove dust and loose mortar

Wet plaster finish applied in two coats: 10mm base and 3mm finishing coat

Skirting ground

Ex 100 x 25mm softwood skirting primed ready for decoration

Sealed access cover

Floor finish

50mm minimum thickness sand/cement screed with trowel smooth finish

Drain to combined sewer

Damp proof membrane and damp proof course sealed together

Back inlet access gully with 50mm minimum water seal

Mass concrete bed

Figure 181   Construction sequence : finishes

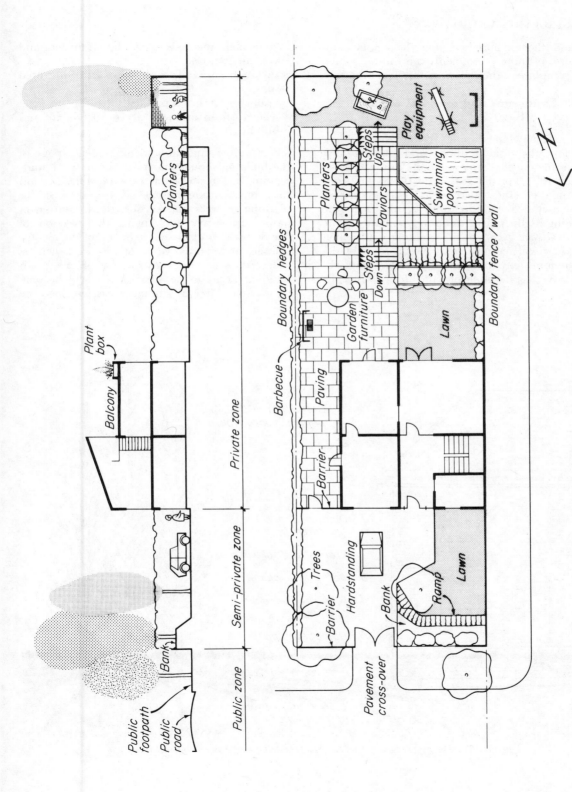

Public footpath

Public road

Bank

Public zone

Semi-private zone

Private zone

Plant box

Balcony

Planters

Trees

Barrier

Hardstanding

Barrier

Bank

Ramp

Lawn

Pavement cross-over

Barrier

Paving

Barbecue

Boundary hedges

Planters

Garden furniture

Steps Down

Steps Up

Paviors

Swimming pool

Play equipment

Lawn

Boundary fence / wall

N

Figure 182   The role of landscaping – see also figure 76 page 93

251

around them before selecting the colours and textures of more permanent wall finishes which will compliment the chosen style of furnishings.

17.12 Landscape and external works
CI/SfB (90)

The external landscaping around the building provides the vital aesthetic and physical link between 'outside' spaces and 'inside' spaces (figure 182). For this reason the design and construction of landscaped areas requires just as much detailed consideration as applies to the building itself. Apart from purely enhancing the *appearance and quality* of the building with which it most closely associates as well as that of the surrounding environment, the landscaping will also serve in providing practical functions relating to:

- convenient methods of circulation for foot and vehicular travel
- shelter from wind, snow, rain, and unwanted sunshine
- visual and/or aural privacy
- security from unlawful access to the site and building.

The required function and subsequent design of the building should maximize the natural landscaping of the site as far as is possible. Beyond this, consideration must be given to the need for planting of trees and shrubs, the formation of paths, driveways and roads, the erection of fences, walls, and the construction of shelters for storage, or amenity. Most of this work will be executed after the construction of the main building because the areas of site will become progressively free as storage areas, etc, become

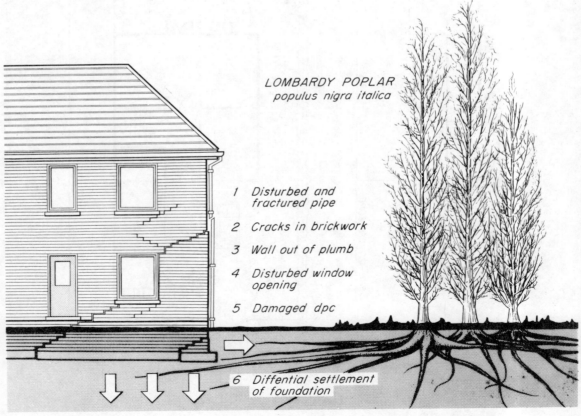

LOMBARDY POPLAR
populus nigra italica

1   Disturbed and fractured pipe

2   Cracks in brickwork

3   Wall out of plumb

4   Disturbed window opening

5   Damaged dpc

6   Diffential settlement of foundation

Figure 183   Danger of planting trees too close to building erected on conventional foundations

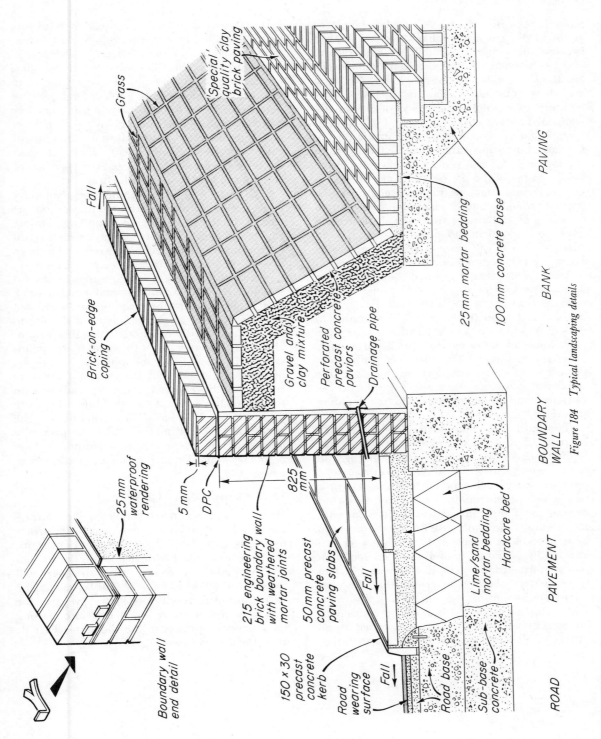

Grass

'Special' quality clay brick paving

Fall

Brick-on-edge coping

25mm waterproof rendering

5mm

DPC

Boundary wall end detail

215 engineering brick boundary wall with weathered mortar joints

50mm precast concrete paving slabs

825 mm

Gravel and clay mixture

Perforated precast concrete paviors

Drainage pipe

25mm mortar bedding

100mm concrete base

150 x 30 precast concrete kerb

Road wearing surface

Fall

Fall

Lime/sand mortar bedding

Hardcore bed

Road base

Sub-base concrete

BOUNDARY WALL

ROAD

PAVEMENT

BANK

PAVING

Figure 184 Typical landscaping details

253

redundant. Also, of course, newly completed landscaped areas are less likely to be damaged because the majority of building construction will be completed. (An exception might occur on large building contracts when the long period of construction can be used to allow planting to mature in advance of the finally completed buildings.)

It would be economically desirable if most of the earth landscape features can be formed from the topsoil saved as a result of excavation of the building area of the site (see 17.3(c)). If there is an insufficient amount, the contractor will be required to import additional quantities from an outside source. Existing trees and shrubs which have been retained and protected during construction work will also provide considerable saving against planting new ones.

It is important that detailed thought is given to the selection of the correct species of new trees or shrubs. These will be small during their infancy and subsequent growth may result in undesirable amount of overshadowing and the curtailment of desirable vistas. It is particularly necessary to analyse the effects of future root growth relative to the type of soil (including the proximity of water sources and the influences of pruning) and their proximity to the building (figure 183).

The hard surface landscaping areas created by paths, driveways and roads on the site must be designed to retain their interest through all the changing seasons of the year. Natural stone pavings and gravels can be incorporated in a way which permits mosses and lichen to spread over areas of their surfaces to unify them with the natural landscape. Alternatively, they can be designed and constructed to create a separation rigidly defining dry and durable circulation paths. Hard surfaces may also require a system of rainwater collection to ensure that they are relatively free from the effects of rainwater. Large expanses can be drained by continuous 'monsoon type' drains which are revealed at the surface only by a continuous gap. Other materials considered suitable for large hard surfaced areas include tarmacadum, hot rolled asphalt, and *in situ* concrete. Care must be taken to ensure they do not give a monotonous appearance or do not crack excessively owing to dimensional movements. Accordingly, these materials can be formed in bays using another material at the joints which helps relieve monotony and create the opportunity for an aesthetically acceptable movement joint. Precast concrete paving slabs, brick pavers, clay tiles or cobbles can also be used with effect.

The provision of screening and boundary fences/walls has an important influence on the overall quality of the site and building. Timber can be used in a variety of ways which result from construction techniques ranging from driving poles for fixing preformed panels, to sophisticated joinery incorporating hardwood sections to give maximum visual appeal. Brick built walls may also be pleasing, and can form an 'introduction' at the perimeter of the site to the aesthetic of the brick faced building within. However, in order for this ideal to be maintained for the life of the building it will be necessary to ensure that the effects of weathering on a fully exposed boundary wall remain similar to the walls of the building which are subjected to less hazardous conditions. The choice of brick and mortar should be primarily influenced by the exposure condition of the boundary wall. Typical details are indicated in figure 184.

17.13 Completion CI/SfB (A8)

Once all work has been completed the contractor will be required to clear away all surplus materials, clear away debris, remove plant/huts, etc, and leave the building and site in a clean and workable condition. The designer will inspect the work, ensure the satisfaction of the local authority inspectors, and prepare the necessary maintenance manual before passing the completed project to the client (see Part B: 16.8 *Maintenance manuals*).

Further specific reading from Mitchell's Building Series and Building Research Establishment Digests which are relevant to each section of this Chapter

17.1 Building team and communication
Structure and Fabric Part 1: Chapter 2 *The production of buildings*
Structure and Fabric Part 2: Chapter 1 *Contract planning and site organization*
                            Chapter 2 *Contractors' mechanical plant*
Components: Chapter 3 *Industrialized system building*

17.2 Site considerations
Structure and Fabric Part 1: Chapter 4 *Foundations and site exploration*
                                    (Soils and soil characteristics)
Structure and Fabric Part 2: Chapter 3 *Foundations* (Soil mechanics)

BRE Digest 63:   *Soils and foundations* Part 1
BRE Digest 64:   *Soils and foundations* Part 2
BRE Digest 67:   *Soils and foundations* Part 3
BRE Digest 202: *Site use of the theodolite and surveyor's level*
BRE Digest 234: *Accuracy in setting-out*
BRE Digest 247: *Waste of building materials*
BRE Digest 283: *The assessment of wind speed over topography*
BRE Digest 298: *The influence of trees on house foundations in clay soils*
BRE Digest 318: *Site investigation for low-rise building: desk studies*
BRE Digest 322: *Site investigation for low-rise building: procurement*

17.3 Initial site works
Structure and Fabric Part 1: Chapter 2 *Contractors' mechancial plant*
                              Chapter 11 *Temporary works* (Timbering for excavations)

BRE Digest 179: *Electricity distribution on sites*
BRE Digest 274: *Fill* Part 1
BRE Digest 275: *Fill* Part 2
BRE Digest 298: *The influence of trees on house foundations in clay soils*
BRE Digest 318: *Site investigation for low-rise building: desk studies*
BRE Digest 322: *Site investigation for low-rise building: procurement*

17.4 Foundation construction
Structure and Fabric Part 1: Chapter 3 *Foundations*
Structure and Fabric Part 2: Chapter 3 *Foundations* (Foundation design and foundation types)

BRE Digest 63:   *Soils and foundations* Part 1
BRE Digest 64:   *Soils and foundations* Part 2
BRE Digest 67:   *Soils and foundations* Part 3
BRE Digest 200: *Repairing brickwork*
BRE Digest 240: *Low-rise buildings on shrinkable clay soils* Part 1
BRE Digest 241: *Low-rise buildings on shrinkable clay soils* Part 2
BRE Digest 242: *Low-rise buildings on shrinkable clay soils* Part 3
BRE Digest 250: *Concrete in sulphate-bearing soils and groundwaters*
BRE Digest 251: *Assessment of damage in low-rise building*
BRE Digest 274: *Fill* Part 1: *Characteristics*
BRE Digest 275: *Fill* Part 2: *Site investigations*
BRE Digest 298: *The influence of trees on house foundations in clay soils*
BRE Digest 313: *Mini-piling for low-rise building*

BRE Digest 315: *Choosing piles for new construction*
BRE Digest 318: *Site investigation for low-rise building: desk studies*
BRE Digest 325: *Concrete* Part 1: *Materials*
BRE Digest 326: *Concrete* Part 2: *Specification, design and quality control*

17.5 Ground floor construction

Structure and Fabric Part 1:
　　　　Chapter 8 *Floor structures* (Ground floor construction)
Components: Chapter 13 *Raised floors*
Finishes: Chapter 3 *Ceramic materials*
　　　　Chapter 4 *Composites*

BRE Digest 18: *Design of timber floors to prevent decay*
BRE Digest 33: *Sheet and tile flooring made from thermoplastic binders*
BRE Digest 54: *Damp-proofing solid floors*
BRE Digest 79: *Clay tile flooring*
BRE Digest 104: *Floor screeds*
BRE Digest 145: *Heat losses through ground floors*
BRE Digest 163: *Drying out buildings*
BRE Digest 201: *Wood preservatives: application methods*
BRE Digest 276: *Hardcore*
BRE Digest 323: *Selecting wood-based panel products*

17.6 External wall construction

Structure and Fabric Part 1:
　　　　Chapter 5 *Walls and piers*
　　　　Chapter 9 *Fireplaces, flues and chimneys*
Structure and Fabric Part 2:
　　　　Chapter 4 *Walls and piers* (Solid masonry walls and cavity loadbearing walls)
Components: Chapter 1 *Component design*
　　　　Chapter 3 *Industrialized system building*
　　　　Chapter 5 *Doors*
　　　　Chapter 6 *Windows*
　　　　Chapter 7 *Glazing*
Finishes: Chapter 2 *Polymeric materials*
　　　　Chapter 3 *Ceramic materials*

BRE Digest 77: *Damp-proof courses*
BRE Digest 108: *Standard 'U' values*
BRE Digest 140: *Double glazing and double windows*
BRE Digest 160: *Mortars for bricklaying*
BRE Digest 190: *Heat loses from dwellings*
BRE Digest 196: *External rendered finishes*
BRE Digest 230: *Fire performance of walls and linings*
BRE Digest 236: *Cavity insulation*
BRE Digest 246: *Strength of brickwork and blockwork walls: design for vertical load*
BRE Digest 262: *Selection of windows by performance*
BRE Digest 273: *Perforated brick*
BRE Digest 277: *Built-in cavity wall insulation for housing*
BRE Digest 281: *Safety of large masonry walls*
BRE Digest 294: *Fire risk from combustible cavity insulation*
BRE Digest 297: *Surface condensation and mould growth in traditionally built dwellings*
BRE Digest 304: *Preventing decay in external joinery*
BRE Digest 338: *Insulation against external noise*

17.7 Internal wall construction

Structure and Fabric Part 1:
　　　　Chapter 5 *Walls and piers* (Partitions)
Components: Chapter 11 *Demountable partitions*

BRE Digest 320: *Fire doors*

17.8 Intermediate floor construction

Structures and Fabric Part 1:
　　　　Chapter 8 *Floor Structures* (Upper floor construction)
　　　　Chapter 10 *Stairs*
Structures and Fabric Part 2:
　　　　Chapter 6 *Floor structures*
　　　　Chapter 8 *Stairs, ramps and ladders* (Stairs)
Components: Chapter 10 *Balustrades and barriers*
Finishes: Chapter 3 *Ceramic materials*
　　　　Chapter 4 *Composites*

BRE Digest 18: *Design for timber floors to prevent decay*
BRE Digest 33: *Sheet and tile flooring made from thermoplastic binders*

BRE Digest 201:	*Wood preservatives: application methods*
BRE Digest 323:	*Selecting wood-based panel products*

17.9 Roof construction

Structures and Fabric Part 1:
        Chapter 7 *Roof structures*
Components: Chapter 14 *Roofings*

BRE Digest 8:	*Built-up felt roofs*
BRE Digest 108	*Standard 'U' valves*
BRE Digest 144:	*Asphalt and built-up felt roofings: durability*
BRE Digest 180:	*Condensation in roofs*
BRE Digest 201:	*Wood preservatives: application methods*
BRE Digest 145:	*Heat losses through ground floors*
BRE Digest 163:	*Drying out buildings*
BRE Digest 201:	*Wood preservatives: application methods*
BRE Digest 233:	*Fire hazard from insulating materials*
BRE Digest 270:	*Condensation in insulated domestic roofs*
BRE Digest 284:	*Wind loads on canopy roofs*
BRE Digest 295:	*Stability under wind loads of loose laid external roof insulation boards*
BRE Digest 311:	*Wind scour of gravel ballasts*
BRE Digest 312:	*Flat roof design: the technical options*
BRE Digest 323:	*Selecting wood-base panel products*
BRE Digest 324:	*Flat roof design: thermal insulation*
BRE Digest 336:	*Swimming pool roofs*
BRE Digest 338:	*Insulation against external noise*

17.10 Services

Environment and Services:
        Chapter 5 *Thermal installations*
        Chapter 8 *Electric lighting*
        Chapter 9 *Water supply*
        Chapter 10 *Sanitary appliances*
        Chapter 11 *Pipes*
        Chapter 12 *Drainage installations*
        Chapter 13 *Sewage disposal*

        Chapter 15 *Electricity and telecommunications*
        Chapter 16 *Gas*
Structure and Fabric Part 1:
        Chapter 9 *Fireplaces, flues and chimneys*
Components: Chapter 11 *Demountable partitions*
        Chapter 12 *Suspended ceilings*
        Chapter 13 *Raised floors*

BRE Digest 83:	*Plumbing with stainless steel*
BRE Digest 248:	*Sanitary pipework* Part 1
BRE Digest 248:	*Sanitary pipework* Part 2
BRE Digest 254:	*Reliability and performance of solar collector systems*
BRE Digest 289:	*Building management systems*
BRE Digest 292:	*Access to underground drainage systems*
BRE Digest 308:	*Unvented domestic hot water system*
BRE Digest 335:	*Electric interferences in buildings*
BRE Digest 339:	*Condensing boilers*

17.11 Finishes

Components: Chapter 5 *Doors*
        Chapter 6 *Windows*
        Chapter 9 *Ironmongery*
        Chapter 11 *Demountable partitions*
        Chapter 12 *Suspended ceilings*
        Chapter 13 *Raised floors*
Finishes: All chapters

BRE Digest 33:	*Sheet and tile flooring made from thermoplastic binders*
BRE Digest 104:	*Floor screeds*
BRE Digest 163:	*Drying out buildings*
BRE Digest 197:	*Painting walls* Part 1
BRE Digest 198:	*Painting walls* Part 2
BRE Digest 213:	*Choosing specifications*
BRE Digest 261:	*Painting woodwork*
BRE Digest 280:	*Cleaning external surfaces of buildings*
BRE Digest 286:	*Natural finishes for exterior timber*
BRE Digest 296:	*Timbers: their natural durability and resistance to preservative treatment*
BRE Digest 301:	*Corrosion of metals by wood*
BRE Digest 323:	*Selecting wood-based panel products*

The following information from the *Construction Indexing Manual 1976* is reproduced by courtesy of RIBA Publications Ltd.

Used sensibly and in appropriate detail, as explained in the manual, the CI/SfB system of classification facilitates filing and retrieval of information. It is useful in technical libraries, in specifications and on working drawings. *The National Building Specification* is based on the system, and BRE Digest 172 describes its use for working drawings.

The CI/SfB system comprises tables 0 to 4, tables 1 and 2/3 being the codes in most common use. For libraries, classifications are built up from:

Table 0	Table 1	Tables 2/3	Table 4
-a number code	-a number code in brackets	-upper and lower case letter codes	-upper case letter code in brackets
eg 6	eg (6)	eg Fg	eg (F)

An example for clay brickwork in walls is: (21) Fg2, which for trade literature, would be shown in a reference box as:

CI/SfB 1976 reference by SfB Agency		
(21)	Fg2	

The lower space is intended for UDC (Universal decimal classification) codes – see BS 100A 1961. Advice in classification can be obtained from the SfB Agency UK Ltd at 66 Portland Place, London W1N 4AD.

In the following summaries of the five tables, chapter references are made to the seven related volumes and chapters of *Mitchell's Building Series* in which aspects of the classifications are dealt with. The following abbreviations are used:

Introduction to Building	*IB*
Environment and Services	*ES*
Materials	*M*

Structure and Fabric, Part 1	*SF (1)*
Structure and Fabric, Part 2	*SF (2)*
Components	*C*
Finishes	*F*

Table 0 **Physical Environment** (main headings only)

Scope: End results of the construction process

0 Planning areas
1 Utilities, civil engineering facilities
2 Industrial facilities
3 Administrative, commercial, protective service facilities
4 Health, welfare facilities
5 Recreational facilities
6 Religious facilities
7 Educational, scientific, information facilities
8 Residential facilities
9 Common facilities, other facilities

Table 1 **Elements**

Scope: Parts with particular functions which combine to make the facilities in table 0

**(0-) Sites, projects
Building plus external works
Building systems** *C* 11

(1-) Ground, substructure
(11) Ground *SF (1)* 4, 8, 11; *SF (2)* 2, 3, 11; *IB* 17.3
(12) Vacant
(13) Floor beds *SF (1)* 4, 8; *SF (2)* 3
(14), (15) Vacant
(16) Retaining walls, foundations *SF (1)* 4; *SF (2)* 3, 4; *IB* 17.4, 17.5
(17) Pile foundations *SF (1)* 4; *SF (2)* 3, 11
(18) Other substructure elements
(19) Parts, accessories, cost summary, etc

(2-) Structure, primary elements, carcass
(21) Walls, external walls *SF (1)* 1, 5; *SF (2)* 4, 5, 10; *IB* 17.6

(22) Internal walls, partitions *SF (1)* 5; *SF (2)* 4, 10; *C* 9; *IB* 17.7

(23) Floors, galleries *SF (1)* 8; *SF (2)* 6, 10; *IB* 17.8

(24) Stairs, ramps *SF (1)* 10; *SF (2)* 8, 10

(25), **(26)** Vacant

(27) Roofs *SF (1)* 1, 7; *SF (2)* 9, 10; *IB* 17.9

(28) Building frames, other primary elements *SF (1)* 1, 6; *SF (2)* 5, 10; Chimneys *SF (1)* 9

(29) Parts, accessories, cost summary, etc

(3-) **Secondary elements, completion of structure**

(31) Secondary elements to external walls, including windows, doors *SF (1)* 5; *SF (2)* 10; *C* 3, 4, 5, 7; *IB* 17.6

(32) Secondary elements to internal walls, partitions including borrowed lights and doors *SF (2)* 10; *C* 3, 7; *IB* 17.7

(33) Secondary elements to floors *SF (2)* 10; *IB* 17.8

(34) Secondary elements to stairs including balustrades *C* 8

(35) Suspended ceilings *C* 10

(36) Vacant

(37) Secondary elements to roofs, including roof lights, dormers *SF (1)* 7; *SF (2)* 10; *C* 6

(38) Other secondary elements

(39) Parts, accessories, cost summary, etc.

(4-) **Finishes to structure**

(41) Wall finishes, external *SF (2)* 4, 10; *F* 3, 4, 5; *IB* 17.6, 17.11

(42) Wall finishes, internal *F* 2, 4, 5; *IB* 17.7, 17.11

(43) Floor finishes *F* 1; *IB* 17.5, 17.8, 17.11

(44) Stair finishes *F* 1; *IB* 17.11

(45) Ceiling finishes *F* 2; *IB* 17.11

(46) Vacant

(47) Roof finishes *SF (2)* 10; *F* 7; *IB* 17.9, 17.11

(48) Other finishes; *IB* 17.11

(49) Parts, accessories, cost summary, etc; *IB* 17.11

(5-) **Services** (mainly piped and ducted)

(51) Vacant

(52) Waste disposal, drainage *ES* 11, 12, 13; *IB* 17.10

(53) Liquids supply *ES* 9, 10; *SF (1)* 9; *SF (2)* 6, 10; *IB* 17.10

(54) Gases supply; *IB* 17.10

(55) Space cooling; *IB* 17.10

(56) Space heating *ES* 7; *SF (1)* 9; *SF (2)* 6, 10; *IB* 17.10

(57) Air conditioning, ventilation *ES* 7; *SF (2)* 10; *IB* 17.10

(58) Other piped, ducted services; *IB* 17.10

(59) Parts, accessories, cost summary, etc Chimney, shafts, flues, ducts independent *SF (2)* 7; *IB* 17.10

(6-) **Services** (mainly electrical)

(61) Electrical supply; *IB* 17.10

(62) Power *ES* 14; *IB* 17.10

(63) Lighting *ES* 8; *IB* 17.10

(64) Communications *ES* 14; *IB* 17.10

(65) Vacant

(66) Transport *ES* 15

(67) Vacant

(68) Security, control, other services; *IB* 17.10

(69) Parts, accessories, cost summary, etc; *IB* 17.10

(7-) **Fittings** with subdivisions (71) to (79)

(74) Sanitary, hygiene fittings *ES* 10

(8-) **Loose furniture, equipment** with subdivisions (81) to (89) Used where the distinction between loose and fixed fittings, furniture and equipment is important.

(9-) **External elements, other elements**

(90) External works, with subdivisions (90.1) to (90.8); *IB* 17.12

(98) Other elements

(99) Parts, accessories etc. common to two or more main element divisions (1-) to (7-) Cost summary

Note: The SfB Agency UK do not use table 1 in classifying manufacturers' literature

---

Table 2 **Constructions, Forms**

Scope: Parts of particular forms which combine to make the elements in table 1. Each is characterised by the main product of which it is made.

A Constructions, forms – used in specification applications for Preliminaries and General conditions

B Vacant – used in specification applications for demolition, underpinning and shoring work

C Excavation and loose fill work

D Vacant

E Cast *in situ* work *M* 8; *SF (1)* 4, 7, 8; *SF (2)* 3, 4, 5, 6, 8, 9

Blocks

F Blockwork, brickwork
Blocks, bricks *M* 6, 12; *SF (1)* 5, 9
SF (2) 4, 6, 7

G Large block, panel work
Large blocks, panels *SF (2)* 4

Sections

H Section work
Sections *M* 9; *SF (1)* 5, 6, 7, 8; *SF (2)* 5, 6

I Pipework
Pipes *SF (1)* 9; *SF (2)* 7

J Wire work, mesh work
Wires, meshes

K Quilt work
Quilts

L Flexible sheet work (proofing)
Flexible sheets (proofing) *M* 9, 11

M Malleable sheet work
Malleable sheets *M* 9

N Rigid sheet overlap work
Rigid sheets for overlappings *SF (2)* 4; *F* 7

P Thick coating work *M* 10, 11; *SF (2)* 4;
F 1, 2, 3, 7

Q Vacant

R Rigid sheet work
Rigid sheets *M* 3, 12, 13, *SF (2)* 4; *C* 5

S Rigid tile work
Rigid tiles *M* 4, 12, 13; *F* 1, 4

T Flexible sheet and tile work
Flexible sheets eg carpets, veneers, papers,
tiles cut from them *M* 3, 9; *F* 1, 6

U Vacant

V Film coating and impregnation work *F* 6;
M 2

W Planting work
Plants

X Work with components
Components *SF (1)* 5, 6, 7, 8, 10; *SF (2)* 4;
C 2, 3, 4, 5, 6, 7, 8

Y Formless work
Products

Z Joints, where described separately

Table 3 **Materials**

Scope: Materials which combine to form the
products in table 2

---

a   **Materials**

b, c, d   Vacant

Formed materials e to o

e   **Natural stone** *M* 4; *SF (1)* 5, 10; *SF (2)*
4

e1 Granite, basalt, other igneous

e2 Marble

e3 Limestone (other than marble)

e4 Sandstone, gritstone

e5 Slate

e9 Other natural stone

f   **Precast with binder** *M* 8; *SF (1)* 5, 7,
8, 9, 10; *SF (2)* 4 to 9; *F* 1

f1 Sand-lime concrete (precast)
Glass fibre reinforced calcium silicate
(gres)

f2 All-in aggregate concrete (precast) *M* 8
Heavy concrete (precast) *M* 8
Glass fibre reinforced cement (gre) *M* 10

f3 Terrazzo (precast) *F* 1
Granolithic (precast)
Cast/artificial/reconstructed stone

f4 Lightweight cellular concrete (precast)
M 8

f5 Lightweight aggregate concrete (precast)
M 8

f6 Asbestos based materials (preformed)
M 10

f7 Gypsum (preformed) *C* 2
Glass fibre reinforced gypsum *M* 10

f8 Magnesia materials (preformed)

f9 Other materials precast with binder

g   **Clay (Dried, Fired)** *M* 5; *SF (1)* 5, 9,
10; *SF (2)* 4, 6, 7

g1 Dried clay eg pisé de terre

g2 Fired clay, vitrified clay, ceramics
Unglazed fired clay eg terra cotta

g3 Glazed fired clay eg vitreous china

g6 Refractory materials eg fireclay

g9 Other dried or fired clays

h   **Metal** *M* 9; *SF (1)* 6, 7, *SF (2)* 4, 5, 7

h1 Cast iron
Wrought iron, malleable iron

h2 Steel, mild steel

h3 Steel alloys eg stainless steel

h4 Aluminium, aluminium alloys

h5 Copper

h6 Copper alloys

h7 Zinc

h8 Lead, white metal

h9 Chromium, nickel, gold, other metals,
metal alloys

i **Wood** including wood laminates *M* 2, 3; *SF (1)* 5 to 8, 10; *SF (2)* 4, 9; *C* 2
i1 timber (unwrot)
i2 Softwood (in general, and wrot)
i3 Hardwood (in general, and wrot)
i4 Wood laminates eg plywood
i5 Wood veneers
i9 Other wood materials, except wood fibre boards, chipboards and wood-wool cement

j **Vegetable and animal materials** – including fibres and particles and materials made from these
j1 Wood fibres eg building board *M* 3
j2 Paper *M* 9, 13
j3 Vegetable fibres other than wood eg flaxboard *M* 3
j5 Bark, cork
j6 Animal fibres eg hair
j7 Wood particles eg chipboard *M* 3
j8 Wood-wool cement *M* 3
j9 Other vegetable and animal materials

k, 1 Vacant

m **Inorganic fibres**
m1 Mineral wool fibres *M* 10; *SF (2)* 4, 7
Glass wool fibres *M* 10, 12
Ceramic wool fibres
m2 Asbestos wool fibres *M* 10
m9 Other inorganic fibrous materials eg carbon fibres *M* 10

n **Rubber, plastics, etc**
n1 Asphalt (preformed) *M* 11; *F* 1
n2 Impregnated fibre and felt eg bituminous felt *M* 11; *F* 7
n4 Linoleum *F* 1

Synthetic resins n5, n6
n5 Rubbers (elastomers) *M* 13
n6 Plastics, including synthetic fibres *M* 13
Thermoplastics
Thermosets
n7 Cellular plastics
n8 Reinforced plastics eg grp, plastics laminates

o **Glass** *M* 12; *SF (1)* 5; *C* 5
o1 Clear, transparent, plain glass
o2 Translucent glass

o3 Opaque, opal glass
o4 Wired glass
o5 Multiple glazing
06 Heat absorbing/rejecting glass
X-ray absorbing/rejecting glass
Solar control glass
o7 Mirrored glass, 'one-way' glass
Anti-glare glass
o8 Safety glass, toughened glass
Laminated glass, security glass, alarm glass
o9 Other glass, including cellular glass

Formless materials p to s
p **Aggregates, loose fills** *M* 8
p1 Natural fills, aggregates
p2 Artificial aggregates in general
p3 Artificial granular aggregates (light) eg foamed blast furnace slag
p4 Ash eg pulverised fuel ash
p5 Shavings
p6 Powder
p7 Fibres
p9 Other aggregates, loose fills

q **Lime and cement binders, mortars, concretes**
q1 Lime (calcined limestones), hydrated lime, lime putty, *M* 7
Lime-sand mix (coarse stuff)
q2 Cement, hydraulic cement eg Portland cement *M* 7
q3 Lime-cement binders *M* 15
q4 Lime-cement-aggregate mixes
Mortars (ie with fine aggregates) *M* 15; *SF (2)* 4
Concretes (ie with fine and/or coarse aggregates) *M* 8
q5 Terrazzo mixes and in general *F* 1
q6 Lightweight, cellular, concrete mixes and in general *M* 8
q9 Other lime-cement-aggregate mixes eg asbestos cement mixes *M* 10

r **Clay, gypsum, magnesia and plastics binders, mortars**
r1 Clay mortar mixes, refractory mortar
r2 Gypsum, gypsum plaster mixes
r3 Magnesia, magnesia mixes *F* 1
r4 Plastics binders
Plastics mortar mixes
r9 Other binders and mortar mixes

s   **Bituminous materials** *M* 11; *SF (2)* 4

s1   Bitumen including natural and petroleum bitumens, tar, pitch, asphalt, lake asphalt

s4   Mastic asphalt (fine or no aggregate), pitch mastic

s5   Clay-bitumen mixes, stone bitumen mixes (coarse aggregate)
Rolled asphalt, macadams

s9   Other bituminous materials

Functional materials t to w

t   **Fixing and jointing materials**

t1   Welding materials *M* 9; *SF (2)* 5

t2   Soldering materials *M* 9

t3   Adhesives, bonding materials *M* 14

t4   Joint fillers eg mastics, gaskets *M* 16 *SF (1)* 2

t6   Fasteners, 'builders ironmongery'
Anchoring devices eg plugs
Attachment devices eg connectors *SF (1)* 6, 7
Fixing devices eg bolts, *SF (1)* 5

t7   'Architectural ironmongery' *C* 7

t9   Other fixing and jointing agents

u   **Protective and Process/property modifying materials**

u1   Anti-corrosive materials, treatments *F* 6
Metallic coatings applied by eg electro-plating *M* 9
Non-metallic coatings applied by eg chemical conversion

u2   Modifying agents, admixtures eg curing agents *M* 8
Workability aids *M* 8

u3   Materials resisting specials forms of attack such as fungus, insects, condensation *M* 2

u4   Flame retardants if described separately *M* 1

u5   Polishes, seals, surface hardeners *F* 1; *M* 8

u6   Water repellants, if described separately

u9   Other protective and process/property modifying agents eg ultra-violet absorbers

v   **Paints** *F* 6

v1   Stopping, fillers, knotting, paint preparation materials including primers

v2   Pigments, dyes, stains

v3   Binders, media eg drying oils

v4   Varnishes, lacquers eg resins
Enamels, glazes

v5   Oil paints, oil-resin paints
Synthetic resin paints
Complete systems including primers

v6   Emulsion paints, where described separately
Synthetic resin-based emulsions
Complete systems including primers

v8   Water paints eg cement paints

v9   Other paints eg metallic paints, paints with aggregates

w   **Ancillary materials**

w1   Rust removing agents

w3   Fuels

w4   Water

w5   Acids, alkalis

w6   Fertilisers

w7   Cleaning materials *F* 1
Abrasives

w8   Explosives

w9   Other ancillary materials eg fungicides

x   **Vacant**

y   **Composite materials**
Composite materials generally *M* 11
See p. 63 *Construction Indexing Manual*

z   **Substances**

z1   By state eg fluids

z2   By chemical composition eg organic

z3   By origin eg naturally occuring or manufactured materials

z9   Other substances

Table 4   **Activities, Requirements** (main headings only)

Scope: Table 4 identifies objects which assist or affect construction but are not incorporated in it, and factors such as activities, requirements, properties, and processes.

Activities, aids

(A)   Administration and management activities, aids *C* 11; *M* Introduction, *SF (1)* 2; *SF (2)* 1, 2, 3; *IB* 14, 17.13

(B)   Construction plant, tools *SF (1) 2; SF (2)* 2, 11

(C)   Vacant

(D)   Construction operations *SF (1)* 2, 11; *SF (2)* 2, 11

Requirements, properties, building science, construction technology
Factors describing buildings, elements, materials, etc

(E) Composition, etc *SF (1)* 1, 2; *SF (2)* 1, 2; *IB* 1

(F) Shape, size, etc *SF (1)* 2; *IB* 4

(G) Appearance, etc *M* 1; *F* 6; *IB* 1

Factors relating to surroundings, occupancy

(H) Context, environment *IB* 6

Performance factors

(J) Mechanics *M* 9; *SF (1)* 3, 4; *SF (2)* 3, 4; *IB* 5

(K) Fire, explosion *M* 1; *SF (2)* 10; *IB* 9

(L) Matter *IB* 10

(M) Heat, cold *ES* 1; *IB* 8

(N) Light, dark *ES* 1; *IB* 10

(P) Sound, quiet *ES* 1; *IB* 7

(Q) Electricity, magnetism, radiation *ES* 14

(R) Energy, other physical factors *ES* 7; *IB* 3

(T) Application

Other factors

(U) Users, resources *IB* 11, 12

(V) Working factors

(W) Operation, maintenance factors

(X) Change, movement, stability factors

(Y) Economic, commercial factors *M* Introduction; *SF (1)* 2; *SF (2)* 3, 4, 5, 6, 9; *IB* 1, 13

(Z) Peripheral subjects, form of presentation, time, place – may be used for subjects taken from the UDC (*Universal decimal classification*), see BS1000A 1961

Subdivision: All table 4 codes are subdivided mainly by numbers

Appendix B

Working drawings and schedules

(a) This list should not be taken as a definitive check list of items to be included on working drawings/schedules. It is only a guide as to the type of information which may be required.

(b) Drawings and schedules may be supplemented by information given in specification, bills of quantities, sample boards and three dimensional representations.

1.00 GENERAL INFORMATION ON EACH DRAWING/SCHEDULE

1.01 Name of project and location
1.02 Project reference and drawing number
1.03 Name of designer/consultant and address of contact including telephone number
1.04 General description of content of drawing
1.05 Cross reference to other relevant drawings, schedules or Bills of Quantities/specification
1.06 Scale(s) employed on drawing(s)
1.07 Date of drawing when completed
1.08 Space of date of issue stamp
1.09 North point when appropriate
1.10 Amendment panel, including date of revision
1.11 General instructions panel
1.12 Sub-titles for each individual item shown
1.13 Key to non-standard conventions

2.00 SITE PLANS

2.01 District and address
2.02 Boundaries and site lines
2.03 Adjoining buildings, road and paths (existing and proposed)
2.04 Rights of way to be maintained/temporarily obstructed
2.05 Existing features (trees, mounds, fences, hedges, ditches, etc)
2.06 Existing datum, levels, and contours
2.07 Existing drainage: sewers, cesspools, septic tanks, soakaways and watercourses including levels
2.08 Existing gas, water, electricity, telecommunication services including levels

2.09 Means of access
2.10 Setting out of proposed building and site including levels
2.11 Proposed routes for rubbish disposal and fire fighting vehicles
2.12 Proposed landscape feature including levels
2.13 Alteration to existing features (boundaries, trees, landscape)
2.14 Alterations to existing drainage, gas, water, electricity and telecommunication services including connection to proposed
2.15 Location proposed drainage, gas, water, electricity and telecommunication services

3.00 FOUNDATION PLANS

3.01 Grid/modular planning lines
3.02 Datum level for excavations
3.03 Indication of existing foundations, earthworks, etc to be removed
3.04 Dimensions of new foundations to define shape including vertical steps
3.05 Levels of top and/or underside of foundations
3.06 Positions of walls relative to foundations
3.07 Positions of services to be installed below ground level
3.08 Location and size of holes left through foundations for service pipes
3.09 Drain and manhole foundations and levels
3.10 Typical details of excavations

4.00 FLOOR PLANS

4.01 Grid/modular planning lines
4.02 Datum level
4.03 Finished floor levels and changes of level
4.04 External dimensions:
  (a) changes of direction, openings, etc
  (b) overall of building
4.05 Internal dimensions:
  (a) changes of direction, openings, etc
  (b) overall of areas
4.06 Room/space names and areas if required
4.07 Wall dimensions and description of materials, thickness and jointing

4.08 Location of movement joints

4.09 Air bricks and ventilation grilles

4.10 Layout of horizontal and vertical service ducting: dimensions and description of services carried (gas, water, electricity, telephone, tv and cable)

4.11 Fireplaces: dimensions and/or description of openings, hearth, projections, etc

4.12 Flues: dimensions and/or description of type, position, size, lining, etc

4.13 Mat wells and other changes in floor surface

4.14 Overhead floor/roof construction: materials, size, spacing and direction of span, etc

4.15 Overhead rooflights, ventilation cowls, etc

4.16 Staircases: dimension and/or description of direction (up/down) treads (and number), rise/going, flight width, handrails, landings, balconies, etc

4.17 Doors and door frames/linings: dimensions and/or description cross referenced to schedules (including ironmongery)

4.18 Windows and window frames: dimensions and/or description cross referenced to schedules (including ironmongery)

4.19 Hatches/access panels and frames: dimensions and/or description cross referenced to schedules (including ironmongery)

4.20 Door (and window) direction of opening, number, and manufacturer's reference if applicable

4.21 Vertical damp proof courses and membranes

4.22 External and internal wall and floor finishes including materials, thickness, jointing, position and techniques

4.23 Internal ceiling finishes where to be linked in with wall components

Depending on the size of the project, the following items may be included either on the main drawings, or on separate 'specialist' drawings *cross referenced* with main drawings

4.24 Built-in furniture, cupboards, units, etc: dimensions and/or description cross referenced to schedule/detail drawings

4.25 Heating units/radiators etc: dimensions and/or description cross referenced to schedule

4.26 Sanitary fittings: dimensions and/or description cross referenced to schedule

4.27 Specialist equipment; fire hoses; alarms; mechanical plant

4.28 Electrical installation: description and dimensioned position of intake, consumer unit (main switch/fuses), power socket outlets, lighting points and switches, emergency lighting, etc

4.29 Gas installation: description and dimensioned position of intake, meter and supply points

4.30 Water installation: description and dimensioned position of intake, valve controls, rising main, supply points; and cold water tank, drain down points, etc

4.31 Heating installation: description and dimensioned position of boiler, (equipment, model, size and rating), hot water cylinder, expansion tank, heating units, drain down points, etc

4.32 Drainage installation: description and dimension of outfall, manholes, drainpipes, soil stacks, wastes, vent pipes, gullies, branch drains from fittings to include details of material, size, gradient and direction of flow when appropriate

4.33 Rainwater drainage installation: description and dimensions of outfall, manholes, drainpipes, rainwater pipes and gullies, surface water gullies and direction of falls, etc

4.34 External works and landscaping: steps, ramps, paving, grass, planting, trees and built features, such as terraces, playgrounds (including equipment), amenities, etc

4.35 Details, dimensions and/or description of external fuel storage, oil tanks and sheds, garages, car ports, etc

5.00 ROOF PLANS

5.01 Grid/modular planning lines

5.02 Datum level

5.03 Dimensions:
   (a) changes of direction, openings, etc
   (b) overall of building

5.04 Description of roof structure: decking, furrings, and integral thermal insulation – vapour barriers and ventilation

5.05 Screeds: composition, maximum and minimum thicknesses

5.06 Roof finishes: materials, sizes, thickness, gauge, pitch, etc

5.07 Direction of falls

5.08 Roof lights and ventilation cowls: description and dimensions cross referenced to schedule if applicable

5.09 Ventilation pipes, rainwater pipes and gutters: description and dimensions

5.10 Parapet, coping, eaves, ridge and upstands: description and dimensions

5.11 Lift motor and mechanical plant rooms: description and dimensions

5.12 Location and description of cantilever beams, cradle fixing position, etc for window cleaning

5.13 Location of areas unsafe for maintenance workers without additional precautions; areas safe for maintenance workers only; areas safe for foot traffic and/or heavier loads

5.14 Access positions: staircases, traps, ladders, etc

5.15 Landscaping, terraces and balconies: description and dimensions including arrangements for waterproofing and drainage

6.00 Sections

6.01 Grid/modular section lines

6.02 Datum levels

6.03 Ground levels: existing and new finished ground and floor levels

6.04 Floor levels and identification

6.05 Room/space names when applicable

6.06 External dimensions:
(a) changes of direction, openings, etc
(b) overall of building

6.07 Internal dimensions:
(a) door/window openings, staircases, ducts, built-in furniture, guard rails, etc
(b) room heights and suspended floors/ceilings

6.08 Foundations: description and dimensions including composition, size and stepping details

6.09 Foundation walls: materials, thickness, bonding, backfilling, damp proof course, holes for drainage and ventilation

6.10 Fill and hardcore material, thickness, and layering

6.11 Solid ground floor construction; thickness, composition, additives, reinforcement, damp proof membrane, insulation, screeds, and finishes

6.12 Suspended ground floor construction: type, materials, dimensions, fixings, insulation, damp proof membrane, ventilation and finishes

6.13 Intermediate floor construction: type, materials, dimensions, fixings, insulation, and finishes including ceilings

6.14 Roof construction: type, materials, dimensions, fixing, insulation and finishes, including soffites and ceilings

6.15 External wall construction: type, material, dimensions, fixings, ties, insulation and finishes

6.16 Internal wall construction: type, material, dimensions, fixings, ties, insulation and finishes

6.17 Staircases: dimension and/or description of direction (up/down) treads (and number), rise/going, flight width, handrails, landings, balconies, etc

6.18 Doors and door frames/linings: dimensions and/or description cross referenced to schedules (including ironmongery)

6.19 Windows and window frames: dimensions and/or description cross referenced to schedules (including ironmongery)

6.21 Lintol types, sizes and materials cross referenced to schedule when appropriate

6.22 Location of movement joints

6.23 Air bricks and ventilation grilles

6.24 Layout of horizontal and vertical service ducting: dimensions and description of services carried (gas, water, electricity, telephone, tv and cable)

6.25 Fireplaces: dimensions and/or description of openings, hearth, projections, etc

6.26 Flues: dimensions and/or description of type, position, size, lining, etc

6.27 Mat wells and other changes in floor surface

Depending on the size of the project, the following items may be included either on the main drawings, or on separate 'specialist' drawings *cross referenced* with main drawings

6.28 Built-in furniture, cupboards, units, etc: dimensions and/or description cross referenced to schedule/detail drawings

6.29 Heating units/radiators etc: dimensions and/or description cross referenced to schedule

6.30 Sanitary fittings: dimensions and/or description cross referenced to schedule

6.31 Specialist equipment; fire hoses; alarms; mechanical plant

6.32 Electrical installation: description and dimensioned position of intake, consumer unit (main switch/fuses), power socket outlets, lighting points and switches, emergency lighting, etc

6.33 Gas installation: description and dimensioned position of intake, meter and supply points

6.34 Water installation: description and dimensioned position of intake, valve controls, rising main, supply points; and cold water tank, drain down points, etc

6.35 Heating installation: description and dimensioned position of boiler, (equipment, model, size and rating), hot water cylinder, expansion tank, heating units, drain down points, etc

6.36 Drainage installation: description and dimension of outfall, manholes, drainpipes, soil stacks, wastes, vent pipes, gullies, branch drains from fittings to include details of material size, gradient and direction of flow when appropriate

6.37 Rainwater drainage installation: description and dimensions of outfall, manholes, drainpipes, rainwater pipes and gullies, surface water gullies and direction of falls, etc

6.38 External works and landscaping: steps, ramps, paving, grass planting, trees and built features, such as terraces, playgrounds (including equipment), amenities, etc

6.39 Details, dimensions and/or description of external fuel storage, oil tanks and sheds, garages, car ports, etc

7.00 ELEVATIONS
7.01 Grid/modular section lines

7.02 Datum levels

7.03 Ground levels: existing and new finished ground and floor levels

7.04 Foundation lines

7.05 General description of facing materials including type, material, texture and colour

7.06 Door positions and cross reference to schedules

7.07 Window positions and cross reference to schedules

7.08 Direction of opening of door and windows

7.09 Air bricks and ventilation cowls

7.10 Description and position of movement joints

7.11 Soil stacks, gutters, rainwater pipes, and other exposed services including alarms, lighting, power points, stand pipes, etc

7.12 Description and position of special features: signs, sculptures, decorative displays, etc

7.13 Description and location of sheds, storage room, etc, detached from main building

7.14 Description and location of landscape features: trees, bushes, plants, earth ramps and mounds, etc

8.00 DETAIL DRAWINGS, AXONOMETRICS AND ISOMETRICS
8.01 Grid/modular planning/section lines

8.02 Location of doors windows and other openings

8.03 Wall, floor, roof thicknesses

8.04 Location of furniture and other fixings

8.05 Details of construction and finishes

9.00 SCHEDULES
9.01 Reference number (from drawings), location, type and size of items

9.02 Manufacturer's reference if applicable

9.03 Ancillary information concerning, colour, ironmongery, methods of fixing, applications and details of immediately surrounding elements or components, etc

9.04 Cross reference with other schedules unless fully described, ie colour to be painted, ironmongery involved, etc

Clerk of works direction

Job reference:
Serial No:
Issue date:

Original to Contractor

Architect:
address:

Employer:
address:

Contractor:
address:

Works:
situated at:

Under the terms of the Contract dated

I issue this day the following direction.

This direction shall be of no effect unless confirmed in writing by the Architect within 2 working days and does not authorize any extra payment.

Direction

Architect's use

Signed

Clerk of works

Copies to:
☐ Architect
☐ File

Covered by AI No:

© RIBA Publications Ltd 1982

Architect's instruction

Job reference:
Serial No:
Issue date:

Original to Contractor

Architect:
address:

Employer:
address:

Contractor:
address:

Works:
situated at:

Under the terms of the Contract dated

I/We issue the following instructions. Where applicable the Contract Sum will be adjusted in accordance with the terms of the relevant Condition.

Instructions

Office use: Approx costs
£ omit   £ add

Signed

Architect

Amount of Contract Sum £
± Approximate value of previous instructions £
± Approximate value of this instruction £
Approximate adjusted total £

Copies to:
— Employer   — Structural consultant   Nominated Sub-Contractors
— Quantity Surveyor — Services consultant   —
— Clerk of works   — Electrical consultant   —
— Site   —   File

© RIBA Publications Ltd 1980

Final certificate

Architect:
address:

Employer:
address:

Contractor:
address:

Works:
situated at:

Job reference:
Serial No:
Issue date:
Valuation date:

Contractor's copy

Under the terms of the Contract dated

for the Works named and situated as stated above

I/We certify that:

1. The Contract Sum adjusted as necessary in accordance with the Contract Conditions is .... £

2. The amount already stated as due in Interim Certificates to the Contractor is .... £

and that .... £

(in words) _____

is a balance due* to the Contractor from the Employer* to the Employer from the Contractor which subject to any deductions authorised by the Contract Conditions shall as from the 14th day after the date of issue of this certificate be a debt payable from the one to the other.

All the above amounts are exclusive of VAT

*Delete as appropriate

Signed _____ Architect

Each Nominated Sub-Contractor shall be notified of the date of issue of this Final Certificate.

© RIBA Publications Ltd 1980

Interim certificate
and Direction

Architect:
address:

Employer:
address:

Contractor:
address:

Works:
situated at:

Job reference:

Interim Certificate No:

*Issue date:

Valuation date:

*This date is to be not later than 7 days from the Valuation date.

Contractor's copy and Direction

Under the terms of the Contract dated

in the sum of £   for the Works named and situated as stated above

I/We certify that the following interim payment is due from the Employer to the Contractor; and

I/We direct the Contractor that the amounts of interim or final payments to Nominated Sub-Contractors included in this Certificate and listed on the attached *Statement of Retention and of Nominated Sub-Contractors' Values* are due to be discharged to those named.

Gross valuation inclusive of the value of Works by Nominated Sub-Contractors .... £

Less Retention which may be retained by the Employer as detailed on the Statement of Retention .... £

.... £

Less total amount stated as due in Interim Certificates previously issued up to and including Interim Certificate No. .... £

Amount due for payment on this Certificate .... £

(in words) _____

All the above amounts are exclusive of VAT

Signed _____ Architect

Contractor's provisional assessment of total amounts included in above certificate on which VAT will be chargeable £   @   %

This is not a Tax Invoice

© RIBA Publications Ltd 1982

Form 1 (upper)

Notification
to Nominated Sub-Contractor concerning amount included in certificates

Job reference:
Interim Certificate No:
Issue date:
Valuation date:

Architect:
address:

Employer:
address:

Contractor:
address:

Works:
situated at:

Nominated Sub-Contractor:
address:

Original to Nominated Sub-Contractor

Under the terms of the Contract dated _____

I/We inform you that the amount of an interim/final* payment due to you has been included in

Interim Certificate No. _____ dated _____

issued to the Employer, in accordance with the attached *Statement of Retention and of Nominated Sub-Contractors' Values* and that the Contractor, named above, has been directed in the said Certificate to discharge his obligation to pay this amount in accordance with the terms of the Contract and the relevant Sub-Contract.

To comply with your obligation to provide written proof of discharge of the certified amount, you should return the acknowledgement slip below to the Contractor immediately upon such discharge.

Signed _____ Architect

*Delete as appropriate

© RIBA Publications Ltd 1982

Nominated Sub-Contractor's acknowledgement of discharge of payment due

Job reference:
Notification date:
Interim Certificate No:

Contractor:
address:

Works:
situated at:

We confirm that we have received from you discharge of the amount included in Certificate No. _____ dated _____

as stated in the Notification dated _____

in accordance with the terms of the relevant Sub-Contract.

Signed _____ Nominated Sub-Contractor

Date _____

Form 2 (lower)

Certificate of completion of **Making good defects**

Job reference:
Serial No:
Issue date:

Architect:
address:

Employer:
address:

Contractor:
address:

Works:
situated at:

Under the terms of the Contract dated _____

I/We hereby certify that the defects, shrinkages and other faults specified in the schedule of defects delivered to the Contractor as an instruction have in my/our opinion been made good.

This Certificate refers to:

*1. The Works described in the Certificate of Practical Completion
Serial No. _____ dated _____

*2. The Works described in the Certificate of Partial Possession of a relevant part of the Works
Serial No. _____ dated _____

*Delete as appropriate

Signed _____ Architect

Date _____

Original to: Employer

Copies to:
Contractor
Quantity Surveyor
Clerk of Works

Structural Consultant
Services Consultant
Electrical Consultant

Nominated Sub-Contractors:

Site

© RIBA Publications Ltd 1982

Certificate of Practical completion

of the Works
OR of works executed
by a Nominated
Sub-Contractor

Job reference:
Serial No:
Issue date:

Architect: address:

Employer: address:

Contractor: address:

Works: situated at:

Nominated Sub-Contract Works: (if applicable)

Under the terms of the Contract dated

I/We certify that:

*1. Practical Completion of the Nominated Sub-Contract Works referred to above, was achieved on:

*2. Practical Completion of the Works was achieved on:

*Delete as appropriate

The Employer should note that as from the date of issue of this Certificate of Practical Completion of the Works the Employer becomes solely responsible for insurance of the Works.

Signed _____ Architect

Original to: ☐ Employer

Copies to:
☐ Contractor
☐ Quantity Surveyor
☐ Clerk of Works
☐ Structural Consultant
☐ Services Consultant
☐ Electrical Consultant

Nominated Sub-Contractors:
☐
☐
☐ Site

© RIBA Publications Ltd 1982

Certificate of Partial possession

by the Employer

Job reference:
Serial No:
Issue date:

Original to Employer

Architect: address:

Employer: address:

Contractor: address:

Works: situated at:

Under the terms of the Contract dated

I/We certify that a part of the Works, referred to as the relevant part, namely:

the approximate value of which I/we estimate for the purposes of this Certificate, but for no other purpose, to be £ ....

was taken into possession by the Employer on: _____ (date)

and that for this relevant part of the Works only the Defects Liability Period will end on: _____ (date)

Signed _____ Architect

The Employer should note that as from the date of possession of the relevant part, he becomes solely responsible for the insurance of the relevant part of the Works.

Copies to:
— Contractor
— Quantity Surveyor
— Clerk of works
— Structural consultant
— Services consultant
— Electrical consultant
— Nominated Sub-Contractors
— File

© RIBA Publications Ltd 1980

INTERIM CERTIFICATE

STANDARD FORM APPROVED BY THE
SOCIETY OF
INDUSTRIAL ARTISTS
& DESIGNERS

SUPERVISING
OFFICER'S
NAME &
ADDRESS

AGREEMENT DATE
CERTIFICATE NO
JOB NO
DATE
CONTRACTOR'S
NAME
EMPLOYER'S
NAME

We certify that under the terms of the above Agreement interim
payment as detailed below is due from the Employer to the
Contractor.

Total Value £
(including the value of works by nominated sub-contractors as
detailed on any accompanying direction form)

Less Retention £
(As applicable)

Balance £
(Cumulative total certified for payment)

Less cumulative total previously certified £

Amount due for payment on this certificate £
(In words)

The above amounts are exclusive of VAT
This is not a Tax Invoice

SIGNED (Supervising Officer)

© 1983 SIAD

INSTRUCTION

STANDARD FORM APPROVED BY THE
SOCIETY OF
INDUSTRIAL ARTISTS
& DESIGNERS

SUPERVISING
OFFICER'S
NAME &
ADDRESS

AGREEMENT TITLE
INSTRUCTION NO
JOB NO
DATE

TO CONTRACTOR

We issue the following instructions under the terms of the Agreement.
Where applicable the Contract Sum will be adjusted in accordance
with the relevant terms of the Agreement.

ITEM	INSTRUCTIONS	£ OMIT	£ ADD

SIGNED (Supervising Officer)

CONTRACT SUM £
VALUE OF PREVIOUS INSTRUCTIONS £
VALUE OF THIS INSTRUCTION £
ADJUSTED TOTAL £

DISTRIBUTED TO
☐ Contractor
☐ Employer
☐ Quantity Surveyor
☐ Structural Engineer
☐ Architect

© 1983 SIAD

FINAL CERTIFICATE

STANDARD FORM APPROVED BY THE
SOCIETY OF
INDUSTRIAL ARTISTS
& DESIGNERS

SUPERVISING
OFFICER'S
NAME &
ADDRESS

AGREEMENT DATE
CERTIFICATE NO
JOB NO
DATE
CONTRACTOR'S
NAME
EMPLOYER'S
NAME

We certify that under the terms of the above Agreement the Defects Period has ended, all/any defects have been made good to our satisfaction and final payment is due as listed below:

Contract Sum
(Adjusted as necessary in accordance with Instructions issued)   £

Less the sum of the amounts paid to the Contractor under Interim Certificates and the amount of Retention monies paid to the Contractor at Practical Completion; and the amounts of any monies paid to other contractors to rectify defects under Clause 2.5 of the Agreement.   £

Balance Now Due
to be paid from/to the Employer to/from the Contractor within 14 days
of the date of issue of this certificate
(In words)   £

The above amounts are exclusive of VAT
This is not a Tax Invoice

SIGNED (Supervising Officer)

© 1983 SIAD

PRACTICAL COMPLETION

STANDARD FORM APPROVED BY THE
SOCIETY OF
INDUSTRIAL ARTISTS
& DESIGNERS

SUPERVISING
OFFICER'S
NAME &
ADDRESS

AGREEMENT DATE
JOB NO
DATE
CONTRACTOR'S
NAME
EMPLOYER'S
NAME

We certify that under the terms of the above Agreement that subject to the completion of any outstanding items and making good of any defects which appear during the Defects Period.

The Works were in our opinion practically completed on

The said Defects Period will end on

One half of the Retention monies deducted under previous Certificates is now due to be paid from the Employer to the Contractor
in the sum of   £

SIGNED (Supervising Officer)

DISTRIBUTED TO
☐ Contractor
☐ Employer
☐ Quantity Surveyor
☐ Structural Engineer
☐ Architect

© 1983 SIAD

Index

INDEX